Games and Dynamics in Economics

Ferenc Szidarovszky · Gian Italo Bischi
Editors

Games and Dynamics in Economics

Essays in Honor of Akio Matsumoto

 Springer

Editors
Ferenc Szidarovszky
Department of Mathematics
Corvinus University of Budapest
Budapest, Hungary

Gian Italo Bischi
Dipartimento di Economia,
Società, Politica (DESP)
Università di Urbino
Urbino, Italy

ISBN 978-981-15-3625-0   ISBN 978-981-15-3623-6   (eBook)
https://doi.org/10.1007/978-981-15-3623-6

This Springer imprint is published by the registered company Springer Nature Singapore Pte Ltd.
The registered company address is: 152 Beach Road, #21-01/04 Gateway East, Singapore 189721, Singapore

Preface 1

This book contains a collection of research papers on Dynamic Economic Systems, Game Theory, and Related Topics with applications and theoretical issues celebrating the academic achievements of Prof. Akio Matsumoto. The authors of this volume are his former collaborators, colleagues, and friends who contributed their newest research results in honor of the well respected scientist, colleague, and also a great friend of all of us who have the privilege of knowing him.

Professor Akio Matsumoto is considered as one of the international leaders in research in dynamic economic models especially dealing with delayed dynamics. His life is a true success story. He was born 70 years ago as the first son of a middle class family. Both parents were educators, from whom he learned honesty, hard work, and demand for high quality intellectual activity. After elementary and junior high schools the years studied in the Keio High School gave him the high quality education which was fundamental for his college education and later research works. During his college years, he started studying Administrative Engineering, and then for a short time he was dedicated to Marxian Economics. After realizing that this was not his primary interest, he turned to General Equilibrium Theory and then became specialized in Economic Dynamics. First linear models were in his focus, where local stability implies global stability, so this type of models were not challenging enough to the young talented researcher. After several shorter periods with different Japanese universities, twenty two years ago he finally joined the Economics Department of Chuo University in Tokyo. The new challenging atmosphere there helped him to further extend his research topics and find his major research interest, Nonlinear Dynamics with special attention to Delayed Systems. The two years of working in the University of Southern California under a Fulbright grant was very influential for his research development, and later the two years in the University of Arizona on a sabbatical leave gave him the opportunity to continue and further improve his research and publishing activity. He is the author and coauthor of several books, book chapters, conference papers in addition to close to 200 journal publications. His two most recent books on Game Theory and Dynamic Oligopolies, as well as two coedited volumes, were published by Springer. His research papers were published in high quality journals, including

Nonlinear Dynamics, Psychology, and Life Sciences; *Nonlinear Analysis: Real World Applications*; *International Game Theory Review*; *Applied Mathematics and Computations*; *Journal of Economic Behavior and Organization*; *Communications in Nonlinear Science and Numerical Simulation*; *CUBO a Mathematical Journal*; *Mathematics and Computers in Simulations*; *Chaos, Solitons & Fractals*; *Economic Modeling*; *Structural Change and Economic Dynamics*; *Metroeconomica*; *Journal of Evolutionary Economics*; *Frontiers in Applied Mathematics* and *Statistics; Environmental Economics and Policy Studies; Focus Issue of Chaos* among many others. He also participated in a great number of international conferences (he also organized a couple of Nonlinear Economic Dynamics conferences), where we had the opportunity to discuss different research topics and results with him. We all learned a lot during these discussions. He was always very helpful, friendly, and all kinds of communications were very pleasant with him. In addition to "dry" scientific works he is a fan of both light and classical music. The community of researchers in Game Theory and Dynamic Economic Systems owes much to Professor Matsumoto. The present volume of the collected papers expresses a modest sign of our gratitude.

Urbino, Italy   Gian Italo Bischi
Budapest, Hungary   Ferenc Szidarovszky
December 2019

Preface 2

Introduction

Summary of the Scientific Works of Prof. Matsumoto

Professor Matsumoto has a large publication record on analyzing a wide variety of static and dynamic economic models and related theoretical issues. In addition to examining dynamic nonlinear monopolies with discrete and continuous time scales and with fixed and distributed delays he also extended these models to duopolies and triopolies with one and two delays. The local asymptotical stability of the equilibrium was studied analytically and global stability was examined with computer simulation. The most important results on these topics are summarized in chapters 2 and 3 of his book: Matsumoto, A. and F. Szidarovszky, *Dynamic Oligopolies with Time Delays*, Springer, 2018. These models and results were later extended to *n*-firm oligopolies: models with and without product differentiation were considered with both discrete and continuous time scales. Cases of fixed and distributed single and multiple delays were analyzed, local stability conditions were derived, and those for global stability were obtained by using numerical simulations. In addition to the classical Cournot competition Bertrand oligopolies were also studied with both linear and hyperbolic price functions. The classical result of Theocharis was generalized for the nonlinear models, as well as special cases with implementation and demand delays were examined. Interesting results were obtained by comparing dynamic Cournot and Bertrand competitions. Several additional extensions of oligopolies were introduced and their dynamic behavior analyzed including models with employee ownership, with advertisements, with flexible and contingent workforce and unemployment insurance, and with consideration to production adjustment and investment costs. Partial cooperation among the firms can model different versions of co-ownership among the firms. This general model contains the classical noncooperative and fully cooperative cases. In addition to equilibrium analysis its stability conditions were derived. As related topics, models with cartelizing groups and antitrust thresholds were introduced and examined. In addition, some difficulties in applying the Herfindahl-Hirschman index were discovered and analyzed. In studying oligopolies with uncertainty, the uncertainty in the price and cost functions were modeled with probabilistic methods. Behavioral uncertainty was modeled in a duopoly, when one firm believed that the competition was à la Cournot, and the other firm thought that it was a Bertrand duopoly. Contest games are generalizations of hyperbolic oligopolies. The stability of such games were examined without and with time delays. A special n-firm oligopoly was introduced with discontinuous profit functions, when the firms could treat their waste until a certain volumes and then contractors were hired to take care of the rest. If the contractors charged setup costs, then the profit functions became discontinuous. This static model was proved to have at least one equilibrium which is not necessarily unique. Based on this work several researchers considered similar models with their dynamic extensions. Models of adaptive learning of the price function were introduced and conditions were derived for the asymptotical stability of the resulted dynamic models guaranteeing successful learning. The dynamism of the models was based on repeated price or output information. Some interesting results on n-firm oligopolies are presented in chapters 4–6 of the book by Matsumoto and Szidarovszky (2018). Several theoretical issues were studied in relation to the above mentioned models. The stability of continuous time dynamics with one or two delays were studies and the stability thresholds and stability switching curves were characterized with elementary methods. Sufficient, and necessary global stability conditions were analyzed for continuous dynamics, as well as stability conditions for discrete and continuous systems were compared. Several classical economic models were reconsidered with dynamic extensions including time delays. Both discrete and continuous time scales were considered with fixed and distributed delays. Goodwin's multiplier-accelerator model was extended to consider consumption, investment, and tax collection delays including piecewise delay investment. The neoclassical Solow model was extended with delays, as well as with physical and human capitals. Delay dynamic Marxian economies were also examined. Stability switching curves were analytically determined in two- and three-delay Lotka-Volterra competition systems and in nonlinear cobweb models. Similar issues were studied in international subsidy games with electronic and traditional traders; in asset price dynamics; in analyzing Hicksian trade cycles and the Kaldor-Kalecki dynamic model, furthermore in IS-LM models with tax collection. Oligopoly models were connected to environmental issues, when point-source and non-point-source emission controls were studied and conditions were derived for the successful control of emission volumes. Both linear and hyperbolic Cournot competitions and linear Bertrand oligopolies were studied in static and dynamic frameworks. The research interest of Prof. Matsumoto was not limited to economics. We can mention here two different areas. One is the application of the Lotka-Volterra-type model into analyzing the dynamic evolution of the devotion of two lovers toward each other. The other area is a very important field of industrial engineering, namely reliability and quality engineering. Models were developed to find optimal imperfect preventive maintenance strategies as a game against nature. Similar models and solution algorithms were suggested to determine optimal scheduling of repairs and preventive replacements of components and subsystems of engineering systems. Since all of these models rely on statistical methods with uncertainties in the obtained decisions, they could be treated by using expected value analysis (as usual) or certainty equivalents.

A Note on Akio's Passion for the Beatles

In 2003 Akio Matsumoto published a paper in *Chaos, Solitons & Fractals, An interdisciplinary journal of nonlinear science*, where he investigated a dynamic model of pure exchange with two goods and two consumers in which chaotic price fluctuations can arise. In order to reveal some statistical properties of such price dynamics, he constructed a density function along a chaotic trajectory, calculated a long-run average utility and then compared it with the utility computed at a stationary state of the model. He essentially obtained two results: he analytically proved and numerically verified that (1) chaotic price dynamics can be beneficial for one consumer and harmful to the other one; (2) in the long-run the whole economy is possibly better off along chaotic fluctuations than at a stationary state. These results imply the possibility that a chaotic variation of price may be preferable respect to a constant price, i.e., chaotic economies may be beneficial. A surprising economic implication, in contrast to many economic policies oriented to the stabilization of chaotic systems. However, in this short note I would better pay attention to the title of the paper proposed by Akio: "Let it be", with subtitle "Chaotic price instability can be beneficial". Of course, "Let it be" is also the title of a famous song by the English rock band "The Beatles", released in March 1970 as a single (and also the title of a famous album) written and sung by Paul McCartney. A song that quite soon became famous all over the world. This may be just a coincidence, but all friends of Akio immediately understood it was not the case, as we all know his passion for the Beatles.

Indeed, some years after Akio wrote another paper, together with his coauthor Ferenc Szidarovszky, where they considered a market with an isoelastic demand function and proposed a three-country model with two active governments and two firms. The purpose of this paper was to study the dynamic behavior of a sequential subsidy game in which the governments determine their optimal trade policies and, accordingly, the firms choose their optimal outputs. The paper was published in the Springer book *Global Analysis of Dynamic Models in Economics and Finance* (edited by G. I. Bischi, C. Chiarella and I. Sushko) with the title "A Little Help from my Friend" and subtitle "International Subsidy Games with Isoelastic Demands". Such title emphasizes the presence of subsidies offered by countries to the firms they are hosting in order to increase their welfare. Of course, "With a Little Help from My Friends" is a song by the Beatles from their 1967 album *Sgt. Pepper's Lonely Hearts Club Band*, written by Paul McCartney and John Lennon, and intended as the album's featured vocal for drummer Ringo Starr. The song became very famous also for a subsequent recording of the track by Joe Cocker, that in 1969 was an anthem for the Woodstock era.

But this is not the end of the story, because in September 2017 one of us (Gian Italo Bischi) received a copy of the Springer book *Optimization and Dynamics with Their Applications* (A. Matsumoto editor) that included a paper by Akio concerning a dynamic model describing love affairs between two people under different conditions in the interaction of the lovers. The paper gave conditions for the existence of steady states and considered the effects of delays in the mutual-reaction process, called the gaining-affection process. Bischi sent a playful email to Akio where he said to be a bit disappointed because after the previous papers with titles inspired to Beatles songs, for that paper on love affairs dynamics he expected the title "All you need is love".

The answer from Akio was almost immediate "Dear Gian Italo, it is a very good idea, 'All you need is love'. Do you mind if I use this title? After finishing the love dynamic paper, I almost lost interest on this issue. However, you encourage me to do it again. In order to entitle 'All you need is love' I will write one more paper on love dynamics".

Any comments are useless. As a conclusion of this short note we can state that, for sure, Akio is really very fond of the Beatles, and this thesis has been supported by himself who recently remarked, during a NED (Nonlinear Economic Dynamics) conference in Kyiv (Ukraine), that in 1968 he was just 18 years old, and that was the time when the success of the Fab Four of Liverpool was worldwide. However, he also said that he was not pretty sure if, in the Sixties and Seventies, he loved more the Beatles or the Rolling Stones. In other words, existence has been proved, but not uniqueness.

Contributions of This Volume

The papers being published in this volume are divided into three groups. Research papers in economic dynamics are in the first group, the second group contains works in game theory, while papers on related areas close the collection.

Papers on Economic Dynamics

The first paper is written by J. Barkley Rosser Jr., who is a mentor and old friend of Prof. Matsumoto. He gives an outline of four dynamic models: A coupled climate-economy model; its variant with flare attractors; a renewable resource market, and complex multi-level hierarchies with bounded rationality. Edgar J. Sanchez Carrera and coauthors introduce and study the co-evolution dynamics of human capital and innovative firms by using evolutionary game theory. They show that a policy to increase the stock of skilled labor can set the economy on a positive path towards technological development. Gian Italo Bischi and coauthors give an excellent survey of different learning schemes focusing mainly on comparing adaptive and statistical learnings. The limit sets and basins of attractions for statistical learning are characterized and the case of Bray learning is described. As examples the cases of unimodal, bimodal, and overlapping generation models with increasing maps are shown in detail. Giovanni Campisi and Fabio Tramontana study a heterogeneous agent-based financial market model and show that the presence of imitators makes the dynamics more complicated but even more realistic. Imitators may stabilize an otherwise unstable market and at the opposite, they can make unstable an otherwise stable market. The paper of Domenico De Giovanni and coauthors is based on the dynamics of compliance and optimal auditing in a population of boundedly rational agents who might decide to engage in tax evasion. The agents have several different ways to do so with different auditing probabilities. The paper studies the intertemporal optimal auditing policy of the tax authority that maximizes revenue with resource constraint. Toshio Inaba and Toichiro Asada examine a three-country Kaldorian model of business cycles where the exchange rates are fixed. Discrete time scales are selected and the authors show the cyclical fluctuations of the real national income and real capital stock of the countries. Numerical simulations verify and illustrate the theoretical findings. Ahmad Naimzada and Marina Pireddu reconsider the Muthian cobweb model where the economy is populated by unbiased fundamentalists and two types of biased fundamentalists, which are optimists and pessimists. Discrete time scales are selected and the stability of the equilibrium is studied. The section is closed by the paper of József Móczár dealing with the characterization of a large class of dynamic models. The generalized Nöther's theorem and the Lie symmetries of the Lagrangian are the theoretical basis of the investigation. Goodwin's non-linear dynamic system illustrates the Lie symmetries, and the author shows that the cyclical trajectories are extremal in the phase space.

Papers on Game Theory

Tamás László Balogh and Attila Tasnádi extend the production-in-advance version of the capacity-constrained Bertrand-Edgeworth mixed duopoly game and prove the existence of the pure-strategy (subgame-perfect) equilibrium for all possible orderings. The authors also analyze the public firm's impact on social surplus and compare the results to the production-to-order models. The famous Nikaido-Isoda theorem gives sufficient conditions for the existence of a Nash equilibrium for n-person concave games. It is also well known for n-person single-product oligopolies without product differentiation that if the firms face capacity limits, the price and cost functions are twice continuously differentiable, the price function is decreasing and concave and the cost functions are increasing and convex, then the Nash equilibrium is unique. The paper by Ferenc Forgó and Zoltán Kánnai shows that under reasonable conditions the concavity of the revenue function or the convexity of the cost functions are necessary for the uniqueness of the equilibrium. Zoltán Kánnai and coauthors give sufficient conditions for the existence of a solution of a large class of optimal control problems. Their analysis is based on the theory of set-valued mappings. The case of a generalized dynamic input-output economic model illustrates the practical application of the theoretical results. Agent based simulation is used in the paper of Ugo Merlone and coauthors to analyze the evolution of networks of interactive agents. N-person social dilemma games are selected with Pavlovian agents and all possible game types of different model parameter values are characterized and the corresponding evolution of the games are analyzed and compared.

Papers on Related Area

The paper by Maryam Hamidi and coauthors extend the earlier models on preventive maintenance policies examined by Prof. Matsumoto and his collaborators. Multi-unit systems are studied with different initial virtual ages where the possible repairable failures, maintenance and repair costs, failure and preventive replacement costs, as well as financial constraints, are included. In addition to developing the associated mathematical model, a numerical case study illustrates the practical application of the methodology. The computation of Nash-equilibria often requires the numerical solution of nonlinear equations with real numbers, vectors or even functions being the unknowns. Therefore the well-known solution techniques, like the Newton method, have to be extended to certain abstract spaces in order to cover most practical cases. The paper of I. K. Argyros and Stepan Shakhno generalizes the Newton-type equation solvers to Banach-space valued operator equations. The convergence analysis of this procedure is more general than that in earlier works making the applicability of the methodology more flexible. The final paper of this volume, by Sándor and Márk Molnár (father and son) considers a generalized form of a linear system, when the derivatives of the input functions show up in the system equations. They give sufficient conditions for the reachability of any final state, which is a nice generalization of the Kalman condition introduced in the classical case. The paper also gives a brief summary of the classical results which makes their comparison to the general case easier. The mathematical methodology is based on the theory of Lee algebras, the fundamentals of which are also discussed in the paper.

Gian Italo Bischi
gian.bischi@uniurb.it

Ferenc Szidarovszky
szidarka@gmail.com

Preface 3

Magical Mystery Tour from Linearity to Nonlinearity

I arrive at an old enough age to look back at my life and work. I have been thinking of my personal past and have a feeling that my life was so simple and straightforward that I could clearly see a straight life trajectory connecting the starting point and the current point. The following is often mentioned. All people's life careers look coherent in retrospect but are likely to have been wild and winding in prospect. You can see whether my life career looks like to be well planed as a linear line or to play it by ear as a chaotic path.

Origins and Boyhood: Now I am going to turn back the clock and talk about my own origins and evolution as a professor of mathematical economics. I was born in 1950 as the first son of a middle-class family in which my father was a mathematics high school teacher and my mother was a teacher of the Japanese language until I was born. My father had grown up on farms and his elder brother was running dairy cattle farm with a lot of cows in the neighborhood of my father's house where my family lives now. So, I had grown up in pastoral surroundings, waking up with cows' morning cry and having fresh milk every day. After spending peaceful days in elementary and junior high schools, I went to Keio high school in 1966 that was an attached school of Keio University, an A-rated historical private university in Tokyo. One good point of this school was that its students could be admitted at Keio University without an entrance examination. Passing the entrance examinations of universities, which was called examination hell, we were forced to have endurance and perseverance. In fact, hundreds of thousands of high school students targeting good schools should devote their young life for hard-studying, followed by mock exams, preparation and review from early morning to mid-night, seven days a week. By contrast, my high school days were very easy-going.

I enrolled at the faculty of engineering of Keio University in 1969 and then joined the department of administrative engineering. This department is one of the pioneering departments focusing on management engineering in Japan. Its selling point then was "enjoy studying on campus and then become a company president".

To become a company president was one of the dreams of the young people those days. However, I did not have any inclination to become a president or even a professor of economics. Around 1970, universities in Japan got rough and student movements were most furious due to domestic and international situations. Japan signed the treaty of mutual cooperation and security with the United States in 1960 in Washington. The treaty was valid for 10 years. Thus, 1970 was the year of revision. Quite a few university students, as well as ordinary people were against the reconfirmation. The Japan Communist Party and Japan Social Party standing behind the scenes, their factions occupied university buildings and halted classes. Internationally the anti-movements against the Vietnam war became more active day by day from the latter half of 1960s. Although I was an apolitical student, I was unavoidably involved in those movements like the main character, Simon, of the *Strawberry Statement* of Warner Brothers Pictures in 1970. At the same time, I was strongly affected by "flower power" that was rooted in the opposition movement of the Vietnam war and peaceful war protest, typically represented by hippies like persons in faded jeans wearing flowers in the long hair. I still have its remnant nowadays.

The Treaty was successfully renewed in 1970. Universities returned to normal and students were gradually back to classes. I encountered with my first bifurcation point of my life in a class of firm research in which we studied secrets of successful firms. I can still remember the following comments by a professor with strong impact on me, "Umami is a type of seasoning discovered in Japan which is used in cooking. It is a white powder stored in a small container with a lid having 10–20 small holes. Umami is sprinkled to add flavor while preparing the food. The company that produced Umami powder tried to enlarge the diameter of the holes of the lid to increase the amount of Umami used while cooking as a way to raise profit". This could be a rare example, but I was amazed how small-minded it was and lost interest all in one gulp on administrative or management engineering. I first turned attention to Marxian economics that was decorated with crisis in capitalism, class struggle, surplus value, labor theory of value, all of which were different from those studied in administrative engineering. Further, they were flying around in student meetings, as well as in class meetings held at the heyday of the movements, and thus I was a bit familiar with them (now I presume that the Communist and Social parties intentionally used these words to dupe naive students and I was almost trapped). Soon after, I found Marxian economics was not what I wanted and then started to study modern economics all by myself.

Graduate Schools and After: Most of my friends took off to society after graduation but I did not feel in that way. To postpone my decision, I studied economics for the time being and enrolled at the graduate school (the master program) of Yokohama City University in 1976. I had a particular interest on the general equilibrium theory, spent a lot of time to read classical monographs, *Theory of Value* of Gerard Debreu (1959) and *General Competitive Analysis* (1971) by K. Arrow and F. Hahn covering the theoretical developments of the general equilibrium theory after Debreu. It was not an easy job to unravel the tight logic of the general equilibrium theory. I wrote a master thesis on synthesizing Keynes and Walras based on the dual decision hypothesis in disequilibrium economy. To continue the doctorate program, I finally decided to study economics in earnest. Gaining a Ph.D. scholarship, I enrolled at the graduate school of Tohoku University in which I completed my doctorate degree in 1980. After graduation, I got a position in a small university in Nagono in a central area of Japan. I taught elementary economics but was not happy with that position. My position-changing journey also started with an aim returning to Tokyo. After spending four years, I moved to the faculty of political science and economics of Takushoku University in Tokyo. I was back to Tokyo, however, it was a non-stationary point something like intermittent chaos that approaches a stationary state and stays in its neighborhood for a long time and finally moves away. Therefore, after spending five years, I moved again to the west coast of Japan, the economic department of Niigata University. It was a national university and the research environment was greatly improved. Although it is located in the west coast, there was a very sharp difference from the west cost of the US with blue sky and brightening sun shine. Niigata is on the Sea of Japan and a snow country having a severe winter that I did not and do not like. In 1998, I finally touched down at the economic department of Chuo University in Tokyo. Teaching basically introductory microeconomics and macroeconomics for those years, I accidentally encountered *Irregular growth cycles*, Richard Day's AER paper, that shifted my research interest definitely to nonlinear dynamics. This was my second bifurcation point.

Richard Day: I got the Fulbright grant on research program in 1990 and chosen the University of Southern California for US affiliation because Prof. Richard Day was there. My first experience of studying abroad, however, started at the Bryn Mawr College in Pennsylvania, the other side of the US. The Japan-US educational commission gave me an opportunity not only to improve English but to obtain the common knowledge of the US for survival in everyday life before starting a life in Los Angeles, California. There were ten other Fulbright grantees. Staying at a college dormitory with full board located in a very beautiful campus outside a major metropolitan area, we studied English from morning to evening. My own first objective was how to correctly pronounce "Bryn Mawr" that includes two "r". Since Japanese language does not have corresponding sound for "r", we (Japanese) are not good at saying it correctly. On the one hand, I had real wonderful time and appreciated warm hospitality provided by people at the college. On the other, I did not have remarkable progress in English, this was because we were back to the dormitory after the classes and spoke Japanese to each other.

At the end of August of that year, I moved to Los Angeles and finally stayed there for two years until 1992 with one-year extension of the grantee. During around the 1990s, Prof. Day was standing at the world center of nonlinear economic dynamic analysis and I was so excited at being closer to him. He organized the economic dynamic seminar once a week and various professors from inside and outside of the US gave one-hour lectures on relevant topics of economic dynamics and exchanged heated arguments with the participants during the seminars. I was immensely lucky to be taught by and to discuss many topics with him. It was also my great pleasure and good memorial that my paper, *Complex dynamics in a simple macro disequilibrium model* was accepted, after a long and hard struggle, by *Journal of Economic Behavior and Organization* (*JEBO*) edited by Professor Day.

Professor Day had many faces, not only a professor of economics but also sailor, motor cyclist, and poet. For several decades, he and his wife lived on a boat anchored at Marina del Ray, North America's largest harbor. They commuted to USC with a Porche. Nevertheless, they called themselves "boat people". They once invited my family to this boat and took us sailing on Los Angeles Sea. Although the sea was not rough, I easily got seasick by the way that the boat was moving. Sailing on the sea sounded very nice but it was pretty hard in fact. All in all, this period should be golden years in my life.

In 2003 a conference honouring Richard Day was held at USC and Barkley Rosser who was the fitting successor for *JEBO* invited me. It was also a great honour for me to attend this conference and to show some results strongly related to his work on statistical dynamics. The revised version of the paper originally presented, *Density function of piecewise linear transformation*, was published in the special issue of *JEBO* in 2005. A few years ago I sent to him my paper, *Delay differential growth model* (*JEBO*, 2011) with a long-time-no-see message since the paper extended the discrete time growth model developed in his AER paper to a delay continuous time growth model and showed the birth of complicated dynamics involving chaos. I received unexpectedly heart-warming response from him and whispered to myself that I could have returned the favor.

Conferences and Friends: In 1994, an international conference on dynamical systems and chaos was held at Tokyo Metropolitan University. One of the organizers of this conference asked me to chair an economic session and to give a talk. There, I met Gustav Feichtinger for the first time. He was a head of the research group ORDYS (Operations Research and Nonlinear Dynamical Systems) at TU Wien and organized a workshop on optimal control, dynamic games, and nonlinear dynamics for every two years. After the session, he came to me and offered me to participate to his next conference, the fifth Viennese workshop on advances in nonlinear economic dynamics. I attended this workshop and made my international debut. At the same time, I fully realized that being international was not an easy job. Carl Chiarella was a discussant of my talk, but I hardly understood his comments because he spoke in very strong Australian accent. It was a small experience but very impressive. Gustav opened the door for the new world of nonlinear economics dynamics and I jumped into it. I accumulated international academic developments by attending various conferences related to nonlinear dynamics, in particular, the conference in Umea, Sweden, organized by Tönu Puu in 1995, the conference in New South Wales in Sydney, Australia, organized by Carl Chiarella in 1996, the Viennese conference again in 1997, the conference in Beer-Sheva in Israel organized by Michael Sonis in 1998 and so on. I made great friends through those conferences: Gian Italo Bischi, Volker Böhm, Herbert Dawid, Christophe Deissenberg, Peter Flashel, Cars Hommes, Steve Keen, Micheal Kopel, Hans-Walter Lorenz, Serena Sordi, to name a only few. This expansion of human network was my third bifurcation point.

Barkley Rosser: I cannot remember when I met Barkley for the first time, but we were at the same conferences and workshops many times. We were getting closer every time we met. I am sure that he was at the Micheal Sonis' conference in Israel in 1998. We have been friends for quite a long time. Soon after he became the editor of *JEBO*, he offered me to become an associate editor. I took this offer and stayed there one more year after Barkley stepped down from *JEBO*. Although time-consuming, I had pretty interesting experiences to learn what have been done after we submitted a paper to a journal. Barkley came to my university more than several times and gave a seminar each time. You all know how his seminar goes. Before the seminar, I always asked him to speak slowly, to avoid slangs, not to call out loud, etc. He nodded and smiled. Actually, the seminar started in peaceful atmospheres in which he slowly and clearly spoke plain English. After ten minutes or so, his engine quietly started and his talk was changed into motion a little but he was in first gear, everything was all right. A few minutes later, unconsciously he shifted the gear to second and then to third. The whole situations changed completely and his solo stage began while many of the participants did not really follow his talk. At some international airport, I coincidentally met students from the economic department of James Madison University and had small chats with them. I asked one of them whether he knew Prof. Barkley Rosser. His answer was yes with "he yelled out in a class". For his honour, I have to quickly add the followings. After his talk at some conference, I asked Carl Chiarella whether he could follow Barkley and he said, "no problem for natives". In 2008, Barkley organized a conference on transdisciplinary perspective on economic complexity at his university and kindly invited me. One of the participants gave a lecture in almost exactly the same way as Barkley, yes, this professor yelled out loud. Although I forgot his name, I was sure he was an American. So I found that Barkley was not alone.

Ferenc Szidarovszky: Among others, Ferenc Szidarovszky influenced me scientifically and socially more than anyone else. The name "Szidarovszky" was well-known among the students at graduate schools of economics in Japan because he coauthored enormous amounts of the papers concerning oligopoly theory with Prof. Koji Okuguchi who was a very well-known professor and his Springer book, *Expectations and Stability in Oligopoly Models* was one of musts for graduate students focusing on microeconomics and game theory. Without having any specific reasons, I thought, on my own, Prof. Szidarovszky was a supervisor of Prof. Okuguchi when he studied in a graduate school in the United States and after graduation, they worked together. I know Prof. Okuguchi only by name and Prof. Szidarovszky were out of my league, a person who lived in another world. When I met him at a small research meeting held in Odense in Denmark in 2002, I got a big surprise to learn that he was still active. As I mentioned earlier, I thought that Prof. Szidarovszky was the supervisor and should be older than Prof. Okuguchi who was old enough then. In consequence, I believed with selfish reasons that Prof. Szidarovszky was already retired (I feel very sorry for having had said this to him. Later I knew that Prof. Szidarovszky was not the supervisor and much younger than Prof. Okuguchi). I remember Gian Italo Bischi, Micheal Kopel, Shahria Yousefi, and some others came to this meeting. In daytime we shared and discussed the recent results of respected researchers while in nighttime, we went to a small restaurant where we raised a glass or two of wine (or three, maybe more) and enjoyed Danish cousin. That just reminds me that this get-together was a starter of the still-continuing NED conference. Professor Szidarovszky was a well-known mathematician and the first-rate game theorist. It was an easy guess that his brain was full of equations and theorems. However, I found in Odense that he divided his brain into two parts, one for mathematics and the other for jokes (later I got to know he was also a classic music lover, so he divided the brain into three at least). He kept presenting a wide variety of jokes ranging from wry smiles or dry laughs to dirty ones and we laughed away the hours. This was my small anecdote with Prof. Szidarovszky whom now I call Szidar. It was definitely sure that this unsuspected encounter was the fourth bifurcation point.

I got a sabbatical leave in 2006 and went to University of Arizona in Tucson where Szidar was. We worked hard together revising, from a delay dynamic viewpoint, the traditional imperfect competition models of monopoly and oligopoly and classical economic dynamic models of Kaldor, Goodwin, Hicks, etc. We published a large number of papers in economic and applied mathematical journals and two monographs from Springer, *Game Theory and its Applications* (2015) and *Dynamic Oligopolies with Time Delays* (2018), as well as editing a conference volume of the 9th International Conference on Nonlinear Economic Dynamics held in Tokyo, *Essays in Economic Dynamics* (Springer, 2016). Further, I edited a book celebrating academic achievements of Szidar, *Optimization and Dynamics with their Applications* (Springer, 2017). He is such a perfectionist about publishing. He usually called himself "picker" when he returned, with a lot of red marks, the draft of a paper I wrote. After repeating this adjustment process many times over more than a decade, I believe that a number of marks is getting a little bit less these days.

From Here: Recently I am still active in writing scientific papers. Having Power Point slides of one or two unfinished papers, I regularly attend the annual conference of the Society for Chaos Theory in Psychology and Life Science (SCTPLS) where Barkley will be the president next term, the conference on Nonlinear Economic Dynamics (NED) in odd years (its 2019 edition organized by Irina Sushko was held in Kyiv, Ukraine) and the workshop Modeli Dinamici in Economia and Finanza (MDEF) organized by Laura Gardini and Gian Italo Bischi in even years. Its 2020 edition will be held in Urbino. I cannot wait for it. After these conferences, I finalize the papers taking account of the comments and criticism obtained. The old professor neither dies nor fades away but plans to be on the Magical Mystery Tour for a few more years.

Akio Matsumoto
Chuo University
Tokyo, Japan
akiom@tamacc.chuo-u.ac.jp

Contents

Games

Related Topics

Contributors

Elvio Accinelli Facultad de Economía, Autonomous University of San Luis Potosi, San Luis Potosi, Mexico

Ioannis K. Argyros Department of Mathematics, Cameron University, Lawton, USA

Toichiro Asada Faculty of Economics, Chuo University, Tokyo, Japan

Tamás László Balogh Mathematics Connects Association, Debrecen, Hungary

J. Barkley Rosser Jr. James Madison University, Harrisonburg, USA

Gian Italo Bischi Department of Economics, Society, Politics (DESP), University of Urbino, Urbino, Italy

Giovanni Campisi Department of Economics Marco Biagi, University of Modena and Reggio Emilia, Modena, Italy

Domenico De Giovanni Department of Economics, Statistics and Finance, University of Calabria, Rende, Italy

Ferenc Forgó Department of Operations Research and Actuarial Sciences, Corvinus University of Budapest, Budapest, Hungary

Armando García Facultad de Economía, Autonomous University of San Luis Potosi, San Luis Potosi, Mexico

Laura Gardini Department of Economics, Society, Politics (DESP), University of Urbino, Urbino, Italy

Maryam Hamidi Department of Industrial Engineering, Lamar University, Beaumont, TX, USA

Toshio Inaba School of Education, Waseda University, Tokyo, Japan

Zoltán Kánnai Department of Mathematics, Corvinus University of Budapest, Budapest, Hungary

Fabio Lamantia Department of Economics, Statistics and Finance, University of Calabria, Rende, Italy;
School of Social Sciences, University of Manchester, Manchester, UK

Reza Maihami School of Business and Leadership, Our Lady of the Lake University, Houston, TX, USA

Ugo Merlone Department of Psychology, Center for Logic, Language, and Cognition, University of Torino, Torino, Italy

József Móczár Department of Mathematical Economics and Economic Analysis, Corvinus University of Budapest, Budapest, Hungary

Mark Molnar Department of Macroeconomics, Institute of Economics, Szent István University, Gödöllő, Hungary

Sandor Molnar Institute of Mechanics and Machinery, Szent István University, Gödöllő, Hungary

Ahmad Naimzada Department of Economics, Management and Statistics, University of Milano-Bicocca, Milano, Italy

Mario Pezzino School of Social Sciences, University of Manchester, Manchester, UK

Marina Pireddu Department of Mathematics and its Applications, University of Milano-Bicocca, Milan, Italy

Laura Policardo Agenzia delle Dogane e dei Monopoli, Sesto Fiorentino, Italy

Behnam Rahimikelarijani Department of Industrial Engineering, Lamar University, Beaumont, TX, USA

Edgar J. Sanchez Carrera Department of Economics, Society and Politics, University of Urbino Carlo Bo, Urbino, Italy;
Research Fellow at CIMA UAdeC, Saltillo Coahuila, Mexico

Daren R. Sandbank Systems and Industrial Engineering Department, University of Arizona, Tucson, AZ, USA

Stepan Shakhno Department of Theory of Optimal Processes, Ivan Franko National University of Lviv, Lviv, Ukraine

Imre Szabó Department of Mathematics, Corvinus University of Budapest, Budapest, Hungary

Ferenc Szidarovszky Department of Mathematics, Corvinus University of Budapest, Budapest, Hungary

Peter Tallos Department of Mathematics, Corvinus University of Budapest, Budapest, Hungary

Attila Tasnádi Department of Mathematics, Corvinus University of Budapest, Budapest, Hungary

Fabio Tramontana Department of Mathematical Disciplines, Mathematical Finance and Econometrics, Catholic University of Milan, Milan, Italy

Dynamics

Coupled Chaotic Systems and Extreme Ecologic-Economic Outcomes

J. Barkley Rosser Jr.

Abstract In sympathy with work of Akio Matsumoto, this essay reviews models that consider how the coupling of systems within ecologic-economic contexts can generate not only chaotic dynamics, but lead to outcomes that exhibit kurototic outcomes rather than reflecting Gaussian distributions. This aligns with arguments made by Martin Weitzmann regarding the global climate system. The models considered included one where climate and economic systems are separately non-chaotic but chaotic when combined and another where the economic system is chaotic and when combined with climate generates kurtotic outcomes through flare attractors. Likewise, similarly coupled models involving fisheries and forestry dynamics are considered where coupling leads to chaotic dynamics. Multi-level systems with such dynamics are then considered with the governance issues involved with such systems are examined.

1 Introduction

Akio Matsumoto has long studied coupled dynamical systems exhibiting various forms of complex dynamics, often involving lags (Matsumoto 1997, 1999; Matsumoto and Szidarovszky 2015). In addition, he has had an interest in implications of such models connecting economics with environmental problems (Matsumoto et al. 2018; Ishikawa et al. 2019). A theme of his work on these topics has indeed been that both coupling and lags tend to increase the complexities arising from such systems. This might appear to run counter to another theme of his work, that sometimes chaotic dynamics "can be beneficial" (Matsumoto 2001, 2003). However, those models involved one-dimensional systems of price dynamics without coupling or lags or other complications that could undermine their relatively sunny outcomes. Nevertheless, this insight of Matsumoto's that chaotic dynamics are not necessarily "bad" has not been fully appreciated.

J. Barkley Rosser Jr. (✉)
James Madison University, Harrisonburg, USA
e-mail: rosserjb@jmu.edu

© Springer Nature Singapore Pte Ltd. 2020
F. Szidarovszky and G. I. Bischi (eds.), *Games and Dynamics in Economics*,
https://doi.org/10.1007/978-981-15-3623-6_1

In appreciation of these themes of Matsumoto's we shall consider how coupled ecologic and economic chaotic systems can generate extreme events, kurtotic "fat tails." While there are various such possible applications, including to fisheries and forests, arguably the most important involves global warming, a more accurate term that this observer prefers to the more anodyne and widely used "climate change." While most of the models underlying official IPCC reports have assumed Gaussian distributions of outcomes, Martin Weitzman (2009, 2011, 2012, 2014) has argued that underlying nonlinear dynamics of the global climate system in interaction with the global economic system is subject to power law or other distributions that exhibit kurtosis and thus a higher probability of extreme outcomes than appearing in the more conventional models. Indeed, Lorenz (1963) first identified a strange attractor associated with sensitive dependence on initial conditions in a chaotic model of climate dynamics. It is thus completely appropriate to consider how such models can bring about these outcomes that Weitzman considered to be so important.

This raises the question of how policy should be carried out in the face of such phenomena, especially as this happens in the context of complications such as the hierarchical complexity of ecologic-economic systems and the bounded rationality of policy makers (Rosser and Rosser 2006, 2015). Such analysis is deeply in synch with the spirit and tradition of the work of Akio Matsumoto.

2   A Coupled Climate-Economy Model

As already noted, Lorenz (1963) modeled climate dynamics as being chaotic, although that term was not yet in use at that time. However, the chaotic nature of climate dynamics is widely accepted, with the "butterfly effect" of sensitive dependence on initial conditions being widely viewed as a reason why weather forecasting has only a fairly short range of reliability, even though longer term averages and trends may be forecasted.

While many theoretical models of chaotic economic dynamics have been proposed (Rosser 2011, Appendix A), solid empirical verification of such dynamics in economic systems has been lacking, although a variety of complex nonlinear dynamics have been accepted as happening in economic systems. However, as studied in Rosser (2002) two systems that by themselves may not exhibit chaotic dynamics can do so when coupled together. This draws on work of Chen (1997), which draws on simple underlying sub-systems.

This simple system has two sectors in its economic part, agricultural and manufacturing. These sectors are each related to global average temperature, T. For agriculture, temperature is a negative input. For manufacturing, it is a positive input to global average temperature. Each sub-system is very simple, but the coupled system can show chaotic dynamics.

On the economic side demand is given by a CES utility function of agriculture, A, and manufacturing, M.

$$U(A, M) = (A^\rho + M^\rho)^{1/\rho}. \tag{1}$$

We are assuming equilibrium on the economic side so that consumption of each good equals its output. This gives the elasticity of substitution as is standard for CES functions to be

$$\sigma = 1/(1-\rho) < 1. \tag{2}$$

Both production functions are linear in labor, L, with total labor normalized to unity, so that

$$L(A) + L(M) = 1. \tag{3}$$

Besides a positive constant and the labor input, agricultural production also includes a negative quadratic term for global average temperature, so that

$$A = (-\alpha T^2 + \beta T + 1)L(A). \tag{4}$$

Manufacturing output is given by

$$M = bL(M). \tag{5}$$

This generates a market clearing manufacturing price of

$$P = (-\alpha T^2 + \beta T + 1)/b. \tag{6}$$

The climate model draws on one due to Henderson-Sellers and McGuffie (1987). This now involves dynamics with time subscripts as temperature in a succeeding time period that is determined by the temperature in the current one along with a long-run normal temperature, T_n, as well as a positive linear function of manufacturing output. With c in the unit interval and $g > 0$, this is given by

$$T_{t+1} = (1-c)(T_t - T_n) + T_n + gM_t. \tag{7}$$

Combining with the economic sub-system generates an equilibrium motion for global temperature that is given by

$$T_{t+1} = (1-c)T_t + g(bp_t^{1-\sigma})/(1 + p_t^{1-\sigma}). \tag{8}$$

Chen simulated this model setting $\sigma = 0.5$, $\alpha = 8$, $\beta = 7$, $b = 1$, and $g = 0.6$. The climatic tuning parameter, c, for this set of other parameter values, generates a unique and stable steady state for values in $(0.233, 1)$. As c declines below 0.233, period-doubling bifurcations appear, and aperiodic chaotic dynamics appear after it goes below $c = 0.209$. The system also exhibits sensitive dependence on initial conditions ("butterfly effect") below this level as well.

3   Flare Attractors and Extreme Ecologic-Economic Outcomes

A related model that can bring about an outcome of a combined ecologic-economic system with a chaotic driver, if differing in important details from the model in the preceding section, involves *flare attractors*. These are key to the not-fully developed *econochemistry* concept. Initially conceived by Otto Rössler and Georg Hartmannn (1995) to study solar flares and various autocatalytic chemical reactions, they came to be applied to economics as well, initially for entrepreneurial activities (Hartmann and Rössler 1998) and then for asset price volatility (Rosser et al. 2003).

This approach differs from that in the previous section by having the underlying fundamental process being chaotic rather than becoming chaotic as a result of the coupling aspect. In the case of this model the "flaring" kurtotic outcomes, sudden bursts coming almost from nowhere, are the result of the coupled second layer deriving from the underlying driving chaotic process. This also involves an introduction of heterogeneous agents into the system. Ironically as one moves from the original model of solar flares to the model of climatic outbursts of extreme temperatures, we see a return to an original physical chemistry application after passing through an economics application that explored financial market dynamics.

The underlying mathematics of this model were developed by Rössler et al. (1995). The attractors involved are extensions of the continuous chaotic attractor model of Rössler (1976) as special cases that are continuous-but-nowhere-differentiable and also exhibit "riddled basins." The full explication of such attractors is due to Milnor (1985).

Here we shall extend this model to an application not previously made, to the problem of global warming, or more generally, extreme outcomes of climate change. The previous model due to Chen (1997), had the ecologic-economic interactions more direct, which arguably reflects a longer run perspective. Here we shall focus more on a shorter-term perspective of economic-to-climate interactions. The coupling aspect involves the second-tier aspect of heterogeneous agents responding to the underlying economic model already assumes an environmental limit on economic growth. This limit is not connected to the higher level global warming issue, but a narrower limit more locally determined. The model is one of the earliest chaotic economic models due to Day (1982). His model involves a logistic equation, which relates to the original model of chaotic application. This was due to May (1976). Such a model depends on a hard upper limit of growth along with a lower bound.

The underlying economic model, due to Day (1982) is a modified Solow growth model. It has the labor exponent as α, and β the capital exponent, y being per capita output, λ being the population growth rate, with m being a "capital-congestion" saturation coefficient, which ultimately drives the logistic formulation that has an upper limit, and which resembles the model of May (1976). The modified production function is given by

$$f(k) = \beta k^{\beta}(m-k)^{y}. \tag{9}$$

Assuming a constant savings rate, the capital-labor ratio implies the following difference growth equation,

$$k_{t+1} = \alpha\beta k_t^{\beta}(m-k_t)^y/(1+\lambda). \tag{10}$$

This formulation coincides with that of May (1976), who made clear the parameter values of this model for which chaotic dynamics will occur. Rosser et al. (2003, p. 80) assumed that

$$A\beta/(1+\lambda) = 3.99 = k_{t+1}/(1-k_t). \tag{11}$$

This formulation provides a chaotic dynamical process as k changes. This process assumes that the capital share remains constant.

Earlier literature has posited at this point that the specification of heterogeneous agents involved human agents responding differently to the underlying system. For this case we follow Hartmann and Rössler (1998) for giving a general form of the agent reaction function. The difference between this formulation and earlier work by these authors in physical chemistry is that while here the agents are heterogeneous individuals or organizations, in this case implicitly the agents are nations or regions of the world subject to climatic variation.

What goes on here is that we have a set of locations that have a varying relation with the exogenous chaotic driving force. In particular there will be a switching value of a, a function of k, beyond which there will be a substantial increase in temperature. Whereas in the asset model of Rosser et al. (2003) these agent reaction functions represent behavior of human agents, including human organizations, in this case these represent locations on the planet with their respective situations that imply heterogeneous behavior. The appearance of an outburst reflects a sufficient number of these agents/locations crossing their critical value of $1 > a > 0$.

The general form of the reaction function for an agent/location of I type out of n, assuming agents/locations, and $c > 0$, and will be given by

$$B_{t+1}^I = b_t^I + b_t^I(a^I - k_t^I) - cb_t^{(I)2} + cs_t. \tag{12}$$

The first term in (12) is an autoregressive component. The second is the switching term. The third provides a stabilizing component. The fourth is the destabilizing element coming from the buildup of previous trends, representing the ongoing overall state of the system determined by overall demand s, and is given by

$$S_{t+1} = b_t^1 + b_t^2 + \cdots + b_t^n. \tag{13}$$

In Rosser et al. (2003), assuming certain values of the parameters allows for a simulation that provides a sequence of outcomes that exhibit scattered kurtotic outbursts consistent with the Weitzman scenario for global warming.

4   Coupled Chaotic Dynamics in Renewable Resource Markets

While we have seen that coupled chaotic dynamics can happen at the global level scale of the climate-economy system, such coupled chaotic dynamics can also happen in lower level ecologic-economic systems. Two examples are in fisheries and also in forestry, although the former have been modeled more clearly, with Conklin and Kohlberg (1994) initially showing the possibility for chaotic dynamics within a fishery in a non-optimizing setting. Central to such dynamics in these systems is when supply curves bend backwards, a result first suggested for fisheries without a formal model by Copes (1970). More complex dynamics for fisheries than those presented below are presented in Foroni et al. (2003).

Hommes and Rosser (2001) have demonstrated the possibility of this for fisheries in what they label a "Gordon-Schaefer-Clark" model of an optimally managed fishery. This assumes on the ecological side a Schaefer (1957) yield function, $f(x)$, with x the fish biomass, which in equilibrium will also be the harvest function, $h(x)$, with r the natural growth rate of the fish and k the carrying capacity of the fishery is given by

$$h(x) = f(x) = rx(1-x/k). \tag{14}$$

Following Clark (1985) the economic side is given by an effort function linear in time fishing, E, with costs $C(E)$, without fixed costs, constant marginal costs, c, and a catchability coefficient, q, with p the price of fish, and R the rent, output Y is given by

$$Y = qEx = h(x). \tag{15}$$

This implies that rent which the present value of which is to be maximized is

$$R(Y) = pqEx - cE. \tag{16}$$

In the optimization non-equilibrium must be allowed where harvest may not equal the yield function. Solving the intertemporal optimal control problem with non-negativity constraints on x and h and a constant discount rate, δ, (Hommes and Rosser 2001) leads to

$$f(x) = \delta = [cf(x)]/(p-c). \tag{19}$$

From this optimal discounted supply curve is given by

$$x(p, \delta) = k/4[1 + (c/pqk) - (\delta/r) + (1 + c/pqk - (\delta/r)^2 + 8c\delta/pqkr)^{1/2}]. \tag{20}$$

The crucial variable determining system dynamics is the discount rate, δ. At zero with no discounting of future rents, the supply curve slopes upwards, but as it increases beyond about 0.02, the supply curve bends backwards, allowing for catastrophic collapses of the fishery. As it goes to infinity implying not counting the future at all, the curve bends backwards the most and becomes identical to that for an open access fishery subject to "tragedy" (Gordon 1954), the problem dealt with by Ostrom (1990) and others. The open access supply curve is given by

$$x(p, \infty) = (rc/pq)(1-c/pqk). \tag{21}$$

Assuming lags in behavior by fishers turns this into a form of a cobweb model. Somewhat similarly to the finding of Matsumoto (1997), Hommes and Rosser (2001) show for an appropriate demand curve and for intermediate values of the discount rate, chaotic dynamics can emerge in this coupled fishery system.

While no one has shown specifically chaotic dynamics in a forest-harvesting model, under certain situations an optimally managed forest can also exhibit backward-bending supply curves for sufficiently high discount rates. This was first proposed by Hyde (1980) with empirical support for backward-bending forestry supply curves found in the Amazon rain forest for certain circumstances (Amacher et al. 2009). Drawing on Colin Clark's fishery model, Binkley (1986) developed a model that formally showed how such a backward-bending supply curve could arise in a forestry model, with this further studied by Rosser (2013). These models are all for a single output, timber from cut trees, with Binkley finding tentative empirical support for the long run supply of loblolly pines in the southeastern US. The basic canonical optimal forestry management model accounting for multiple uses and infinite time horizon is given by Hartman (1976).

Letting most variables be identical to the above fishery model, the main new variable that appears in the system is T, the optimal rotation age for the forest, the time that trees should be cut and then replanting of them occurs. This T depends on the discount rate and also p, the price of timber, and unlike the fishery, the yield function is a function of time since the last replanting, $f(t)$, with the growth at optimal rotation age given by $f(T(p))$. From all this an optimal inverse supply function for p as a function of T and δ is given by

$$p = c/[f(T) - f'(t)(1 - e^{-\delta t})/\delta)]. \tag{22}$$

This is consistent with the possibility of a backward-bending supply curve for certain parameter values. Binkley (1986, p. 173) provides an intuitive explanation of what is happening in such situations.

High stumpage prices imply not only that the output from the forest has high value, but also that the capital in the form of growing stock has a high opportunity cost. At high prices, it is optimal to conserve on the use of capital and therefore to reduce the stock inventory by reducing the rotation age.

While it has not been shown explicitly that this model can generate chaotic dynamics, I am reasonably certain that with appropriate lags for forester behavior, such will occur for certain situations. I close this discussion by observing that chaotic dynamics have been found for a variety of both harvested biological populations (Sakai 2001) as well as non-harvested ones (Zimmer 1999; Turchin 2003; Solé and Bascompte 2006).

5   Policy in Complex Multi-level Hierarchies with Bounded Rationality

The difficulty of managing such dynamically complex coupled systems is complicated when they exist within hierarchical ecologic-economic contexts (Radner 1992; Rosser 1995, 2001). This complexity enforces the necessary reliance on bounded rationality as posed by Simon (1957, 1962; Rosser and Rosser 2015). It also involves positing the appropriate level of the system as the locus of such policymaking in order to overcome the difficulties of common property resources that arise in such situations (Netting 1976; Ostrom 1990; Bromley 1991; Rosser and Rosser 2006; Rosser 2016).

While Simon (1962) formalized the discussion of hierarchy in complex systems, his arguments for dynamical systems were prefigured in general systems theory (von Bertalanffy 1962) and its predecessor, tektology (Bogdanov 1925–1929). These have been more fully generalized for ecological systems by Holling (1992). A deep issue is the relation between higher and lower levels of such systems. While it is generally argued that higher levels dominate or at least constrain lower levels (Radner 1992), it may be possible for changes in lower levels to lead to changes in higher levels, or even the complex emergence of higher levels through hypercyclic morphogenesis (Rosser 1991).

We can consider such systems that allow for ultimately flexible relations with both fast and slow dynamics in the formalization of synergetics as developed by Haken (1977). Let there be a well-defined hierarchy with n levels. Higher levels constrain more rapidly oscillating lower levels under normal conditions. Thus fast dynamics operate at lower levels and slow dynamics operate at higher levels.

At a given level let q be the vector of fast variable dynamics and F the vector of slow variable dynamics, with A, B, and C being matrices and $\varepsilon(t)$ be an i.i.d. stochastic fluctuations term. Then the fast dynamics are given by

$$dq/dt = Aq + B(F)q + C(F) + \varepsilon(t). \tag{23}$$

Haken argues that such a system can be simplified by rearranging this equation in order to exhibit adiabatic approximation in which the fast dynamics are shown to depend solely upon the slow dynamics based on order parameters in F. This is given by

$$dq/dt = -(A + B(F))^{-1}C(F). \tag{24}$$

The order parameters are the variables in F and can be ranked in inverse order of the absolute values of the variables in $A + B(F)$. Curiously these order parameters are unstable in the sense that they possess positive real parts of their eigenvalues. Other variables are the slaved variables and have negative real parts of their eigenvalues.

Structural changes in the sense of Holling can come from either the bottom or the top in such a system. Bottom-up changes can come about through a slaved variable destabilizing by having the real part of its eigenvalue going positive in a process known as "the revolt of the slaved variables" (Diener and Poston 1984). Haken saw this as a key to the emergence of chaotic dynamics in a structured system. An example might be the outbreak of the Great Plague in Europe in the mid-14th century as an accumulation of malnutrition weakening population immune systems reached a critical mass such that the plague could sweep through the population (Braudel 1967).

The top-down mechanism can happen through the emergence of a new constraining higher level of the system, such as the emergence of a city in an urban hierarchy much larger than previous ones that dominates them through the appearance of new economic activities (Rosser 1994). The mechanisms for such anagenetic moments of hypercyclic morphogenesis can arise from frequency entrainment as modeled by Nicolis (1986). Another way may be through the appearance of cooperative forms leading to multi-level evolution in an evolutionary process (Crow 1955).

The policy problem must confront this hierarchical complexity. This is an issue that Ostrom (1990) and others have tried to confront. A clear outcome is that governance should operate at the most crucial level that determines the crucial dynamics of the system. In light of the analysis above of synergetic systems, it may not always be obvious what that level is (Wilson et al. 1999). A system apparently dominated by the highest level may actually be dependent on dynamics at the bottom and vice versa. More generally, focusing policy on an ecologic-economic hierarchy level that is not crucial to the system dynamics can lead to worse outcomes than doing nothing.

Indeed, for the most difficult problems the complex links mean that actions may need to be taken at several levels. This would seem to be especially the case for the global climate system, where in fact given the coupled nonlinear dynamics involved it would seem that multiple levels are involved. Global agreements are necessary for setting overall goals. But individual nations must set goals and establish specific policies. But many of these policies end up being carried out at lower levels. Likewise it is not just the political and economic elements that have this multi-level aspect, but also the ecological and climatological. The ecologic-economic system functions at levels ranging from almost minutely local to the totally global.

A further complication due to the complexities associated especially with chaotic dynamics is that when a system is decomposed from the global to the regional level, it may be subject to severe effects due to sensitive dependence on initial conditions. Thus Massetti and Di Lorenzo (2019) have considered in detail the regional level forecasts from simulations of global level climate models used by the United Nations IPCC for projecting possible future climate outcomes. In particular they ran simulations slightly varying initial starting values for certain variables and indeed found substantial sensitive dependence for regional level projections. Thus for the west-central portion of the United States some projections would have substantial warming while others actually found cooling happening even as the global average was for warming, again for starting values only slightly apart, thus replicating the old result found by Lorenz (1963) for climate models. Needless to say, this seriously complicates knowing what to do at more local levels for such situations.

These multi-layered complexities involve deep uncertainties about all the matters noted above and more. These include ongoing debates about underlying science issues, as well as the full nature of the interactions between the economic and climatological aspects. That the elements of this involve chaotic dynamics subject to sensitive dependence on initial conditions makes the whole matter that much more difficult to understand. All this leads to the inability of any observer or agent to reliably understand in full detail how it works. This means that inevitably bounded rationality is the best that can be hoped for to be used in analyzing such a system.

6   Conclusions

The coupled global ecologic-economic system deeply involves chaotic dynamics. This means the system is subject to sensitive dependence on initial conditions. Also it may be subject to flare phenomena. These involve kurtotic outbursts that increase the dangers involved in understanding the system and increase the risks involved in the analysis. These issues extend to other kinds of coupled dynamical ecologic-economic systems such as those involving fisheries and forests. As a multi-layered complex system, where management must apply at the appropriate level, decision makers are limited to bounded rationality in dealing with it. The ideas involved in these matters are deeply linked to ideas that Akio Matsumoto has studied in his lifetime of research.

References

Amacher, G. S., Merry, F. D., & Bowman, M. S. (2009). Smallholder timber sale decisions on the Amazon frontier. *Ecological Economics, 68,* 1787–1796.
von Bertalanffy, L. (1962). *General systems theory.* New York: George Braziller.
Binkley, C. S. (1986). Long-run timber supply, price elasticity, inventory elasticity, and the use of capital in timber production. *Natural Resource Modeling, 7,* 163–181.
Bogdanov, A. A. (1925–1929). *Tektologia: Vseobschaya Organizatsionnaya Nauka,* 3 vols. Berlin: Z.I. Grschbein. (English translation, G. Gorelik (1980) *Essays in Tektology: The general science of organizations.* Seaside: Intersystem Publications).
Braudel, F. (1967). *Civilization Matérielle et Capitalisme.* Paris: Librairie Armand Colin. (English translation: K. Miriam. (1973) *Capitalism and material life: 1400–1800.* New York: Harper and Row).

Bromley, D. W. (1991). *Environment and economy: Property rights and public policy*. Oxford: Basil Blackwell.

Chen, Z. (1997). Can economic activity lead to climate chaos? *Canadian Journal of Economics, 30*, 349–366.

Clark, C. W. (1985). *Bioeconomic modelling and fisheries management*. New York: Wiley Interscience.

Copes, P. (1970). The backward-bending supply curve of the fishing industry. *Scottish Journal of Political Economy, 17*, 69–77.

Conklin, J. E., & Kohlberg, W. C. (1994). Chaos for the halibut? *Marine Resource Economics, 9*, 153–182.

Crow, J. F. (1955). General theory of population genetics: Synthesis. *Cold Spring Harbor Symposia on Quantitative Biology, 20*, 54–59.

Day, R. H. (1982). Irregular growth cycles. *American Economic Review, 72*, 406–414.

Diener, M., & Poston, T. (1984). The perfect delay convention, or the revolt of the slaved variables. In H. Haken (Ed.), *Chaos and order in nature* (2nd ed., pp. 249–268). Berlin: Springer.

Foroni, I., Gardini, L., & Rosser, J. B., Jr. (2003). Adaptive statistical expectations in a renewable resource market. *Mathematics and Computers in Simulation, 63*, 541–567.

Gordon, H. S. (1954). Economic theory of a common-property resource: The fishery. *Journal of Political Economy, 62*, 124–142.

Haken, H. (1977). *"Synergetics", nonequilibrium phase transitions and social measurement*. Berlin: Springer.

Hartman, R. (1976). The harvesting decision when a standing forest has value. *Economic Inquiry, 14*, 52–58.

Hartmann, G. C., & Rössler, O. E. (1998). Coupled flare attractors—A discrete prototype for economic modelling. *Discrete Dynamics in Nature and Society, 2*, 153–159.

Henderson-Sellers, A., & McGuffie, K. (1987). *A climate modeling primer*. New York: Wiley.

Holling, C. S. (1992). Cross-scale morphology, geometry, and dynamics of ecosystems. *Ecological Monographs, 62*, 447–502.

Hommes, C. H., & Rosser, J. B., Jr. (2001). Consistent expectations equilibria and complex dynamics in renewable resource markets. *Macroeconomic Dynamics, 5*, 180–203.

Hyde, W. F. (1980). *Timber supply, land allocation, and economic efficiency*. Baltimore: Johns Hopkins University Press.

Ishikawa, T., Matsumoto, A., & Szidarovszky, F. (2019). Regulation of non-point source pollution under n-firm Bertrand competition. *Environmental Economics and Policy Studies, 21*(4). https://doi.org/10.1007/1008-09-0243-9.

Lorenz, E. N. (1963). Deterministic non-periodic flow. *Journal of Atmospheric Science, 20*, 130–141.

Massetti, E., & Di Lorenzo, E. (2019). Chaos in climate change impacts estimates. Working Paper, Georgia Institute of Technology.

Matsumoto, A. (1997). Ergodic cobweb chaos. *Discrete Dynamics in Nature and Society, 1*, 135–146.

Matsumoto, A. (1999). Preferable disequilibrium behavior in a nonlinear cobweb model. *Annals of Operations Research, 89*, 101–123.

Matsumoto, A. (2001). Can inventory chaos be welfare improving? *International Journal of Production Economics, 71*, 31–43.

Matssumoto, A. (2003). Let it be: Chaotic price instability can be beneficial. *Chaos, Solitons & Fractals, 18*, 745–758.

Matsumoto, A., & Szidarovszky, F. (2015). The asymptotic behavior in a nonlinear cobweb model with time delays. *Discrete Dynamics in Nature and Society, 15*, 1–14.

Matsumoto, A., Szidarovskzy, F., & Yabuta, M. (2018). Environmental effects of ambient change in Cournot oligopoly. *Journal of Environmental Economics and Policy, 7*(1). https://doi.org/10.1080/21606544,2017.1347527.

May, R. M. (1976). Simple mathematical models with very complicated dynamics. *Nature, 261,* 471–477.

Milnor, J. (1985). On the concept of an attractor. *Communications in Mathematical Physics, 102,* 517–519.

Netting, R. (1976). What alpine peasants have in common: Observations on communal tenure in a Swiss village. *Human Ecology, 4,* 134–146.

Nicolis, J. S. (1986). *Dynamics of hierarchical systems: An evolutionary approach.* Berlin: Springer.

Ostrom, E. (1990). *Governing the commons.* Cambridge, UK: Cambridge University Press.

Radner, R. S. (1992). Hierarchy: The economics of managing. *Journal of Economic Literature, 30,* 1382–1415.

Rosser, J. B., Jr. (1991). *From catastrophe to chaos: A general theory of economic discontinuities.* Boston: Kluwer.

Rosser, J. B., Jr. (1994). Dynamics of emergent urban hierarchy. *Chaos, Solitons & Fractals, 4,* 553–562.

Rosser, J. B., Jr. (1995). Systemic crises in hierarchical ecological economies. *Land Economics, 71,* 163–172.

Rosser, J. B., Jr. (2001). Complex ecologic-economic dynamics and environmental policy. *Ecological Economics, 37,* 23–37.

Rosser, J. B., Jr. (2002). Complex coupled system dynamics and the global warming policy problem. *Discrete Dynamics in Nature and Society, 7,* 93–100.

Rosser, J. B., Jr. (2011). *Complex evolutionary dynamics in urban-regional and ecologic-economic systems: From catastrophe to chaos and beyond.* New York: Springer.

Rosser, J. B., Jr. (2013). Special problems of forests as ecologic-economics systems. *Forest Policy and Economics, 35,* 31–38.

Rosser, J. B., Jr. (2016). Governance issues in complex ecologic-economic systems. *Review of Behavioral Economics, 3,* 335–357.

Rosser, J. B., Jr., Ahmed, E., & Hartmann, G. C. (2003). Volatility via social flaring. *Journal of Economic Behavior & Organization, 50,* 77–87.

Rosser, J. B., Jr., & Rosser, M. V. (2006). Institutional evolution of environmental management under economic growth. *Journal of Economic Issues, 40,* 421–429.

Rosser, J. B., Jr., & Rosser, M. V. (2015). Complexity and behavioral economics. *Nonlinear Dynamics, Psychology, and Life Sciences, 19,* 201–226.

Rössler, O. E. (1976). An equation for continuous chaos. *Physics Letters A, 57,* 397–398.

Rössler, O. E., & Hartman, G. (1995). Attractors with flares. *Fractals, 3,* 285–286.

Rössler, O. E., Knudsen, C., Hudson, J. L., & Tsuda, I. (1995). Nowhere differentiable attractors. *International Journal of Intelligent Systems, 10,* 15–23.

Sakai, K. (2001). *Nonlinear dynamics in agricultural systems.* Amsterdam: Elsevier.

Schaefer, M. B. (1957). Some considerations of population dynamics and economics in relation to the management of marine fisheries. *Journal of the Fisheries Research Board of Canada, 14,* 669–681.

Simon, H. A. (1957). *Models of man.* New York: Wiley.

Simon, H. A. (1962). The architecture of complexity. *Proceedings of the American Philosophical Society, 106,* 467–482.

Solé, R. V., & Bascompte, J. (2006). *Self-organization in complex ecosystems.* Princeton: Princeton University Press.

Turchin, P. (2003). *Complex population dynamics: A theoretical/empirical synthesis.* Princeton: Princeton University Press.

Weitzman, M. L. (2009). On modeling and interpreting the economics of catastrophic climate change. *Review of Economics and Statistics, 91,* 1–19.

Weitzman, M. L. (2011). Fat-tailed uncertainty in the economics of catastrophic climate change. *Review of Environmental Economics and Policy, 5,* 275–292.

Weitzman, M. L. (2012). GHG targets as insurance against catastrophic climate change. *Journal of Public Economic Theory, 14,* 221–244.

Weitzman, M. L. (2014). Fat tails and the social cost of carbon. *American Economic Review: Papers and Proceedings, 104,* 544–546.

Wilson, J., Low, B., Costanza, R., & Ostrom, E. (1999). Scale misperception and the spatial dynamics of a social-ecological system. *Ecological Economics, 31,* 243–257.

Zimmer, C. (1999). Life after chaos. *Science, 284,* 83–86.

A Co-evolutionary Model for Human Capital and Innovative Firms

Edgar J. Sanchez Carrera, Laura Policardo, Armando García
and Elvio Accinelli

Abstract The paper aims to study the co-evolution dynamics of human capital and innovative firms by means of an evolutionary game theory model. We analyze the properties of the model, showing that if the demand for skilled labor is higher than its supply, then innovative firms may have an incentive to become non-innovative and stop hiring skilled workers. If, by contrast, the supply of skilled labor is higher than its demand, then there could be incentives for non-innovative firms to become innovative. Then, we introduce the dynamic extension of the model, applying a replicator dynamics equation for the fraction of innovative firms and the fraction of skilled workers. The steady states of the system are identified and as the most interesting one, the interior steady state, is discussed. Subsequently some simplified versions of the model are proposed and studied. By means of such analysis, we claim that a policy oriented to increasing the stock of skilled labor can set the economy on a positive path towards technological development.

Keywords Behavioral macroeconomics · Economic growth · Evolutionary dynamics · Innovative firms · Skilled labor

JEL Codes: C72 · C73 · O11 · O55 · K42

Laura Policardo: Opinions expressed in this publication are those of the authors and do not necessarily reflect the official opinion of the Italian Agenzia delle Dogane e dei Monopoli.

E. J. Sanchez Carrera (✉)
Department of Economics, Society and Politics, University of Urbino Carlo Bo, Urbino, Italy
e-mail: edgar.sanchezcarrera@uniurb.it

Research Fellow at CIMA UAdeC, Saltillo Coahuila, Mexico

L. Policardo
Agenzia delle Dogane e dei Monopoli, via Santa Croce dell'Osmannoro 24, Sesto Fiorentino, Italy
e-mail: laura.policardo@gmail.com

A. García · E. Accinelli
Facultad de Economía, Autonomous University of San Luis Potosi, San Luis Potosi, Mexico
e-mail: gama.slp@gmail.com

E. Accinelli
e-mail: elvio.accinelli@eco.uaslp.mx

© Springer Nature Singapore Pte Ltd. 2020
F. Szidarovszky and G. I. Bischi (eds.), *Games and Dynamics in Economics*,
https://doi.org/10.1007/978-981-15-3623-6_2

1 Introduction

The effects of advanced technological firms on the labor market has been deeply studied in economics (see, for example, Guerrini et al. 2019; Matsumoto and Szidarovszky 2020). Seminal papers by Griliches (1969) and Welch (1970), pointed out that the implementation of new technologies in an effective and efficient way, requires adequate skills on the part of workers. A direct consequence of this hypothesis is that an insufficient number of skilled workers is a limitation to the adoption and diffusion of new technologies. In the literature this restriction is known as 'restriction of human resources', which has been widely discusssed discussed by Amendola and Gaffard (1988), among others.

Due to the complementarity between skilled workers and firms' investments in R&D, and any additional cost spent on innovation, the demand for skilled workers increases, and with it, the differential between wages paid to skilled workers compared to respect to the unskilled workers. When resources spent on innovation by firms become substantial, a virtuous cycle of economic performance characterized by high wages and high firms' productivity, and therefore growth, emerges (see Krusell et al. 2000).[1] Seminal papers have developed models to demonstrate that skilled workers and innovative firms complement each other, giving rise to the conformation of a high level equilibrium or economic growth, where new technologies reduce the demand for unskilled workers and increase the demand for those who are qualified, while their training allows them to adapt more easily to technological changes (see Acemoglu 1997, 1998, 2002; Aghion 2006; Hornstein et al. 2005).

As it is well known, the technological progress requires a constant update of the educational and training system. Education can be considered as a productive input that can be accumulated and that can be transferred through different economic sectors and production processes,[2] and therefore, the creation and adoption of new technologies and the accumulation of human capital are interdependent factors that are endogenously determined in an economy (Caselli 1999; Gould 2002).

As it is widely recognized, technological progress has a favorable effect on social well-being (Acemoglu and Restrepo 2018). Differences in countries' economic growth rates are mainly due to differences in the fraction of skilled workers and innovative firms that are capable of adapting to (or producing) technological progress. Invention, automation, adoption of new technology and accumulation of human capital are interdependent factors that, in the common view, generate growth and prosperity.

It is difficult to determine if the accumulation of human capital is what drives the process of economic growth or technological progress, based on theoretically and

---

[1]In the sophisticated context of firms' evolutionary theory the idea of Nelson and Phelps (1966) is referred to, created in the macroeconomic sphere, that "internal" knowledge is needed to absorb new knowledge produced outside, showing a kind of inverse causal process between intellectual capital and innovation.

[2]The accumulation of human capital and the improvement of its management are the sources of sustainable competitiveness (Barney 1991; Hitt et al. 2001).

empirically robust results (see Risso and Sanchez Carrera 2019; Sanchez Carrera 2019). We understand that the increase in the quality of human capital acts as an incentive for investment in technology, in the same way that the increase in investment in technology by firms encourages the decision of workers in to train and get up-to-date. The reverse process is also verified. The non-linear evolutionary model we present corresponds to this self-replicating process.[3] The existence of incentives in one way or another does not imply the immediate change in the decisions of each firm or each worker. It simply indicates a possible direction of evolution, or more precisely, a trend. In our model, if the trend continues in the long term, it will lead to a technologically advanced economy or, conversely, will lead to a poorly performing economy. It will create a situation of conformity in a high social or technological equilibrium or in a low one dominated in the Pareto sense.

In our model we consider, in contrast to the standard endogenous growth theory (Grossman and Helpman 1991; Romer 1990) where agents maximize their direct utility, that agents imitating the more successful strategy, given the current state of the economy. Utility maximization indeed requires perfect information, which is an assumption that in the real world is not always realized. People indeed make choices based on what their neighbours do, and choose to act like them if they believe that the results are better than those they attain, in terms of economic performance, lifestyles and so on.

The economic behaviour driven by imitation is not a novelty in the economic literature, as it is based on the idea posited by Accinelli and Sánchez Carrera (2011, 2012) and Sanchez Carrera (2019). Through their contributions, they explore the foundations of strategic complementarities between firms and workers by means of the evolutionary game theory, where firms and agents can respectively decide to invest in R&D and education (bearing the associated costs) or not, given the current state of the economy, that is to say, the fraction of high technological firms and the fraction of skilled worker over the total workforce.

Based on the above theoretical motivations, this research paper aims to study the co-evolution dynamics of human capital and innovative firms by means of an evolutionary game theory model. More specifically, our model encompasses two types of firms (innovative and non-innovative) and two types of workers (skilled and unskilled). The main difference between the two firms is that the innovative firms employ skilled workers as production input, while non-innovative firms do not. The main difference between skilled and unskilled workers is that the former are more productive than the latter. Moreover, unskilled labor is absolutely elastic, while skilled labor is not. Firms maximize their profits given the equilibrium price in the consumption good market (price-takers) and input prices are fixed to marginal productivity (Matsumoto and Szidarovszky 2018). From these strategic interactions, under some reasonable hypothesis that will be introduced and justified later, two evolutionary stable strategies emerge: for the firms, to invest in research and development and for the workers, to invest in education, giving rise to prosperity and

---

[3]For a broad overview of the topic on nonlinear economic dynamics, see Matsumoto et al. (2018).

economic growth, or not to invest in R&D nor in education, hence giving rise to a poverty trap.

The remainder of the paper is organized as follows. In Sect. 2, we develop the model by explaining how firms maximize their benefits given the distribution of workers types (i.e. skilled or unskilled). In Sect. 2.1 we study the relationship between firms' profit rates and the labor market. The rates of profits, workers' wages and education costs are considered as the main variables for the decision making of firms and workers respectively. Section 2.2 develops an evolutionary dynamics for the firms' decisions and the evolutionary dynamics for workers' decisions. Section 3 develops the evolutionary dynamics of the economy, analyzing the main characteristics of the possible stationary states and some particular transitions path. Section 4 presents the conclusions from the model and offers further research directions.

2 Setup of the Model

Let us consider an economy where there are two types of firms (which differ in the level of technology of their production process) producing an homogeneous product in a perfect competitive market where the equilibrium price of the product is determined as a market clearing price that here is assumed given to $p > 0$. A group of firms, namely innovative firms denoted by the subscript i, uses state-of-the-art technology, while the remaining ones, non-innovative firms denoted by n, use "traditional" technology. Factors of production used by both types of firms are generically represented by physical capital, $K > 0$, and labor, $L > 0$. Labor is differentiated between skilled workers, s, and unskilled workers, $\bar{s}$.

The amount of capital used by innovative firms is denoted by $K_i > 0$. Innovative firms employ skilled workers, and the amount of skilled workers employed is denoted by $L_{is} \geq \bar{L}_{is} > 0$, where $\bar{L}_{is}$ represents the 'reference amount' of skilled workers which is necessary for production. Innovative firms demand also unskilled workers, and $L_{i\bar{s}} > 0$ represents such amount of hired unskilled workers. Non-innovative firms use capital $K_n > 0$ and only demand unskilled labor $L_{n\bar{s}} > 0$. Firms' production functions (satisfying the usual convexity assumptions) are represented by:

$$f_i : \mathbb{R}^3_+ \to \mathbb{R},$$

$$f_n : \mathbb{R}^2_+ \to \mathbb{R}$$

where the function $f_{j \in \{i,n\}}$ a strictly concave, differentiable and not decreasing in each of its variables. That is, firms' production functions are, respectively:

$$y_i = f_i(K_i, L_{is}, L_{i\bar{s}}) \text{ and } y_n = f_n(K_n, L_{\bar{s}}) \tag{1}$$

Optimal output supply of innovative firms is denoted by $y_i^*(p)$ and that of non-innovative firms is $y_n^*(p)$, at prices p. We also denote by K_i^*, L_{is}^*, $L_{i\bar{s}}^*$ and K_n^*, $L_{n\bar{s}}^*$ the optimal inputs demand of each type of firms, at prices p.

At any moment of time, t, the market demand is satisfied by $N_i(t) \geq 0$ total number of innovative firms and by $N_n(t) \geq 0$ total number of non-innovative firms, where $N = N_i(t) + N_n(t)$ is the whole population of firms. At prices p the market demand is $y(p)$, then:

$$y(p) = N_i(t)y_i(p) + N_n(t)y_n(p)$$

where $y_j(p)$, $j \in \{i, n\}$ corresponds to individual supply of each firms' type. In addition, we assume that total demand cannot be covered only by innovative firms, that is to say $y(p) > N_i(t_0)y_i(p)$. The remainder $\bar{y} = y(p) - N_i(t_0)y(p)_i$ is covered by non-innovative firms, in such a way that the relation $\bar{y} = N_n(t_0)y_n(p)$ is verified.

Let us assume that the marginal costs of innovative firms are smaller than those of the non-innovative firms, that is:

$$\frac{dc_i(y)}{dt} < \frac{dc_n(y)}{dt}$$

where c_j, $j \in \{i, n\}$ is the respective cost function, we also assume that:

$$\frac{dc_i(0)}{dt} < \frac{dc_n(0)}{dt} < p$$

meaning that both types of firms have incentives to produce positive quantities for the market.

Firms are maximizing their profits and we consider that input prices are determined by their marginal productivity, i.e.

$$w_s^* = \frac{df_i}{dL_s}(K_i^*, L_{is}^*, L_{i\bar{s}}^*)$$

$$w_{\bar{s}}^* = \frac{df_i}{dL_{n\bar{s}}}(K_i^*, L_{is}^*, L_{i\bar{s}}^*) = \frac{df_n}{dL_{\bar{s}}}(K_n^*, L_{n\bar{s}}^*)$$

$$r_i^* = \frac{df_n}{dK_i}(K_i^*, L_{is}^*, L_{i\bar{s}}^*)$$

$$r_n^* = \frac{df_n}{dK_n}(K_n^*, L_{n\bar{s}}^*)$$

where w_j is labor wage and r_j^* is the capital price, for $j \in \{i, n\}$.

Total labor available at each time t is denoted by $H(t) = H_s(t) + H_{\bar{s}}(t) > 0$, where $H_s(t)$ and $H_{\bar{s}}(t)$ correspond, respectively, to the amounts of skilled and unskilled labor, at time t. We also assume that the supply of unskilled labor is absolutely elastic, i.e. it is always possible to find in the market the amount required by the firms. This does not necessarily happen with skilled labor.

Firms are competitive and therefore seek to maximize their profit functions:

$$\Pi_i(K_i, L_{is}, L_{i\bar{s}}) = pf_i(K_i, L_{is}, L_{i\bar{s}}) - (r_i K_i + w_s L_{is} + w_{\bar{s}} L_{i\bar{s}})$$

$$\Pi_n(K_n, L_{n\bar{s}}) = pf_n(K_n, L_{n\bar{s}}) - (r_n K_n + w_{\bar{s}} L_{\bar{s}})$$

$$(2)$$

The equilibrium values of quantities K_i^*, K_n^*, L_{is}^*, $L_{i\bar{s}}^*$, $L_{n\bar{s}}^*$, satisfy:

$$\Pi_i(K_i^*, L_{is}^*, L_{i\bar{s}}^*) \geq \Pi_i(K_i, L_{is}, L_{i\bar{s}}) \ \forall \ (K_i, L_{is}, L_{i\bar{s}}), \quad \text{and}$$

$$\Pi_n(K_n^*, L_{n\bar{s}}^*) \geq \Pi_n(K_n, L_{n\bar{s}}) \ \forall \ (K_n, L_{n\bar{s}}).$$

Remark 1 If $\partial f_i/\partial L_{i\bar{s}} \neq 0$, the function $L_{i\bar{s}}; \Re_+ \to \Re$ exists by *the global implicit function theorem* satisfying $L_{i\bar{s}}(L_{is}) = L_{i\bar{s}}$ for all $L_{is} \geq \bar{L}_{is} > 0$. As a consequence of the implicit function theorem the following identity holds:

$$\frac{dL_{i\bar{s}}}{dL_{is}}(L_{is}) = -\frac{\frac{\partial f_i}{\partial L_{is}}(K_i, L_{is}, L_{i\bar{s}})}{\frac{\partial f_i}{\partial L_{i\bar{s}}}(K_i, L_{is}, L_{i\bar{s}})}. \tag{3}$$

The equation corresponds to the marginal rate of substitution of unskilled by skilled labor. The curve determined by $(L_{is}, L_{i\bar{s}}(L_{is}))$ corresponds, with fixed K_i, to an iso-product curve. Operating outside the optimum implies an additional cost for the firm.

2.1 Firms and Labor Market

Consider that in period $t = t_0$, K_i^* and K_n^* are the optimal amounts of capital required by each type of firms, and the optimal amounts of labor are L_{is}^*, $L_{i\bar{s}}^*$ for innovative firms and $L_{n\bar{s}}^*$ for non-innovative firms. Hence, the following relationship can be verified:

$$\frac{\Pi_i(K_i^*, L_{is}^*, L_{i\bar{s}}^*)}{c_i(y^*)} = \frac{\Pi_n(K_n^*, L_{n\bar{s}}^*)}{c_n(y^*)} \tag{4}$$

That is with optimally production the profit rates of the two types of firms, is identical. However, the above relationship (2) is no longer true if the required optimal labor quantities are not acquired in the labor market because supply of skilled labor does not satisfy its demand. In that case, firms must operate outside the optimum. So we can state the following result.

Proposition 1 *If in the labor market, the skilled workers demand is greater than its supply, then innovative firms produce in a suboptimal situation. Consequently, incentives may arise to modify strategies, i.e. firms prefer being non-innovative and workers prefer being unskilled.*

Proof Suppose that in period $t = t_0$, the following holds:

$$N_i(t_0)L_i^* > H_i(t_0), \quad N_i(t_0)L_{i\bar{s}}^* + N_n(t_0)L_{n\bar{s}}^* < H_{\bar{s}}(t_0).$$

This leads to some, if not all, innovative firms producing in a suboptimal situation. This can happen in three different ways:

1. Innovative firms must keep the production at the optimum level $y_i^*(p)$ for which they must demand unskilled labor to replace the skilled labor in the iso-product curve determined by the relationship: $L_{i\bar{s}} = L_{i\bar{s}}(L_{is})$, see Remark 1. The identity $f_i\left(K_i^*, L_{is}, L_{i\bar{s}}(L_{is})\right) = y_i^*$ must be valid.
2. Some innovative firms produce less than the optimal amount: $y_i(p) < y^*(p)$.
3. Some innovative firms pay higher wages than equilibrium wages to the hired skilled workers i.e.;

$$w_s(t_0) > w_s^* = \frac{\partial f_i}{\partial L_{is}}(K_i^*, L_s^*).$$

Any of these alternatives is considering that some innovative firms, momentarily at least, are getting lower profit rates than those corresponding to non-innovative firms. So firms have incentives to stop being innovative, with the consequent fall in the demand for skilled labor and an increase in the demand for unskilled labor. This situation is maintained until the profit rates are equalized, which happens in some time $t = t_1$ when:

$$N_i(t_1)L_{is}^* = H_i(t_1),$$

and

$$N_i(t_1)L_{i\bar{s}}^* + N_n(t_1)L_{n\bar{s}}^* = H_{\bar{s}}(t_1),$$

where

$$N_i(t_1) < N_i(t_0), \quad N_n(t_1) > N_n(t_0),$$

consequently

$$H_s(t_1) < H_s(t_0), \quad H_{\bar{s}}(t_1) > H_{\bar{s}}(t_0).$$

If additionally we assume that for an unskilled worker to become a skilled worker, he or she has to face the cost of education (or training cost), then when the number of innovative firms is low enough, that brings a disincentive for the qualification of workers. Wages of skilled workers can never be lower than wages of unskilled workers, since there are costs associated with the qualification, the wages difference should cover such costs, otherwise there would be no incentives for the qualification.

Corollary 1 *If there is an excess supply of skilled labor, then innovative firms could lower their production costs by offering to the hired skilled workers salaries w_s lower than those of equilibrium, but slightly higher than the wages of unskilled workers, that is, $0 < w_{\bar{s}} \leq w_s < w_s^*$. Profit rates of innovative firms grow, and there are incentives for non-innovative firms to become innovative.*

From Corollary 1, It follows that $r_i^* K_i + w_s L_{is}^* + w_{\bar{s}}^* L_{i\bar{s}}^* \leq c_i(y^*(p))$ and $r_n^* K_n + w_{\bar{s}}^* L_{n\bar{s}}^* \leq c_n(y^*)$ and so we have that:

$$\frac{\Pi_i(K_i^*, L_{is}^*, L_{i\bar{s}}^*)}{r_i^* K_i^* + w_s L_{is}^* + w_{\bar{s}}^* L_{i\bar{s}}^*} > \frac{\Pi_i(K_i^*, L_{is}^*, L_{i\bar{s}}^*)}{c_i(y^*)} = \frac{\Pi_n(K_n^*, L_{n\bar{s}}^*)}{c_n(y^*)}. \tag{5}$$

Since we consider that firms (business owners) seek to invest in the production sectors with higher profit rates, equality in (5) assumes that there will be no changes in technology, that is, there are no incentives to change from being a non-innovative type, or the other way around. Nevertheless note that, if the economy is able to create incentives for workers to qualify themselves, then firms owners will have incentives to transform non-innovative firms into innovative firms, thus making an improvement in productive technology.

Next, we analyze the dynamic interdependence between innovative firms and skilled labor, by applying replicator dynamics.

2.2 Firms and Workers Evolutionary Dynamics

Let us consider that firms remain in the market as long as their respective profits are positive and the firms' owners' aim is to maximize the profit rate. Managers and owners have the same interests, that is, there are no agency problems in this model. We denote by $\bar{B}_i(H_s(t)) > 0$ the average profit rate of the innovative firms when the amount of skilled labor is given by $H_s(t) > 0$. In this case, it results that:

$$\bar{B}_i(H_s(t)) \begin{cases} \leq B_i(y_i^*) \ \forall t : \ H_s(t) \leq N_i(t)L_{is}^* \\ \geq B_i(y_i^*) \ \forall t : \ H_s(t) \geq N_i(t)L_{is}^* \\ = B_i(y_i^*) \Leftrightarrow H_s(t) = N_i(t)L_{is}^* \end{cases} \tag{6}$$

Notice that, in (6) the first inequality is the result of having to replace skilled workers work by unskilled workers. Therefore some innovative firms should produce outside the optimum, as a consequence of which the benefits decrease as costs increase or because of the suboptimal level of production. The second inequality corresponds to an excess of skilled labor supply. Therefore innovative firms will be able to offer salaries less than those of equilibrium to skilled workers without decreasing the optimal level of production. Equality corresponds to the fact that it is produced according to an optimal plan with equilibrium prices.

Similarly for non-innovative firms, we will denote by $\bar{B}_n(H_{\bar{s}}(t))$ the average profit rate of the non-innovative firms. Observe that under the assumptions of our model $\bar{B}_n(H_{\bar{s}}(t)) = B_n(y_n^*)$, given that the supply of unskilled workers is undoubtedly supposed to be absolutely elastic.

Consider that the rate of innovative firms existing in the economy, at any time t, is given by $n_s = N_s/N$, and by applying the replicator's dynamics, such rate will grow according to the differential equation:

$$\dot{n}_i = n_i(1 - n_s)[\bar{B}_i(H_s(t)) - \bar{B}_n(H_{\bar{s}}(t))], \tag{7}$$

where $n_s = N_s/N$ corresponds to the rate of non-innovative firms.

Expected payoffs of workers' type s is given by the expression:

$$E(s) = \text{prob}_s(I)[w_s - CE] + \text{prob}_s(NI)[w_{\bar{s}} - CE] \tag{8}$$

where $\text{prob}_s(I)$ and $\text{prob}_s(NI)$ denote the probabilities that skilled workers will be hired by innovative and non-innovative firms, $\text{prob}_s(I) + \text{prob}_s(NI) = 1$. $CE > 0$ represents the discounted education costs to unit time period faced by skilled workers.

While expected payoffs of worker's type $\bar{s}$, is given by:

$$E(\bar{s}) = \text{prob}_{\bar{s}}(I)[w_{\bar{s}}] + \text{prob}_{\bar{s}}(NI)[w_{\bar{s}}]. \tag{9}$$

where

$$\text{prob}_{\bar{s}}(I) = \begin{cases} 0 \ \ if \ \ H_i(t) \geq N_i L_s^* \\ P_{\bar{s}}(I) \ \ if \ \ H_s(t) < N_i L_{is}^* \end{cases}$$

Claim 1 *There are strategic complementarities, meaning that: $\text{prob}_s(I) > \text{prob}_s(NI)$, and $\text{prob}_{\bar{s}}(NI) > \text{prob}_{\bar{s}}(NI)$. That is, innovative firms and skilled workers are complementary and the same holds for non-innovative firms and unskilled workers.*

Let us denote by $w_s(t) = w_s(H_i(t))$ the wage of the skilled workers when the supply of skilled labor is $H_i(t)$. We then have:

$$w_s(H_s(t)) = \begin{cases} > w_s^* \ \text{for all } t : N_i(t)L - is^* > H_s(t) \\ < w_s^* \ \text{for all } t : N_i(t)L_{is}^* < H_s(t) \\ = w_s^* \ \text{for all } t : N_i(t)L_{is}^* = H_s(t) \end{cases} \tag{10}$$

Workers are indifferent between being skilled or not if, $E(s) = E(\bar{s})$, i.e. the probability $\text{prob}_s(I)$ for a qualified worker to be hired by an innovative firm is given by:

$$\text{prob}_s^*(I) = \frac{w_{\bar{s}} + CE}{w_s - w_{\bar{s}}}. \tag{11}$$

By $h_s = H_s/H$ we denote the rate of qualified labor at each moment and then, for $h_{\bar{s}} = H_{\bar{s}}/H$ we denote the rate of unskilled labor, then $h_s + h_{\bar{s}} = 1$. According to the dynamics of the replicator, the growth rate of skilled labor grows proportionally to the net income difference $w_s(t) - w_{\bar{s}}(t) - CE$, as:

$$\dot{h}_s = h_s(1 - h_s)(w_s - w_{\bar{s}} - CE) \tag{12}$$

where CE is the discounted cost of education or qualification that a worker must deal with to become a skilled worker. We consider that this cost is exogenously determined.

3 Evolutionary Dynamics of the Economy

Remember that $h_s(t) = \frac{H_s(t)}{H}$ is the rate of qualified workers existing in the market, and $n_i(t) = \frac{N_i(t)}{N}$ is the rate of innovative firms in the economy at every moment. From differential equations (7) and (12) and taking into account that for all t, equalities $1 = h_s + h_{\bar{s}}$ and $1 = n_i + n_n$ hold, we obtain the result that the evolution of the economy is represented by the following dynamical system:

$$\begin{aligned}
\dot{h}_s &= h_s(1 - h_s)\left[(w_s(t) - w_{\bar{s}} - CE)\right] \\
\dot{n}_i &= n_i(1 - n_i)[\bar{B}_i(h_s(t)) - \bar{B}_n(h_{\bar{s}})]
\end{aligned} \tag{13}$$

Note that the profit rate of non-innovative firms, as well as wages of unskilled workers are, according to the assumptions of the model, constant. To facilitate the resolution of the system, we assume that the costs of education are also constant over time. The dynamic system turns out to be coupled, which shows the interdependence between the variables considered, in this case the interdependence between innovative firms that use advanced technology and skilled workers or qualified human capital.

Observe that the system:

$$\begin{aligned}
h_s(1 - h_s)(w_s(t) - w_{\bar{s}} - CE)) &= 0 \\
n_i(1 - n_i)[\bar{B}_i(h_s(t)) - \bar{B}_n(h_{\bar{s}})] &= 0
\end{aligned} \tag{14}$$

defines the stationary states of the economy, meaning that:

- The dynamic equilibria represented by $(h_s^1, n_i^1) = (1, 0)$ and $(h_s^2, n_i^2) = (0, 1)$ have no meaning in our model. This is because they represent an economy composed exclusively either of skilled workers and non-innovative firms, or on an economy composed exclusively of unskilled workers and all technologically advanced firms.
- The dynamic equilibria represented by $(h_s^3, n_i^3) = (0, 0)$ and $(h_s^4, n_i^4) = (1, 1)$ correspond to very special situations. The first one $((h_s^3, n_i^3) = (0, 0))$ corresponds to an economy in which all firms produce with low-technology, and there is no skilled labor. The second one $((h_s^4, n_i^4) = (1, 1))$ is the counterpart and corresponds to an advanced economy where all firms are technologically advanced, and the workforce is exclusively qualified. These equilibria are analyzed in Accinelli and Sánchez Carrera (2012); Sanchez Carrera (2019), where it is shown that the low equilibrium corresponds precisely to a poverty trap.

- The most interesting equilibrium is that in which the following equalities hold:

$$w_s(t) - w_{\bar{s}} - CE = 0$$
$$\bar{B}_i(h_s(t)) - \bar{B}_n(h_{\bar{s}}) = 0 \tag{15}$$

This state corresponds to the situation represented by (h_s^5, n_i^5) in which the efficiency wage, w_s^*, is reached and equal to $w_{\bar{s}} - CE$ and at the same time that, at the optimum level of production, the profit rates of innovative and non-innovative firms are equal, that is, $\bar{B}_i(h_s^5(t)) = \bar{B}_n(h_{\bar{s}})$. Therefore they are non existing incentives so that neither the workers nor the firms modify the way they are acting. In this steady state, the equality will prevail $H_s = Hh_s^5 = N_i L^* = Nn_i^5 L_{is}^*$ which supposes supply of skilled labor equal to its demand. We conclude that the rate n_i^5 of the firms acting in the economy will be innovative and the rest will be with rate $(1 - n_i^5)$ and are non-innovative firms. Finally, the rate of workers h_s^5 choosing to be qualified and the rest $(1 - h_s^5)$ remaining as non-qualified workers. This equilibrium is determined by the cost of education and the speed with which changes in technology will have an impact on the determinant variables of the model: the supply of skilled workers and the rates of benefit of the firms. The equilibrium values (h_s^5, n_i^5) will be different depending on the economic conditions of the different countries. This result is consistent with the considerations made in Acemoglu 1997, 1998, 2002; Aghion 2006; Hornstein et al. 2005, among others that show the existence of considerable differences in the use and implementation of the new technologies in different countries.

3.1 The Feasible Transition Paths

The seminal papers by Acemoglu (2002), Autor et al. (1998), and Bowles et al. (2001), pointed out that the growing adoption of new technologies in the work centers modifies, at the time of their adoption, the demand for workers, demanding from them the adoption of necessary capabilities for the better use of the new technology. This process is accompanied by a growth in the wage difference between skilled and unskilled workers. Whichever way, the question remains the same: who determines who? Does the increase of qualified workers in the market imply an increase in investment by firms in technology, or is it the other way round?

Although, in general, it is impossible to obtain an analogous solution for the system of equations that determines the evolution of the economy just described above, it is possible to consider a particularly relevant case. Assume a linear relationship: $w_s(t) - [w_{\bar{s}} - CE] = \beta \dot{N}_i$ where β is a positive constant. If N is fixed, it will result in $\dot{N}_i = N\dot{n}_i$. This hypothesis reflects the fact that the wage difference between skilled and unskilled labor depends on the rate increase in technologically advanced firms, which are precisely those which demand skilled workers. Substituting into system (14), we get:

$$\dot{h}_s = h_s(1 - h_s)\beta N \dot{n}_i \tag{16a}$$

$$\dot{n}_i = n_i(1 - n_i)[\bar{B}_i(h_s(t)) - \bar{B}_n(h_{\bar{s}})]. \tag{16b}$$

Being $0 < h_s(t_0) < 1$ the rate of skilled workers existing at the initial moment and $n_i(t_0)$ the rate of innovative firms existing at that time. Obtaining then for $0 < h_s(t) < 1$

$$h_s(t) = \frac{e^{\beta N[n_i(t) - n_i(t_0)]}}{\frac{1 - h_s(t_0)}{h_s(t_0)} + e^{N\beta[n_i(t_0) - n_i(t)]}}.$$

In order to better discuss the meaning of this solution, consider: $A = \frac{1 - h_s(t_0)}{h_s(t_0)}$ and $C = e^{\beta N_i(t_0)}$. It turns out then

$$h_s(t) = \frac{1}{ACe^{-\beta N n_i(t)} + 1}. \tag{17}$$

The trajectories are determined once the initial conditions are known $(h_s(t_0), n_i(t_0))$. The phase diagram corresponds to points in the square $(0, 1) \times (0, 1)$ determined by the trajectories $(h_s(t), n_i(t))$, that result from solving the system (16). The next proposition states our main result.

Proposition 2 *A long-run co-evolutionary dynamics between types of workers and firms holds. That is, if the rate of skilled workers is high enough, then the rate of innovative firms increases, but also in other way around, i.e. when the initial rate of innovative firms is high enough, then the rate of skilled workers increases as well.*

Proof According to the second equation of the dynamical system (16), it follows that, this change in the rates of profits becomes a new incentive for the transformation of non-innovative firms into technologically advanced firms, and therefore, the subsequent increase in the demand for skilled labor results in the increase of wages of skilled workers. The duration of the virtuous cycle will depend on the time that the new technologies take to expand and the workers will be trained in the use of new technologies. An interesting and exhaustive discussion of this topic can be found in Card and DiNardo (2002).

It is important to highlight that, if $N_i(t) \to N$, that is, if all firms become innovative, when $t \to \infty$ then, $h_s(t) \to \frac{1}{ACe^{\beta N} + 1} = h_\infty$. As a result, in the long term, if the rate of innovative firms increase along the time, the rate of skilled labor will converge to the value

$$h_\infty(h_s(t_0), n_i(t_0)) = \frac{1}{\frac{1 - h_s(t_0)}{h_s(t_0)} e^{\beta(N_i(t_0) - N)} + 1}$$

which corresponds to different values according to the initial conditions of the model measured at $t = t_0$. That is, the long term rate of skilled workers will depend on the initial rate of skilled labor and the number of existing innovative firms at $t = t_0$ (see Fig. 1).

Fig. 1 Case in which $\dot{n}_i > 0$, and $h_s(t) \to h_F = \frac{1}{ACe^{-\beta N}+1}$

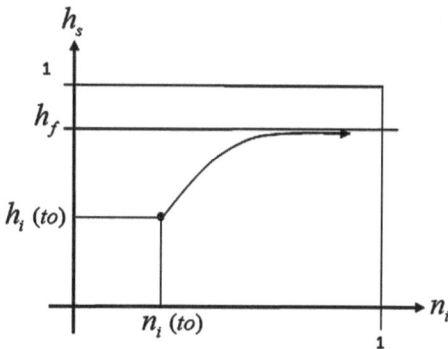

This shows that, for a certain period of time, those countries in which the initial rate of skilled labor and the number of innovative firms is higher, then there are at the end of the period higher values of skilled workers. The fact that when all firms are innovative then there are unskilled workers as well because they use skilled and unskilled workers as inputs for their production.

Note that the Eq. (16b) can be rewritten as:

$$\dot{n}_i = n_i(1 - n_i)[(\bar{B}_i + \bar{B}_n)h_s(t) - \bar{B}_n]$$

so if $[\bar{B}_i + \bar{B}_n]h_s(t) - \bar{B}_n > 0$ or equivalently if

$$h_s(t) > \frac{\bar{B}_n}{\bar{B}_i + \bar{B}_n} \tag{18}$$

then $\dot{n} > 0$ and if $\dot{n} > 0$ then $\dot{h}_s > 0$.

By contrast, if the rate of innovative firm decreases (see Fig. 2), then in the long run, i.e. if $\dot{n}_i < 0$ we get:

$$h_f(n_i(t_0)) = h_f = \frac{1}{Ae^{\beta N_i(t_0)} + 1}.$$

This is the co-evolutionary dynamics for human capital and innovation. If the rate of skilled workers is high enough, the rate of innovative firms will increase, and then the rate of skilled workers increases, too. The fact that the initial rate of skilled workers in the country is sufficient to start a technological growth trajectory, it will depend on the relationship between the expected profits for innovative and non-innovative firms. That is, the smaller the quotient $\bar{B}_i/\bar{B}_n$, the higher the initial rate of workers needed to start a technological development path. In this way, a policy maker interested in the technological development of a country, may implement a policy of incentives suitable for the establishment of technologically advanced firms and/or through incentives for workers to become skilled, i.e. increasing the quality of human capital.

Fig. 2 Case in which
$\dot{n}_i < 0$, and $h_s(t) \to \frac{1}{AC+1}$

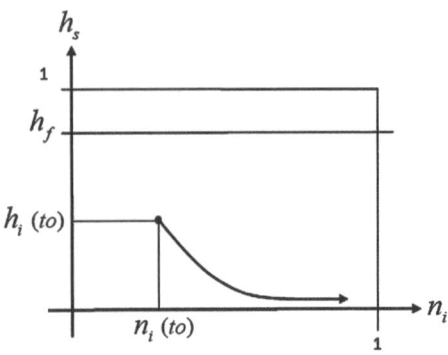

4 Conclusion

The model we have developed highlights the fact that the technological change and
the development of skilled labor are coevolutionary processes complementing and
determining each other. The dynamic system that reflects the evolution of technology
is a system coupled with that of skilled workers. Hence, the relative weights of
the variables in the future evolution may be varied, but both will jointly determine
the long-term evolution. The economic policy maker must act on the political and
economic elements that may delay the evolution of the variables, assuring adequate
costs of education or opportunities of implementing new technologies without costs
aggravated by excessive bureaucracy, for example.

From our results, further research should focus on the following. For example, a
policy maker interested in the technological development of the national economy
must address favourable incentives for workers to aspire to become skilled and firms
to become innovative. The existence of negative incentives either for the introduction
of new production technologies or for the permanent training of workers, will result
in slowing or stopping the process of technological change in the economy as a
whole. For instance, high costs of education or high taxes on technology, will have
negative results, and both will lead to the technological stagnation of the economy.

Education, training and innovation must be considered as a social organic structure interrelated with the productive structure. That is to say, economies or economic
policymakers should implement strategies that result in the combination of education at all levels: competitiveness, the technology level, the training and skills of
the workforce, the population of scientists and development of the country. A qualified education is an important instrument to overcome poverty, improve equality of
opportunities and productivity, and move towards development.

Acknowledgements We benefitted from discussions and comments from Costas Azariadis, Gian
Italo Bischi, Sebastian Ille, Akio Matsumoto, Lionello F. Punzo, Ferenc Szidarovszky and Laura
Veldkamp. We thank the anonymous reviewers for their constructive comments, which helped
us to improve the manuscript. We would like to thank the Research Workgroup GEU at DESP-
University of Urbino Carlo Bo, Italy. We would also like to thank the Research Workgroup on

Seminario de Investigacion y Estudio en Teoria Economica at the Department of Mathematical Economics, University of San Luis Potosi, and SNI-CONACYT Mexico. This research work has been developed in the framework of the research project on "Models of behavioral economics for sustainable development" financed by DESP- University of Urbino Carlo Bo. The usual disclaimer applies.

References

Acemoglu, D. (1997). Training and innovation in an imperfect labor market. *Review of Economic Studies, 64*, 445–64.

Acemoglu, D. (1998). Why do new technologies complement skills? Directed technical change and wage inequality. *Quarterly Journal of Economics, 113*(4), 1055–1089.

Acemoglu, D. (2002). Technical change, inequality and the labor market. *Journal of Economic Literature, 40*, 7–72.

Acemoglu, D., & Restrepo, P. (2018). Artificial intelligence, automation and work (NBER Working Paper No. 24196). National Bureau of Economic Research, Inc.

Accinelli, E., & Sánchez Carrera, E. (2011). Strategic complementarities between innovative firms and skilled workers: The poverty trap and the policymaker's intervention. *Structural Change and Economic Dynamics, 22*(1), 30–40.

Accinelli, E., & Sánchez Carrera, E. (2012). The evolutionary game of poverty traps. *The Manchester School, 80*(4), 381–400.

Aghion, P. (2006). On institutions and growth. In T. S. Eicher & C. García-Peñalosa (Eds.). *Institutions, development, and economics growth.* Cambridge: The MIT Press.

Amendola, M., & Gaffard, J. L. (1988). *The innovative choice.* Oxford: Basil Blackwell.

Autor, D., Katz, L., & Krueger, A. (1998). Computing inequality: Have computers changed the labor market?. *Quarterly Journal of Economics, 113*(4), 1169–1213.

Barney, J. (1991). Firm resources and sustained competitive advantage. *Journal of Management, 17*, 99–120.

Bowles, S., Gintis, H., & Osborne, M. (2001). The determinants of earnings: A behavioral approach. *Journal of Economics Literature, 39*(4), 1137–1176.

Caselli, F. (1999). Technological revolutions. *American Economic Review, 89*, 78–102.

Card, D., & DiNardo, J. (2002). Skill-biased technological change and rising wage inequality: Some problems and puzzles. *Journal of Labor Economics, 20*(4), 733–783.

Gould, E. (2002). Rising wage inequality, comparative advantage, and the growing importance of general skills in the United States. *Journal of Labor Economics, 20*, 105–47.

Griliches, Z. (1969). Capital-skill complementarity. *Review of Economics and Statistics, 51*(4), 465–468.

Grossman, G. M., & Helpman, E. (1991). *Innovation and growth in the global economy.* Cambridge, MA: MIT Press.

Guerrini, L., Matsumoto, A., & Szidarovszky, F. (2019). Neoclassical growth model with two fixed delays. *Metroeconomica, 70*, 423–441.

Hitt, M. A., Biermant, L., Shimizu, K., & Kochhar, R. (2001). Direct and moderating effects of human capital on strategy and performance in professional service firms: A resource-based perspective. *Academy of Management Journal, 44*(1), 13–28.

Hornstein, A., Krusell, P., & Violante, G. L. (2005). The effects of technical change on labor market inequalities. In P. Aghion & S. Durlauf (eds.). Handbook of economic growth, vol. 1, Part B (pp. 1275–1370). Elsevier.

Krusell, P., Ohanian, L., Rios-Rull, J., & Violante, G. (2000). Capital-skill complementarity and inequality: A macroeconomic analysis. *Econometrica, 68*(5), 1029–1053.

Matsumoto, A., & Szidarovszky, F. (2020). Delay growth model augmented with physical and human capitals. *Chaos Solitons & Fractals, 130.* https://doi.org/10.1016/j.chaos.2019.109452

Matsumoto, A., & Szidarovszky, F. (2018). *Dynamic oligopolies with time delays.* Singapore: Springer.

Matsumoto, A., Szidarovsky, F., & Asada, T. (2018). *Essays in economic dynamics: Theory, simulation analysis, and methodological studies.* Singapore: Springer.

Nelson, R., & Phelps, E. (1966). Investments in human capital, technological diffusion, and economic growth. *American Economic Review, 56,* 69–75.

Risso, A., & Sanchez Carrera, E. (2019). On the impact of innovation and inequality in economic growth. *Economics of Innovation and New Technology, 28*(1), 64–81.

Romer, P. (1990). Endogenous technological change. *Journal of Political Economy, 98,* S71–S102.

Sanchez Carrera, E. (2019). Evolutionary dynamics of poverty traps. *Journal of Evolutionary Economics, 29*(2), 611–630.

Welch, F. (1970). Education in production. *Journal of Political Economy, 78,* 3–59.

Path Dependence in Models with Fading Memory or Adaptive Learning

Gian Italo Bischi, Laura Gardini and Ahmad Naimzada

Abstract We consider a learning mechanism where expected values of an economic variable in discrete time are computed in the form of a weighted average that exponentially discounts older data. Also adaptive expectations can be expressed as weighted sums of infinitely many past states, with exponentially decreasing weights, but these are not averages since the weights do not sum up to one for any given initial time. These two different kinds of learning, which are often considered as equivalent in the literature, are compared in this paper. The statistical learning dynamics with exponentially decreasing weights can be reduced to the study of a two-dimensional autonomous dynamical system, whose limiting sets are the same as those obtained with adaptive expectations. However, starting from a given initial condition, different transient dynamics are obtained, and consequently convergence to different attracting sets may occur. In other words, even if the two different kinds of learning dynamics have the same attracting sets, they may have different basins of attraction. This implies that local stability results are not sufficient to select the kind of long-run dynamics since this may crucially depend on the initial conditions. We show that the two-dimensional discrete dynamical system equivalent to the statistical learning with fading memory is represented by a triangular map with denominator which vanishes along a line, and this gives rise to particular structures of their basins of attraction, whose study requires a global analysis of the map. We discuss some examples motivated by the economic literature.

Keywords Economic dynamics · Learning · Expectations · Attractors · Basins · Bifurcations

G. I. Bischi (✉) · L. Gardini
Dipartimento di Economia, Società, Politica (DESP), Università di Urbino, Urbino, Italy
e-mail: gian.bischi@uniurb.it

L. Gardini
e-mail: laura.gardini@uniurb.it

A. Naimzada
Dipartimento di Economia, Management e Statistica, Università Bicocca, Milano, Italy
e-mail: ahmad.naimzada@unimib.it

F. Szidarovszky and G. I. Bischi (eds.), *Games and Dynamics in Economics*,
https://doi.org/10.1007/978-981-15-3623-6_3

1 Introduction

Many dynamic models involve memory of past states to determine the future time evolution of systems in physics, engineering, natural sciences and economics. The inclusion of past history in the time evolution adds nontrivial complexities, balancing the advantage of dealing with more realistic models. In economics the inclusion of memory in modeling human decisions may be considered as a method to represent learning processes (see e.g. Hommes et al. 2012). The effects of memory in continuous time models of oligopoly markets has recently been analyzed by Matsumoto and Szidarovszky in a series of papers dealing with problems of stability of equilibrium points as time lags are varied, as well as bifurcations leading to dynamic complexities whose consequences are studied both by analytical and numerical methods, see e.g. Matsumoto and Szidarovszky (2018), and Matsumoto and Szidarovszky (2015), as well as Matsumoto (2017). Dynamic models involving delays often generate dynamical systems of infinite dimension. However, some particular kinds of distributed delays have been introduced, expressed by integral terms with kernels (denoted as gamma functions) characterized by an exponential decay going back in the past, that allow to transform an integrodifferential equation into an expanded set of ordinary differential equations of finite dimension (see e.g. Cushing 1978; MacDonald 1978; Chiarella 1991).

In this paper we consider discrete time dynamic models, often used to describe social and economic systems characterized by event-driven time, simulate to describe agents that take decisions by considering past information with exponentially distributed weights, i.e. an exponentially fading memory. In particular, we consider economic models that involve agents' expectations about the future states of the system, and are formulated as mappings from beliefs to realizations, such as $x_t = F\left(x_t^{(e)}\right)$ or $x_t = F\left(x_{t+1}^{(e)}\right)$, where $x_t^{(e)}$ and $x_{t+1}^{(e)}$ represent agents' expectations about current or future states respectively. In order to close the model one must introduce a learning mechanism by which agents make forecastings on the basis of the past history of the system. In this paper we consider one-dimensional models with expectations endowed with two kinds of learning: The first is known as *adaptive learning* (see e.g. Hommes 2013 and references therein) where expectations are obtained by assuming that at each time the expected value is a weighted average of the previous forecast and the previous observed value; the second, obtained by assuming that at each time period the agents compute the expected value as an average of the past realized values, starting from a given initial time $t = 0$, is sometimes called *statistical learning* (see e.g. Guesnerie and Woodford 1992).

Both learning mechanisms share the same equilibrium points of the corresponding model with rational expectations (or perfect foresight) $x_t^{(e)} = x_t$ for each t, that is, assuming that agents are able to anticipate the future outcomes, so that expectations are fulfilled at each time. So, it is interesting to consider the problem of stability of such "rational equilibria" under these learning mechanisms. The *Rational Expectations* (RE) hypothesis, based on the assumptions that agents have complete knowledge of the economic model and fully exploit all the available pieces of information, has been criticized from many points of view, mainly because the assumptions behind the RE paradigm seem to require too much agents' rationality. So, models with boundedly rational agents that converge in the long run to a rational equilibrium may be seen as an evolutionary interpretation of rationality, and some authors say that in this case the boundedly rational agents are able to learn, in the long run, what rational agents already know under very pretentious rationality assumptions (see e.g. Fudenberg and Levine 1998). However, it may happen that under different starting conditions (or as a consequence of exogenous perturbations) the same adaptive process leads to non-rational equilibria as well, i.e. equilibrium situations which are different from the ones forecasted under the assumption of full rationality, as well as to dynamic attractors characterized by endless asymptotic fluctuations or unfeasible evolutions. The coexistence of several attracting sets, each with its own basin of attraction, gives rise to path dependence, irreversibility, hysteresis and other nonlinear and complex phenomena commonly observed in real systems as well as in laboratory experiments. So, stability arguments under some learning dynamics are often used as *equilibrium selection* criteria.

In this paper we consider a particular statistical learning in which the agents discount older data by making weighted averages with exponentially decreasing weights (see Bischi and Gardini 1996; Bischi and Naimzada 1997), so it is the analogous of an exponentially decreasing gamma kernel often used in continuous time dynamic models with distributed delays. Moreover, the discrete fading memory analyzed in this paper includes, as a limiting case, the learning process proposed by Bray (1983). Even adaptive expectations can be expressed as weighted sums of infinitely many past states, with exponentially decreasing weights, but these are not averages since the weights do not sum up to one for any finite initial time. These two different kinds of learning are often considered as equivalent in the literature, because they assume the same form as $t \to +\infty$. Indeed, statistical learning dynamics with exponentially decreasing weights can be reduced to the study of a two-dimensional autonomous dynamical system, whose limiting sets are the same as those obtained with adaptive expectations. However, starting from a given initial condition, different transient dynamics are obtained, and consequently convergence to different attracting sets may occur. In other words, even if the two different kinds of learning dynamics have the same attracting sets, they may have different basins of attraction. In situations of *multistability*, i.e. when several coexisting attractors are present, local stability results are not sufficient to provide selection criteria since this may crucially depend on the initial conditions. Hence, adaptive and statistical learning may give different results when the problem of equilibrium selection arises. Moreover, we show that the two-dimensional discrete dynamical system equivalent to the statistical learning with fading memory is represented by an iterated two-dimensional triangular map with denominator which vanishes along a line, and this gives rise to particular structures of their basins of attraction, whose study requires a global analysis of the map following a stream of literature dealing with maps which are not defined in the whole phase space due to the presence of vanishing denominators, see Bischi et al. (1999), Bischi et al. (2003), and Bischi et al. (2005). In particular, we show that the structure of the basis is strongly influenced by the presence of particular points, called *focal points* in Bischi and Gardini (1997) and Bischi et al. (1999), whose existence, in the case of models with expectations, is related to the presence of fixed points of the map F, which are rational expectations equilibria. These global properties are specific to discrete time models, in the sense that they cannot be observed in continuous time models with delays.

The plan of the paper is the following. In Sect. 2 we compare the mathematical form of models with adaptive expectations and those with statistical learning. In Sect. 3 their dynamical properties are studied, in particular those of statistical learning with fading memory analyzed through the study of an equivalent two-dimensional iterated map, with particular emphasis on the study of the basins of attraction and their global bifurcations specific to maps with a vanishing denominator. In Sect. 4 some examples are discussed, Sect. 5 concludes and suggests some possible extensions.

2   From Beliefs to Realizations: Rational Expectations and Learning Dynamics

Let us consider one-dimensional discrete time economic models represented by mappings from expected values to realized values of the same period, i.e. the outcome of the state variable x_t at time t is a function of the value $x_t^{(e)}$ which agents expect, at the same time t, for the state variable, computed by the agents on the basis of the information held at the previous time $(t - 1)$

$$x_t = F\left(x_t^{(e)}\right) \tag{1}$$

If the agents have *Rational Expectations* (RE), which in a deterministic framework means that they are endowed by *Perfect Foresight* (PF), the expected values coincide with the realized values at each time

$$x_t^{(e)} = x_t \quad \forall t \tag{2}$$

If (2) is inserted into (1) we get
$$x_t = F(x_t)$$

which means that only a *Rational Expectations Equilibrium* (REE) is a fixed point of the map F. It is often argued that the assumption of rational expectations is too strong, since economic models should take into account human limited ability to make forecastings. This leads to the weaker assumption of *Bounded Rationality* (BR) which assumes that agents compute the expected values $x_t^{(e)}$ by some *learning mechanism* based on past experience, i.e.

$$x_t^{(e)} = \Psi\left(x_{t-1}, x_{t-2}, ..., x_{t-1}^{(e)}, x_{t-2}^{(e)}, ...\right) \tag{3}$$

This assumption is not only introduced by claiming that BR is more realistic than RE, but it is often used as a *REE justification* or as a dynamic mechanism for *equilibrium selection* when several RE equilibria exist (see e.g. Marimon 1997). Of course, this requires that a REE must also be an equilibrium for the model with learning. In this context, the local stability of a REE with respect to the dynamics induced by bounded rationality learning is commonly referred to as an *evolutive* explanation of the RE solution, see Guesnerie and Woodford (1992). Moreover, some REEs may be more likely to be reached than other ones when some learning mechanism is introduced (some may be not reached at all if they are unstable under the chosen learning process). Even more interesting situations of equilibrium selection arise when there are attractors of the dynamics with learning which do not exist with RE. This leads, for the dynamics with bounded rationality, to situations of coexistence of attractors which are rational, i.e. also exist for the model with RE assumption, with attractors which are non rational, i.e. asymptotic evolutions which do not exist under the assumption of RE, so that the long-run behavior is characterized by agents which continue to make wrong forecastings.

The selection of the attractor, in particular the convergence to a rational or a non rational attractor, may depend on the initial condition, i.e. from the boundaries among the different basins of attraction. This aspect has been rather neglected in the literature because it requires a global analysis of the dynamics with learning.

2.1 Adaptive Learning and Reduction to One-Dimensional Dynamics

A simple and frequently used learning mechanism is given by the *adaptive expectations*, expressed by

$$x_{t+1}^{(e)} = x_t^{(e)} + \alpha \left(x_t - x_t^{(e)} \right) \quad 0 \leq \alpha \leq 1. \tag{4}$$

i.e. for each time $t = 0, 1, \ldots$ the value $x_{t+1}^{(e)}$ expected for the next period $(t+1)$ is obtained by "adapting" the previous forecasting $x_t^{(e)}$ in the direction of the corresponding observed value x_t, with a speed of adjustment α. Rearranging (4) the new expected value $x_{t+1}^{(e)}$ can be expressed as a convex combination (i.e. a weighted average) of the previous expected value $x_t^{(e)}$ and the currently observed value x_t

$$x_{t+1}^{(e)} = (1 - \alpha) x_t^{(e)} + \alpha x_t \quad 0 < \alpha < 1 \tag{5}$$

We can observe that the limiting case $\alpha = 1$ corresponds to static (or naive) expectations

$$x_{t+1}^{(e)} = x_t \tag{6}$$

and decreasing values of α correspond to higher inertia in updating the previously expected value according to the more recent observation.

Using (5) repeatedly, the adaptively expected value can be expressed as

$$x_{t+1}^{(e)} = \alpha \sum_{k=0}^{\infty} (1 - \alpha)^k x_{t-k} \tag{7}$$

i.e. infinitely many past realizations are considered, with weights exponentially decreasing as more remote past values are considered (decreasing as the terms of a geometric sequence of ratio $(1 - \alpha)$). Some authors call (7) a weighted average of the values observed in the past, but it is important to remark that (7) *cannot be considered as an average*, since the weights do not sum up to one for any finite initial time. Indeed, the weighted sum (7) involves infinitely many "realized" values x_t, even with $t < 0$. The model (1), endowed with adaptive learning, can be reduced to a one-dimensional dynamical system in the space of expected values by inserting (1) inside (5)

$$x_{t+1}^{(e)} = (1 - \alpha) x_t^{(e)} + \alpha F\left(x_t^{(e)}\right) \tag{8}$$

This means that, given an initial expectation $x_0^{(e)}$, the whole time evolution (or trajectory) of expected values is obtained by the iteration of the one dimensional map

$$g_\alpha(z) = (1 - \alpha) z + \alpha F(z). \tag{9}$$

Of course, the corresponding time evolution (or trajectory) of realized values x_t, $t \geq 0$, is simply obtained by (1), i.e. by taking the images by F of the expected values, $x_t = F(x_t^{(e)})$, $t \geq 0$.

The properties of the map (9) are well known (see e.g. Hommes 1994; Chiarella 1988). It is a convex combination of the map F and the identity map, so its graph is included inside the region between the graph of F and the diagonal. This implies that the map g_α and the map F have the same fixed points, i.e. the REEs are fixed points of g_α as well. It is immediate to realize that adaptive expectations are fulfilled for each t, i.e. $x_t^{(e)} = x_t$ $\forall t$, if and only if $x_t = F(x_t)$, i.e. at the REE. Instead, the cycles of g_α are in general different from those of F, and the adaptive forecastings are always wrong along invariant sets that are not fixed points of $F(x)$.

Let us assume that the functions are smooth enough on a compact interval of interest, i.e. $F : I \to I$, F of class $\mathcal{C}^{(1)}$. It is worth to note that the graph of g_α approaches the graph of F as $\alpha \to 1$, i.e. in the limiting case of naive expectations, whereas the graph of the map g_α approaches the diagonal as $\alpha \to 0$. This implies that for each F a value $\overline{\alpha} \in (0, 1)$ exists such that g_α is an increasing function for any $\alpha \in (0, \overline{\alpha})$ and, as it is well known, an increasing map cannot have cycles of period $k > 1$. In other words, an adaptive learning, with sufficiently low values of α, rules out any dynamic behavior which is more complex than convergence to a REE. However, not all the REEs are stable under adaptive learning. From the properties of the map

g_α, the following well known results follow, which are immediate consequences of the fact that the first derivative of $g_\alpha(z)$ is a convex combination of 1 and $F'(z)$.

(i) If α is sufficiently small $\left(\text{i.e. } \alpha < \min \left\{ \frac{1}{1-F'(x)}, \ x \in I, \ F'(x) \neq 1 \right\} \right)$ then g_α is increasing, so that only REEs exist.

(ii) If a REE x^* is unstable with $F'(x^*) > 1$ then it is also unstable for g_α being $g'_\alpha(x^*) > 1$.

(iii) If a REE x^* is unstable with $F'(x^*) < -1$ then it is stable for g_α for a sufficiently small value of α.

The properties listed above suggest a stabilizing role of adaptive expectations with respect to naive expectations. However, cases in which, for intermediate values of α, dynamic behaviors of the map g_α can be obtained which are more complex than the dynamics of the map F, have been given in the literature (Chiarella 1988; Hommes 1991; Hommes 1994). This happens, for example, with decreasing functions F. In these cases the iteration of F can only exhibit convergence to a fixed point or to cycles of period 2, whereas the corresponding map g_α which governs the time evolution of adaptive expectations, may be noninvertible (a bimodal map) for intermediate values of α, so that cycles of any period and chaotic dynamics can be observed, and even distinct coexisting attractors.

2.2 Statistical Learning and Reduction to Two Dimensional Dynamics

Another frequently used learning mechanism is obtained by assuming that, at any time period $t = 0, 1, \ldots$ the agents compute the expected value at the next time period $(t + 1)$ as a weighted arithmetic mean of past realized values

$$x_{t+1}^{(e)} = \sum_{k=0}^{t} a_{tk} x_k \tag{10}$$

with weights

$$a_{tk} \geq 0, \ k = 0, \ldots, t, \ \text{normalized to 1, i.e.} \ \sum_{k=0}^{t} a_{tk} = 1 \tag{11}$$

Some authors call *statistical learning* this method to obtain expected values (see e.g. Guesnerie and Woodford 1992). The learning mechanism as suggested by Bray (1983) in the form of a simple arithmetic mean

$$x_{t+1}^{(e)} = \frac{1}{t+1} \sum_{k=0}^{t} x_k \tag{12}$$

is a particular case. More general distributions of weights can be proposed, which reflect the different methods that the agents use to exploit information contained in the past observations. These can be obtained by defining, for each time $t \geq 0$, a $(t + 1)$-dimensional vector of *relative weights*

$$\omega^{(t)} = \left\{ \omega_0^{(t)}, \omega_1^{(t)}, ..., \omega_t^{(t)} \right\} \tag{13}$$

with $\omega_k^{(t)} \geq 0$, from which the weights (11) are computed as

$$a_{tk} = \frac{\omega_k^{(t)}}{W_t}, k = 0, ..., t \quad \text{with} \quad W_t = \sum_{k=0}^{t} \omega_k^{(t)} \tag{14}$$

A reasonable assumption for the computation of the relative weights is that old observations are less considered by economic agents, i.e. they use decreasing weights which discount older data (see e.g. Friedman 1979; Radner 1983; Lucas 1986). A simple method to obtain this consist in assigning a fixed value to the weight of the last observed value, say $\omega_t^{(t)} = 1, t \geq 0$, and then the other weights are computed so that the ratio between two successive weights is fixed, that is, $\omega_{k-1}^{(t)}/\omega_k^{(t)} = \rho, \rho \in [0, 1]$. With this assumption (13) becomes

$$\omega^{(t)} = \left\{ \rho^t, \rho^{t-1}, ..., \rho, 1 \right\} \tag{15}$$

i.e. $\omega_k^{(t)} = \rho^{t-k}$, and consequently

$$a_{tk} = \frac{\rho^{t-k}}{W_t} \tag{16}$$

where W_t is the (t-th) partial sum of a geometric series

$$W_t = \sum_{k=0}^{t} \rho^{t-k} = \begin{cases} \frac{1-\rho^{t+1}}{1-\rho} & if \ \ 0 \leq \rho < 1 \\ t+1 & if \ \ \rho = 1 \end{cases} \tag{17}$$

Statistical learning with "geometrically" distributed weights (16) have been used in Bischi and Naimzada (1997) as a generalization of that proposed by Bray: in fact, for $\rho = 1$ it gives the Bray's average (12). In the other limiting case $\rho = 0$ it reduces to naive expectations $x_{t+1}^{(e)} = x_t$, whereas for intermediate values of the *memory ratio* ρ this learning rule describes agents which, at each time period t, compute their expectations according to a weighted estimation procedure which "exponentially discounts older observations" (see Friedman 1979), that is, an *exponentially fading memory*, see also Foroni et al. (2003), Naimzada and Tramontana (2009), Pecora and Tramontana (2016), Tramontana (2016), and Cavalli and Naimzada (2015), as well as a further generalization with power means in Bischi et al. (2015).

The learning rule (10) with "geometric weights" (16)

$$x_{t+1}^{(e)} = \sum_{k=0}^{t} \frac{\rho^{t-k}}{W_t} x_k \qquad (18)$$

can be written as a generalized "adaptive rule" with nonautonomous (i.e. "time-dependent") adjustment speed. In fact,

$$x_{t+1}^{(e)} = \frac{\rho W_{t-1}}{W_t} \sum_{k=0}^{t-1} \frac{\rho^{t-1-k}}{W_{t-1}} x_k + \frac{1}{W_t} x_t = \frac{W_t - 1}{W_t} x_t^{(e)} + \frac{1}{W_t} x_t$$

where the recursive relation

$$W_{t+1} = 1 + \rho W_t, \qquad W_0 = 1. \qquad (19)$$

has been used. So, if we define

$$\alpha_t = \frac{1}{W_t}. \qquad (20)$$

we get

$$x_{t+1}^{(e)} = (1 - \alpha_t) x_t^{(e)} + \alpha_t x_t = (1 - \alpha_t) x_t^{(e)} + \alpha_t F\left(x_t^{(e)}\right) \qquad (21)$$

which is very similar to an "adaptive rule" (4) except for the fact that the constant speed α is replaced by a time-dependent speed of adjustment given by a decreasing sequence $\{\alpha_t\}$ with $\alpha_t \in (0, 1)$ for each t and $\alpha_t \to (1 - \rho)$ as $t \to +\infty$. Hence, for $t \to +\infty$, the nonautonomous recurrence (21) tends to the *limiting form*

$$x_{t+1}^{(e)} = g_{1-\rho}(x_t^{(e)}) = \rho x_t^{(e)} + (1 - \rho) F(x_t^{(e)}). \qquad (22)$$

i.e., in the long run it behaves like a model with a standard adaptive rule, with speed of adjustment $\alpha = 1 - \rho$. This fact led many authors to consider the two learning rules, the adaptive rule (4) and the statistical rule (18), as practically equivalent, and justify this equivalence statement by the property that the dynamics of the expected values under both the learning rules are governed, in the long run, by the a one-dimensional map which has the same form $g_{1-\rho}(z)$, given by (9). However, even if the limiting sets are the same, their time evolutions are different, because starting from the same initial conditions, the two learning mechanisms exhibit different transient dynamics due to the fact that during the early iterates the dynamics with statistical learning (18), governed by (21), is different from the one governed by (22), and in the presence of several attractors this may be crucial to decide which one will be reached in the long run. In particular, if several attractors are present, convergence to different attractors under the two learning mechanisms may be observed even starting from the same initial condition, thus giving different equilibrium selection results.

This can be equivalently stated by saying that *even if the two different kinds of learning dynamics have the same attracting sets, their basins of attraction may be different.* This is also true in the particular case of Bray learning (12), corresponding to the limiting case $\rho = 1$, as we shall see in the following through some examples.

Some remarks on the crucial role of initial conditions in models with Bray's learning can be found in the literature. For example, in Holmes and Manning (1988) the learning rule (12) is used in a nonlinear cobweb model with decreasing F, and it is stressed that such type of learning has a stabilizing effect on the long run dynamics. However, the authors remark that the short and intermediate run dynamics can be rather complex and of considerable interest. A similar argument is given in Dimitri (1988) where a quadratic map F is proposed as a modification of a linear model of price dynamics with $p_t^{(e)}$ computed according to (12) as proposed in Bray (1983). On the basis of numerical results Dimitri writes "...the evolution of the model is indeed very much dependent upon the starting position..." as a comment to the fact that even if a REE is locally stable, divergent price sequences are obtained even if initial conditions are taken rather close to the REE. These considerations lead us to face the problem of the basins of attraction. This is not, in general, an easy task for nonautonomous recurrences like (21), because for nonautonomous recurrences the ω-limit sets are not invariant sets, due to the fact that the iterated map changes as t varies. However, a global characterization of the basins is possible for (21) since it can be reduced to an autonomous two-dimensional map. This is easily obtained by noticing that, from (19), the sequence $\{\alpha_t\}$ defined in (20) can be defined recursively as

$$\alpha_{t+1} = \frac{\alpha_t}{\alpha_t + \rho}, \quad \alpha_0 = 1$$

So, the model (1) endowed with learning (18) can be written as

$$\begin{cases} x_t = F\left(x_t^{(e)}\right) \\ x_{t+1}^{(e)} = (1 - \alpha_t) x_t^{(e)} + \alpha_t x_t \\ \alpha_{t+1} = \frac{\alpha_t}{\alpha_t + \rho} \end{cases}$$

with initial conditions x_0 (the initial realized value) and $\alpha_0 = 1$. This recursive relation is already known in the literature, at least for the limiting case of Bray learning, i.e. for $\rho = 1$ (see e.g. Marimon 1997). Following the same procedure as in the case of adaptive expectations, we can use (1) to obtain a mapping (which is two-dimensional in this case) which defines the time evolution of the expected values. However, it is important to remark that in this case the iteration procedure starts with the value observed in $t = 0$, given by x_0, and this implies that the first expected value used to start (23) is given, according to (10), by $x_1^{(e)} = x_0$. So, the sequences of expected values generated by (21) can be obtained from the iteration of (23) starting from $\alpha_0 = 1$ and $x_1^{(e)} = x_0$. This means that the difference equation by which $x_t^{(e)}$ is recursively computed is shifted of one period with respect to the other one:

$$\begin{cases} x_{t+2}^{(e)} = (1 - \alpha_{t+1}) x_{t+1}^{(e)} + \alpha_{t+1} F\left(x_{t+1}^{(e)}\right) \\ \alpha_{t+1} = \frac{\alpha_t}{\alpha_t + \rho} \end{cases} \tag{23}$$

Following Bischi and Naimzada (1997) and Bischi and Gardini (1997), in order to study the general properties of the two-dimensional map (23) we rewrite it in the equivalent form

$$T : \begin{cases} z_{t+1} = \frac{\rho W_t}{1 + \rho W_t} z_t + \frac{1}{1 + \rho W_t} F(z_t) \\ W_{t+1} = 1 + \rho W_t \end{cases} \tag{24}$$

where $z_t = x_{t+1}^{(e)}$ and W_t is defined in (17). The sequence of expected values of the model (1) with learning (18) are obtained from the trajectories of the two-dimensional recurrence (24) provided that the conditions are chosen as:

$$z_0 = x_1^{(e)} = x_0 \quad \text{and} \quad W_0 = 1 \tag{25}$$

Starting from a given (z_0, W_0) the iterations of the map T uniquely defines the trajectory $\tau = \{(z_t, W_t) = T^t (z_0, W_0), t \geq 0\}$ and if $(z_0, W_0) = (x_0, 1)$ then the sequence $\{z_t, t \geq 0\}$ represents the time evolution of the expected variables $\{x_t^{(e)}, t \geq 1\}$ from which the sequence of realized values $\{x_t\}$ starting with the given x_0 is simply obtained as the images under the function F:

$$x_t = F\left(x_t^{(e)}\right) \quad t \geq 1 \tag{26}$$

In other words, if $\{(z_0, W_0), (z_1, W_1), \dots, (z_t, W_t), \dots\}$ is the sequence generated by the map T starting from the initial condition $(z_0, W_0) = (x_0, 1)$, then $\{x_1^{(e)} = z_0,$ $x_2^{(e)} = z_1, \dots, x_{t+1}^{(e)} = z_t, \dots\}$ is the sequence of expected values, and $\{x_0, x_1 = F(z_0), \dots, x_t = F(z_{t-1}), \dots\}$ is the corresponding sequence of realized values. Thus the study of the general model (1) with learning rule (18) is reduced to that of a two-dimensional map with initial conditions constrained on the line $W = 1$ (*line of initial conditions*). The class of maps (24) has been initially studied in Bischi and Gardini (1997), which inspired a stream of literature on maps with vanishing denominator, see Bischi et al. (1999), Bischi et al. (2003), and Bischi et al. (2005), from which several applications followed, e.g. Tramontana (2016) and Gu and Hao (2007).

3 Limit Sets and Basins of Attraction for Statistical Learning With fading Memory

Any trajectory of (24) starting from initial conditions on the line $W = 1$ is confined in the strip $0 < W < \frac{1}{1-\rho}$. In fact, this strip is mapped into itself by T because the second difference equation in (24), which gives the dynamics of the variable W, is independent of z and gives a monotonically increasing sequence (the partial sums of a geometric series of ratio ρ) and if $0 \leq \rho < 1$ such sequence $\{W_t\}$ converges to the sum of the geometric series

$$W^* = \frac{1}{1 - \rho} \,. \tag{27}$$

For $0 \leq \rho < 1$ the line $W = W^*$ is an invariant and globally attracting line for the map T, on which the ω-limit sets of all its trajectories are located. For this reason we shall call it *line of ω-limit sets* . The restriction of T to this line is given by the one-dimensional map

$$g(z) = \rho z + (1 - \rho) F(z) \,, \tag{28}$$

already obtained in (22) as the limiting form of the nonautonomous recurrence (21). The map (28) will be called *limiting map*, since it governs the asymptotic behavior of the map T. This implies, as proved in Bischi and Gardini (1996), that any k-cycle $A = \left\{z_1^*, \ldots, z_k^*\right\}$ of the map $g_\rho(z)$ is in one-to-one correspondence with a k-cycle $\mathcal{A} = A \times \{W^*\} = \{(z_1^*, W^*), \ldots, (z_k^*, W^*)\}$ of the map T, located on the line of ω-limit sets. Moreover, the attractors of the model (1) with learning scheme (18), as well as their basins of attraction, can be studied on the basis of the following proposition, given in Bischi and Gardini (1996) (see also Bischi and Naimzada 1997):

Proposition 1 *Let A be a k-cycle, $k \geq 1$, of the map $g_{1-\rho}(z)$, $0 \leq \rho < 1$. Then*
   (i) if A is attracting for the limiting map $g_\rho(z)$, then the set $\mathcal{A} = A \times \{W^*\}$ is an attracting cycle of the map T, and $F(A)$ is an attracting cycle of the model (1) with learning scheme (18);
   (ii) the basin of attraction D_1 of the attractor $F(A)$ of the model (1) with learning scheme (18) is given by the intersection of the two-dimensional basin $\mathcal{B}$ of the cycle $\mathcal{A}$ of the map T with the line of initial conditions $W = 1$.

We recall that the case $k = 1$ corresponds to a fixed point z^* of $g(z)$, and $F(A) = F(z^*) = z^*$ is a REE, since the fixed points of $g(z)$ are also fixed points of $F(z)$.

The part (i) of the Proposition 1 confirms that the asymptotic behavior, i.e. the kinds of attractors and their stability properties, are the same as those of a standard adaptive learning rule with adaptive coefficient $\alpha = 1 - \rho$. For example, a sufficient condition for the attractivity of a REE z^*, under learning (18) with $\rho < 1$, is given by $\left|g'(z^*)\right| < 1$, that is,

$$-\frac{\rho + 1}{1 - \rho} < F'(z^*) < 1 \,. \tag{29}$$

However, the most important implications of Proposition 1 are due to part (ii), since it suggests a general procedure to obtain the boundaries of the basins of attraction when two or more coexisting attractors are present, as often occurs in the case of nonlinear models. This is an important issue that cannot be studied on the basis of the limiting map g_ρ, because the initial conditions are to be taken on the line $W = 1$, whereas g_ρ only governs the dynamics near the line of ω-limit sets $W = W^*$. This means that only a global knowledge of the two-dimensional map T allows one to follow the short-run behavior, during which the dynamics is not governed by the limiting map g.

Moreover, as outlined in the Proposition 1, the basins of attraction of the two-dimensional map T, whose intersection with the line of initial conditions $W = 1$ gives the basins of the model with statistical learning, are obtained by considering the preimages of proper neighborhoods of the attracting sets located along the line of ω-limit sets. We recall that the two dimensional *basin of attraction* of an attractor A of the map T is the open set of points which generate trajectories converging to A:

$$\mathcal{B}(A) = \left\{ (z, W) \mid T^t(z, W) \to A \text{ as } t \to +\infty \right\} \quad . \tag{30}$$

A closed invariant set $A \subset \{W = W^*\}$ is called *asymptotically stable* (or *attracting*) if a neighborhood U of A exists such that $T(U) \subseteq U$ and $T^n(x) \to A$ as $n \to +\infty$ for each $x \in U$. Then, the *basin* of A is obtained by taking all the preimages of the points of U

$$\mathcal{B}(A) = \bigcup_{n=0}^{\infty} T^{-n}(U)$$

where $T^{-n}(x)$ denotes the set of all the preimages of x of rank n, i.e. the set of all the points which are mapped into x after n iterations of T. So, the study of the two-dimensional basin is based on the study of the inverses of T. In the case of the map (24) we have that the properties and the qualitative changes of its basins are strongly influenced by the presence of the denominator which can vanish along the line $W = -\frac{1}{\rho}$ and, in particular, by the points in which the first component of T assumes the form 0/0, see Bischi et al. (1999, 2003, 2005), Gardini et al. (2007), and Bischi and Gardini (1997) for the particular class of triangular maps (24). In these papers it is proved that the presence of points where a component of the map assumes the form 0/0, called *focal points*, may have important consequences on the structure of the basins and their global bifurcations, because fans of basins boundaries arise from them giving peculiar finger-shaped structures called *lobes*. The existence of lobes, originating from the focal points, may have important consequences on the structure of the basins of attraction of the model with learning (18) whenever they intersect the line of initial conditions $W = 1$. This occurrence causes the creation of basins with a complicated topological structure, such as basins formed by many disjoint intervals, as we shall see in the next section.

3.1 Global Properties and Structure of the Basins of The two-Dimensional Triangular Map

In the following we briefly recall some definitions and properties specific to maps with vanishing denominator (see Bischi et al. 1999, 2003, 2005 for a more complete treatment). Let us consider a map $(x, y) \rightarrow (x', y') = T(x, y)$ of the form

$$T : \begin{cases} x' = F(x, y) \\ y' = G(x, y) \end{cases} \tag{31}$$

where x and y are real variables and at least one of the components has the form of a fractional rational function, i.e.

$$F(x, y) = \frac{N_1(x, y)}{D_1(x, y)} \quad \text{and/or} \quad G(x, y) = \frac{N_2(x, y)}{D_2(x, y)} \tag{32}$$

The functions $N_i(x, y)$ and $D_i(x, y)$, $i = 1, 2$, are defined in the whole plane $\mathbb{R}^2$, so that the set with no definition δ_s of the map T coincides with the locus of points in which at least one denominator vanishes:

$$\delta_s = \left\{ (x, y) \in \mathbb{R}^2 | D_1(x, y) = 0 \text{ or } D_2(x, y) = 0 \right\} \tag{33}$$

The two dimensional recurrence obtained by the iteration of T is well defined, i.e. it generates not terminating trajectories, provided that the initial condition belongs to the set E given by

$$E = \mathbb{R}^2 \setminus \bigcup_{k=0}^{\infty} T^{-k}(\delta_s) \tag{34}$$

so that $T : E \rightarrow E$. We recall here the following definition

Definition *A point $Q = (x_Q, y_Q)$ is a focal point if at least one component of the map T takes the form 0/0 in Q and there exist smooth simple arcs $\gamma(\tau)$, with $\gamma(0) = Q$, such that $\lim_{\tau \to 0} T(\gamma(\tau))$ is finite. The set of all such finite values, obtained by taking all the arcs $\gamma(\tau)$ through Q, is the prefocal set δ_Q.*

Roughly speaking, a *prefocal curve* is a set of points for which at least one inverse exists which maps (or "focalizes") the whole set into a single point, called *focal point*. For maps with a vanishing denominator, new kinds of contact bifurcations have been evidenced which involve the singularities defined above. In particular, contacts between basin boundaries and prefocal curves may cause the creation of particular structures of the basin boundaries, denoted as *lobes* and *crescents*. These particular structures have been observed in the study of discrete dynamical systems of the plane which arise in some different contexts, see e.g. Billings and Curry (1996), Billings et al. (1997), Foroni et al. (2003), Gardini et al. (1999), Tramontana (2016),

Pecora and Tramontana (2016), Cavalli and Naimzada (2015), and Yee and Sweby (1994). As already mentioned, the existence of lobes, originating from the focal points, has important consequences on the structure of the basins of attraction of the model with statistical learning whenever they intersect the line of initial conditions $W = 1$, causing the creation of one dimensional basins, of the model (1) with learning (18) with a complicated topological structure.

We now briefly describe the basic mechanism leading to the formation of lobes and crescents. In order to do this, let us consider the map T given in (24), which we rewrite as $T : (z, W) \to (z', W')$, i.e.

$$T : \begin{cases} z' = \frac{\rho W}{1+\rho W} z + \frac{1}{1+\rho W} F(z) \\ W' = 1 + \rho W \end{cases} \tag{35}$$

and we consider the image of an arc crossing through a *focal point*. We shall see that, according to the general results given in Bischi et al. (1999), a one-to-one correspondence is obtained between the slopes of the arcs through a focal point and the points in which their images cross the corresponding prefocal curve.

We first notice that the map (35) is not defined in the whole plane, because the denominator of the first component vanishes on the points of the line δ_s of equation $W = -\frac{1}{\rho}$. So, in order to have a well defined recurrence we must exclude from the phase plane of T the singular line as well as all its preimages of any rank δ_s^{-n} for each $n \geq 1$, belonging to a sequence of lines located below the singular line obtained by backward iteration of the second component of T, i.e.

$$W = \frac{W' - 1}{\rho}. \tag{36}$$

So, δ_s^{-1} has equation $W = -\frac{1+\rho}{\rho^2}$ and is located below δ_s, and analogously δ_s^{-n}, the set of points which are mapped in the singular line after n iterations of T, are located on the line of equation

$$W = -\frac{\sum_{k=0}^n \rho^k}{\rho^{n+1}} = -\frac{1 - \rho^{n+1}}{\rho^{n+1} - \rho^{n+2}}. \tag{37}$$

and the phase space of the recurrence defined by the map T is given by

$$E = \mathbb{R}^2 \setminus \bigcup_{n=0}^{\infty} \delta_s^{-n}. \tag{38}$$

where $\delta_s^{-(n+1)}$ is below δ_s^{-n} for each $n \geq 1$. In the map (35) only the first component has denominator, which vanishes in the points of the singular line $y = -\frac{1}{\rho}$, where the numerator becomes $F(x) - x$, hence it vanishes at every fixed point of the function F

(and thus also of the limiting map (22)). It follows that the a focal point is necessarily of type $\left(x^*, -\frac{1}{\rho}\right)$, where x^* is a fixed point of $F(x)$.

In order to explain the role of a *focal point* and the related *prefocal set* in the geometric and dynamic properties of the map T, following the arguments given in Bischi et al. (1999) we consider a smooth simple arc γ transverse to δ_s and how it is transformed by T. Let $(z_0, -1/\rho)$ be the point where γ intersects δ_s and assume that the arc γ is deprived of $(z_0, -1/\rho)$. If $z_0 \neq x^*$, i.e. $F(z_0) \neq z_0$, then the image $T(\gamma)$ is made up of two disjoint unbounded arcs asymptotic to the line of equation $y = 0$, as qualitatively shown in Fig. 1. A different situation may occur if $z_0 = x^*$, i.e. $F(z_0) = z_0$, because in this case the numerator of the first component also vanishes, and the limit of $T(\gamma)$ may take finite values as $(z, W) \rightarrow \left(x^*, -\frac{1}{\rho}\right)$, so that $T(\gamma)$ is a bounded arc (as qualitatively sketched in Fig. 1 for the arc γ_2). If m is the slope of the tangent to the smooth arc γ in the focal point $Q = \left(x^*, -\frac{1}{\rho}\right)$ then in Bischi et al. (1999) it is proved that the image $T(\gamma)$ crosses the line $W = 0$ in the point $(u_m, 0)$ with

$$u_m(x^*) = x^* + \frac{F'(x^*) - 1}{\rho m}. \tag{39}$$

This means that the images of the arcs crossing through $\left(x^*, -\frac{1}{\rho}\right)$ with slope $m \neq 0$ are bounded arcs (as qualitatively shown in the right panel of Fig. 1), and as m varies in $\mathbb{R}$ all the points of the line $W = 0$ are obtained, provided that $F'(x^*) \neq 1$. Thus the line of equation $W = 0$ represents the *prefocal set* δ_Q for the map (35). The situation in which $F'(x^*) = 1$ can be considered as a bifurcation case (see Bischi et al. 2005).

This suggests some consequences when we consider the preimages. The map (35) may be a noninvertible map, because the number of distinct inverses of T depends on the function $F(x)$. In fact, even if a point (z, W) has a unique image under the

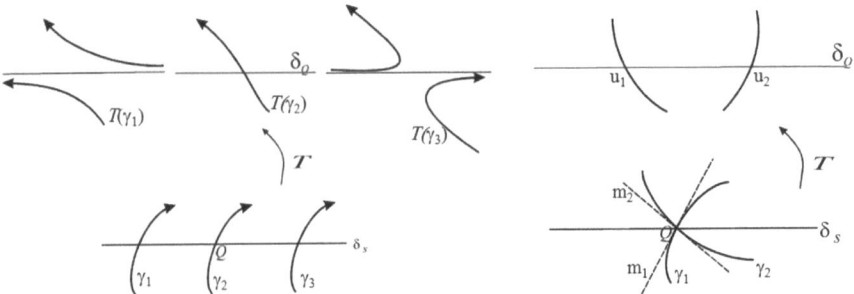

Fig. 1 Schematic picture of the action of a two-dimensional map on an arc crossing a singular curve δ_S along wich a denominator vanishes. Left: The arcs γ_1 and γ_3 cross the singular curve in a generic point of δ_S whereas γ_2 crosses it through a focal point. Right: Two arcs crossing δ_S through a focal point with different slopes are mapped into finite arcs crossing the prefocal curve δ_Q in different points

application of T, given by $(z', W') = T(z, W)$, the backward iteration of T may not be uniquely defined, since given a point (z', W') its preimages (z, W) are obtained by solving, with respect to the unknowns z and W, a system which may have several real solutions, i.e. several inverses. If n is the number of distinct inverses we denote them by $T_i^{-1}(z', W')$ for $i = 1, ..., n$ and $T^{-1}(z', W') = \bigcup_{i=1}^n T_i^{-1}(z', W')$. Moreover, if $F(x)$ has N fixed points (hence also T has N fixed points) then the prefocal line must belong to a region, say Z_N, whose points have N distinct rank-1 preimages.

To sum up, *for each focal point* $Q_i = \left(x_i^*, -\frac{1}{\rho}\right)$ *the map T in (35) defines a one-to-one correspondence between the slope m of an arc γ through Q_i and the point $(u, 0)$ in which the image $T(\gamma)$ crosses the prefocal curve δ_Q, given by*

$$
\begin{array}{lll}
m \to (u, 0) & : & u = x_i^* + \frac{F'(x_i^*)-1}{\rho m} \\[2mm]
(u, 0) \to m & : & m = \frac{F'(x_i^*)-1}{\rho(u-x_i^*)}
\end{array}
\tag{40}
$$

Some consequences of this correspondence, important for the characterization of the basins' boundaries and their bifurcations, are deduced by considering a smooth arc ω that intersects the prefocal line in two points. In this case, the N rank-1 preimages of ω, say $T_i^{-1}(\omega)$, $i = 1, ..., N$, are arcs such that each $T_i^{-1}(\omega)$ has a loop with knot in the focal point $Q_i = (x_i^*, -\frac{1}{\rho})$. This implies that a remarkable contact bifurcation occurs when a smooth curve segment ω moves towards the prefocal curve δ_Q until it has a contact and then crosses δ_Q (as qualitatively shown in Fig. 2). As ω moves toward δ_Q, its preimages move towards Q_i, and when ω becomes tangent to δ_Q then each preimage $\omega_{-1}^i = T_i^{-1}(\omega)$ has a cusp point at Q_i. The slope of the common tangent line to the two arcs that join in Q_i is given by $m_i(u_c)$, according to (40). If the curve segment ω moves further, so that it crosses δ_Q in two points, say $(u_1, 0)$ and $(u_2, 0)$, then its preimages form loops with double points at the focal points Q_i.

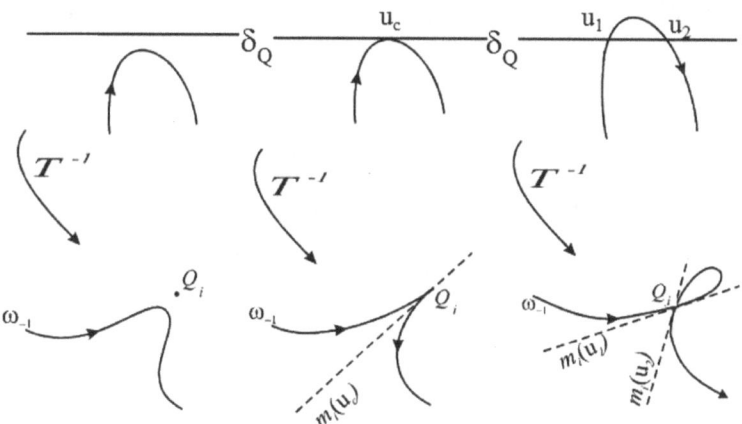

Fig. 2 Qualitative picture of a preimage of an arc moving towards a prefocal line until crossing it

This kind of contact bifurcation is important in the study of the boundaries of the basins of attraction, because if ω is a portion of a basin boundary, a contact between ω and δ_Q implies that a loop is created along the basin boundary, because a basin boundary is backward invariant, i.e. it includes all the preimages of any portion of it, and the portion of the basin inside the loop is a lobe, as we shall discuss in the next sections. Moreover, in the case of noninvertible maps the creation of *crescents* can be obtained as well, obtained from the merging of lobes as qualitatively shown in Fig. 3. It is caused by contacts of a critical curve LC with a basin boundary which already includes lobes which merge along LC_{-1} after the contact (see e.g. Mira et al. 1996 for a definition of critical curves).

Now, let us consider the forward iteration of T. It is easy to see that the image of rank-n of the prefocal line $W = 0$ belongs to the line of equation $W = W_n$ where

$$W_n = \frac{1 - \rho^{n+1}}{1 - \rho} \tag{41}$$

i.e. a sequence of lines parallel to the prefocal line δ_Q and convergent to the line of the ω-limit sets $W = W^*$. This implies that any cycle belonging to the ω-limit set $W = W^*$ is transversely attracting. This property is important in order to study the boundaries of the basins. In fact, recall that, in general, the boundaries of a basin are obtained by taking the stable sets of some cycles on it. In the case of maps (35) such cycles can only be located on the line of ω-limit sets. To get the stable set W^s of a

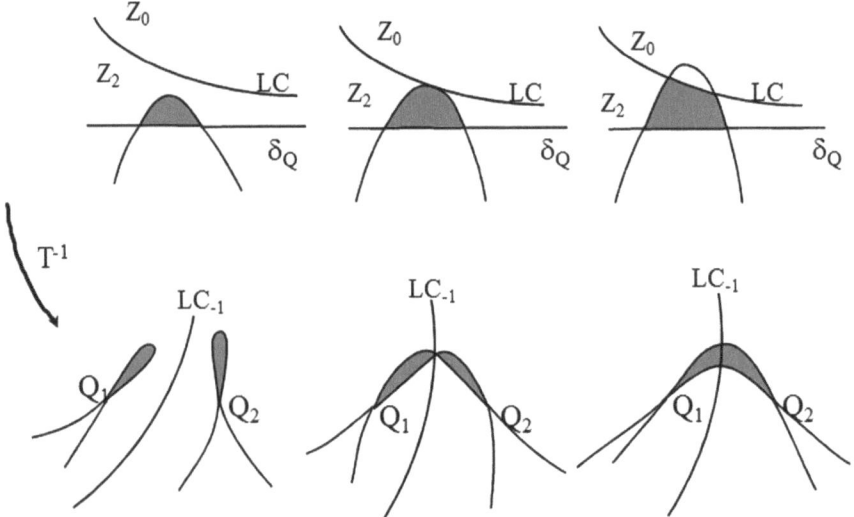

Fig. 3 Qualitative sketch to describe the formation of a *crescent* obtained by the merging of two *lobes* when a portion of a basin of attraction crossing the prefocal line δ_Q has a contact with a critical curve LC

saddle it is enough to take the preimages of any rank of a local stable set W^s_{loc}, that is $W^s = \bigcup\limits_{n=0}^{\infty} T^{-n}(W^s_{loc})$, where W^s_{loc} is transverse to the line $W = W^*$. Due to the expansive character of T^{-1} along the W direction, as defined in (36), such preimages must necessarily reach, in a finite number of steps, the prefocal line $W = 0$. So, all these preimages must necessarily cross the singular line $W = -\frac{1}{\rho}$ through focal points Q_i.

From this observation it follows that the stable set of any saddle cycle of T, obtained by taking the preimages of a local stable set, is made up of branches issuing from the focal points. In fact, the preimages of any local stable set, transverse to the line of ω-limit sets $W = W^*$, necessarily go back to the prefocal line $W = 0$ in a finite number of steps. Thus any stable set must be made up of branches which "cross" the singular line in the focal points, i.e. are "focalized" through the focal points. This argument, applied to the stable set of some saddle cycle belonging to the line of ω-limit sets, constitutes the global mechanism which causes the formation of the particular structures of the basins which will be shown in the examples.

3.2 Increasing Maps

We have seen that even if the law of motion (1) with the two different kinds of learning (4) and (18) has the same attracting sets, the corresponding basins of attraction are generally different. This is related to the fact that the asymptotic dynamics obtained with both the learning mechanisms are governed by the map $g_\rho = g_{1-\alpha}$, whereas the basins are obtained by different procedures.

However, things are simpler when F is an increasing map. In this case the limiting map g is also increasing. As it is well known, for a continuous and increasing map the only invariant sets are the fixed points, and when several fixed points exist, say $x^*_1 < x^*_2 < ... < x^*_k$ they are alternatingly stable and unstable: the unstable fixed points are the boundaries that separate the basins of the stable ones. Indeed, in this case the same situation also holds for the basins of the stable fixed points under the statistical learning.

Proposition *If F is increasing then the basins with adaptive learning and speed α are the same as those for the statistical learning with fading memory and ratio $\rho = 1 - \alpha$.*

Proof Let F be a continuous and increasing function and let x^* be an unstable fixed point of F. The stable set of the saddle $S = (x^*, 1/(1 - \rho))$ is obtained by taking the preimages of any rank of the local stable set of S, which is included in the line $z = x^*$. The rank-1 preimages of a point (x', W') are obtained by

$$\begin{cases} F(z) + z\left(W' - 1\right) = x'W' \\ W = \frac{W'-1}{\rho} \end{cases}$$

and for any $W' \geq 1$ this has only one solution, because the left hand side of the first equation is increasing if F is increasing and $W' \geq 1$. In particular, the only rank-1 preimage of a point (x^*, W') belongs to the line $z = x^*$, so that the projections of the unstable fixed points on the line of initial conditions $W = 1$ are the only boundaries which separate the basins, like in the one-dimensional map g □.

We remark that two dimensional basins of the attracting fixed points of the map T, located on the line $W = W^*$, may include portions which do no belong to the vertical lines through the saddle points, but these are necessarily confined in the region $W < 1$, so that they have no influence on the basins of the model (1) with learning (18). An example will be shown below.

3.3   The Particular Case of Bray Learning

We have seen that the case of the statistical learning (12) proposed by Bray (1983) can be obtained from the statistical learning with fading memory (18) in the limiting case $\rho \to 1^-$. It can be noticed that in this case any REE x^* with $F'(x^*) < 1$ is stable, i.e. it can be "learned" by the agents, because the stability condition (29) is always satisfied as $\rho \to 1^-$. In other words, for the general model (1) with Bray's learning (12), the steady states x^* characterized by $F'(x^*) < 1$ are locally attracting equilibria, whereas those with $F'(x^*) > 1$ are unstable saddles. This confirms, and extends, the stability results obtained, for particular models, by Bray (1983), Dimitri (1988), and Holmes and Manning (1988). That is: *in the case of Bray's learning* (12) *any complexity is lost, and any trajectory is either divergent or convergent to a stable REE x^**.

However, besides divergent trajectories there may be two or more coexisting stable REEs, and the arguments given above about the complexity in the basins also hold in this case. In fact, even if $W^* = 1/(1 - \rho) \to +\infty$, for each REE x^* with $F'(z^*) < 1$ the invariant lines $z = x^*$ are attracting sets whose basins can be obtained following the same procedure outlined in the previous sections. The triangular map (35) becomes

$$
T : \begin{cases} z' = z + \dfrac{F(z) - z}{1 + W} \\[2mm] W' = 1 + W \end{cases}
$$

with initial condition taken on the line $W = 1$. The line with no definition is $W = -1$, on it a focal point is associated with each REE and the prefocal line is $W = 0$.

Except for the uninteresting case in which a unique unstable REE exists (and all the trajectories are divergent), or the simple case in which only one globally attracting REE exists, there are both attracting lines $z = x_i^*$ and repelling lines $z = z_j^*$, associated with stable REE x_i^* and unstable REE z_j^* respectively. Then, any repelling invariant line $z = z_j^*$ has a stable set (made up of all the preimages of any rank of the line $z = z_j^*$) which separates different basins. Such preimages may intersect the prefocal line of T. Thus, depending on the topological structure of these preimages, associated with the inverses of T, the basins on the line $W = 1$ may have a simple or complex geometrical structure.

4 Examples

4.1 Unimodal Maps: A Cobweb Model

One of the simplest models expressed by a law of motion of the form $x_t = F\left(x_t^{(e)}\right)$ is the *cobweb model* (see e.g. Nerlove 1958, Hommes 1991, Chiarella 1988, Jensen and Urban 1984). In the market of a given good, let $q_D = D(p)$ and $q_S = S(p)$ be the demand and supply functions respectively. At the time t, q_D depends on the current price p_t, whereas q_S depends of the price $p_t^{(e)}$ expected by producers at the previous time in which they decided their production. If the production delay is taken as the time unit, the market clearing condition becomes $D(p_t) = S\left(p_t^{(e)}\right)$, and assuming that $D(p)$ is a continuous and decreasing function (hence invertible) the law of motion of the market clearing price is $p_t = D^{-1}S(p_t^{(e)})$ at which the adaptive learning (4) or the statistical learning (18) can be applied. As an exercise to illustrate the results of the previous sections, and to compare the two kinds of expectations, we consider a cobweb model where $F(x) = D^{-1}S(x)$ is a quadratic map, like in Jensen and Urban (1984) or Dimitri (1988), where a linear demand function is considered together with a backward bending supply curve, expressed by a quadratic function, so that $F(x)$ is conjugate to the standard logistic map

$$f(x) = \mu x (1 - x), \ \mu > 1 \tag{42}$$

So, in the following we consider a model $x_t = f\left(x_t^{(e)}\right)$ with f given by (42). In this case, the dynamics of the expected prices under the assumption of adaptive expectations (4) is governed by a quadratic map as well, given by

$$z' = g_\alpha(z) = (1 - \alpha) z + \alpha \mu z (1 - z) \tag{43}$$

whereas under the assumption of statistical learning (18) the dynamics of the expected prices is obtained by the two-dimensional map (24)

$$T : \begin{cases} z_{t+1} = \frac{\rho W_t z_t + \mu z_t (1 - z_t)}{1 + \rho W_t} \\ W_{t+1} = 1 + \rho W_t \end{cases} \tag{44}$$

and the limiting map which governs the asymptotic behavior is:

$$z' = g_{1-\rho}(z) = \rho z + (1 - \rho)\mu z (1 - z) \tag{45}$$

For each $\mu > 1$ there are two non negative REEs, given by

$$s^* = 0 \quad \text{and} \quad p^* = \frac{\mu - 1}{\mu}$$

where s^* is an unstable fixed point of f, with $f'(s^*) > 1$, hence it is also an unstable fixed point of g for each $\rho \in [0, 1)$, whereas p^* is stable for $1 < \mu \leq 3$, and unstable for $\mu > 3$ with $f'(p^*) < -1$. This means that the REE p^* is stable under the assumption of adaptive expectations provided that α is sufficiently small (and the same is true for the statistical learning with ρ sufficiently close to 1). In the following we shall consider values of α or ρ such that the REE p^* is stable, however even in this case, p^* is not globally stable since for each $\alpha \geq 0$ ($\rho \leq 1$) diverging sequences of expected values can be obtained as well. This raises the question of the study of the basins of attraction, i.e. the delimitation of the boundary which separates the set of initial conditions that generate trajectories converging to the REE (i.e. the basin of attraction of p^*) from the set of initial conditions that generate unbounded trajectories (i.e. the basin of attraction of infinity). This question is very easily solved for the case of adaptive learning, whose study requires a simple analysis of the one-dimensional quadratic map (43), for which the basin of p^* is given by the interval $]0, g_\alpha^{-1}(0)[=]0, 1[$. Instead, for the statistical learning a global analysis of the two dimensional map (44) requires more advanced methods. In fact, (44) is a noninvertible map with two focal points,

$$Q_1 = \left(0, -\frac{1}{\rho}\right) \quad \text{and} \quad Q_2 = \left(\frac{\mu - 1}{\mu}, -\frac{1}{\rho}\right). \tag{46}$$

In Fig. 4a, obtained with $\rho = 0.75$ and $\mu = 6$, the basins of the two-dimensional map (44) are shown: The white region represents the set of points converging to the fixed point $P = (p^*, W^*) = \left(\frac{\mu-1}{\mu}, \frac{1}{1-\rho}\right)$, and the grey region represents the set of points which generate diverging trajectories. The intersections with the line of initial conditions $W = 1$ represent the respective one-dimensional basins of the cobweb model with statistical learning (18) given by the interval $(0, \bar{z})$ with $\bar{z} = 1.125$. Instead, if we consider the adaptive learning (4) with $\alpha = 0.25$, so that the dynamics of expected prices are governed by the same one-dimensional map, the basin of the REE p^*, is $\left(0, g_{0.25}^{-1}(0)\right) = (0, 1.5)$, where $g_{0.25}^{-1}(0)$ is the preimage of the unstable fixed point s^* different from itself. This basin coincides with the portion of the line of ω-limit sets $W = W^*$ included in the white region of Fig. 4. A trajectory starting from $x_0^{(e)} = 1.2$ converges to p^* under adaptive expectations, whereas for the model with statistical learning (18) with $\rho = 0.75$, for which the asymptotic dynamics of the expected prices are governed by he same limiting map, the trajectory starting

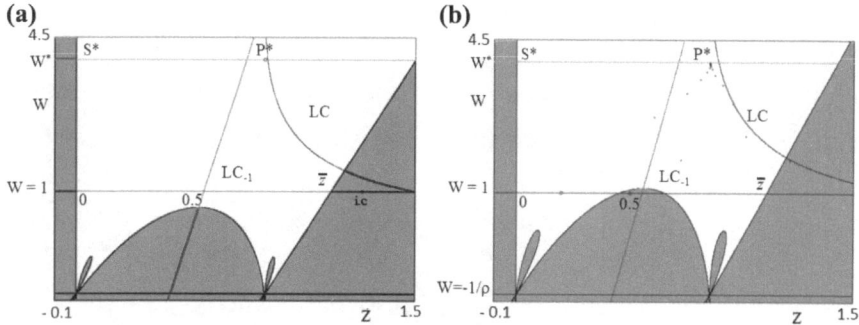

Fig. 4 Numerically generated basins of attraction of the two-dimensional map (44): The white region represents the set of points converging to the fixed point P^* and the grey region represents the set of points which generate diverging trajectories. **a** $\rho = 0.75$ and $\mu = 6$; **b** $\rho = 0.75$ and $\mu = 7$

from $z_0 = p_1^{(e)} = 1.2$ diverges. So, with the same starting condition, the trajectory obtained with the standard adaptive learning converges to the REE p^*, whereas the model with statistical learning (18) with $\rho = 1 - \alpha$ does not converge.

An even more evident difference is obtained in the situation shown in Fig. 4b, obtained with $\rho = 0.75$ and $\mu = 7$. Now the basin $B(p^*)$ is formed by two disjoint intervals, because a "hole" formed by points which generate diverging trajectories is nested inside $B(p^*)$.

It can be noticed that the size, in the z direction, of $\mathcal{B}(\mathcal{P})$, increases for higher values of W, so that stronger shocks are necessary to bring the phase point inside $\mathcal{B}(\infty)$, i.e. the system is less vulnerable with respect to exogenous perturbations as time goes on. Loosely speaking we may say that as the amount of information (i.e. the number of observed realized values) increases the system has a greater probability to converge, because agents learn to behave more and more rationally as time goes on.

The above considerations are even more evident when applied to situations like the one shown in Fig. 5, obtained for $\rho = 0.75$ and $\mu = 9$. In this case the basin of p^* is formed by 4 disjoint intervals, due to the presence of lobes of $\mathcal{B}(\infty)$ intersecting the line of initial conditions $W = 1$.

We now describe a procedure to obtain the exact delimitation of the boundary $\mathcal{F}$ that separates the two basins. In fact, the complementary set of $\mathcal{B}(\infty)$ is the set of bounded trajectories which converge to invariant sets of the limiting map g, on the line $W = W^*$. As remarked above, the attractor always coincides with the REE if the memory ratio ρ is sufficiently close to 1, whereas for lower values of ρ the bounded attractor of the map $g_\rho(z)$ may be a cycle or even a chaotic set. Here we are only interested in parameters' values for which the REE p^* is locally stable, but it is clear that arguments similar to those used below hold independently of the topological structure of the attracting set of $g_\rho(z)$. Thus $P = (p^*, W^*)$ denotes the attractor of T located on the line $W = W^*$, whose basin will be denoted by $\mathcal{B}(\mathcal{P})$. The frontier $\mathcal{F}$ behaves as a repelling set for the points near it, since it acts as a watershed for

Fig. 5 The same as Fig. 4, with $\rho = 0.75$ and $\mu = 9$

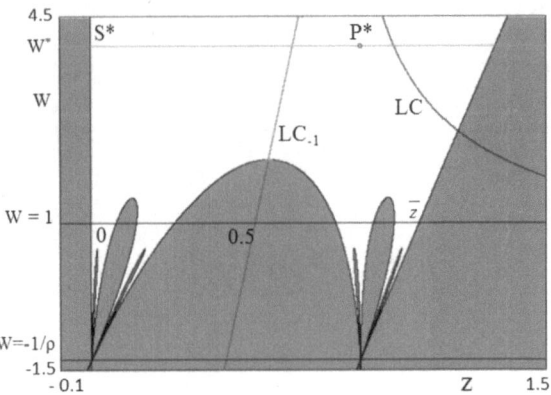

the trajectories of the map T. Points belonging to $\mathcal{F}$ are mapped into $\mathcal{F}$ both under forward and backward iteration of T: more exactly $T(\mathcal{F}) \subseteq \mathcal{F}$, $T^{-1}(\mathcal{F}) = \mathcal{F}$ (see Mira et al. 1996 Chap. 5). This implies that if a saddle-point belongs to $\mathcal{F}$, then $\mathcal{F}$ must also contain the whole stable manifold (see Gumowski and Mira 1980; Mira et al. 1996). In our case, for each $\mu > 1$ and $0 \leq \rho < 1$ the point $S = (0, W^*)$, located on the line of ω-limit sets $W = W^* = 1/(1 - \rho)$, is a saddle point for the map T, with local stable manifold along the invariant line $z = 0$ and unstable set along the invariant line $W = W^*$. The local stable set of S belongs to $\mathcal{F}$ because the unstable manifold, along the line $W = W^*$, has a branch pointing toward the attractor P, and the opposite branch going to infinity (see Fig. 6). Then $\mathcal{F}$ includes the whole stable set of S, i.e.

$$\mathcal{F} \supseteq W^s(S) = \bigcup_{n \geq 0} T^{-n}(W^s_{loc}(S)) \tag{47}$$

where $W^s_{loc}(S)$ is given by the portion of the W axis with $W \in (0, W^*)$, denoted by ω_0 in Fig. 6, and $T^{-n}(z, W)$ denotes the set of all the rank-n preimages of the point (z, W), i.e. the set of points which are mapped into (z, W) after n applications of T. In our case, the map (44) is a noninvertible map of $Z_0 - Z_2$ type, i.e. a point (z', W') has no rank-1 preimages or two preimages, given by $T^{-1}(z', W') = T_1^{-1}(z', W') \cup T_2^{-1}(z', W')$, where

$$T_1^{-1} : \begin{cases} z = \dfrac{(W'+\mu-1) - \sqrt{(W'+\mu-1)^2 - 4\mu z' W'}}{2\mu} \\ W = \dfrac{W'-1}{\rho} \end{cases} \qquad T_2^{-1} : \begin{cases} z = \dfrac{(W'+\mu-1) + \sqrt{(W'+\mu-1)^2 - 4\mu z' W'}}{2\mu} \\ W = \dfrac{W'-1}{\rho} \end{cases}$$

$$\tag{48}$$

if $\Delta(z, W) = (W' + \mu - 1)^2 - 4\mu z' W' > 0$.

We say that a point (z, W) has two preimages, given by (48), if $\Delta (z, W) > 0$, and that no inverses are defined in the points (z, W) when $\Delta (z, W) < 0$. The curve defined by the equation

$$\Delta(z, W) = (W + \mu - 1)^2 - 4\mu z W = 0, \tag{49}$$

is called critical curve LC (from the french "Ligne Critique"). Its points have two coincident preimages located on the line LC_{-1} given by

$$LC_{-1} = \{(x, y) \,|\, \rho W - 2\mu z + \mu = 0\} \tag{50}$$

obtained from T_i^{-1} with $\Delta = 0$. It can also be obtained as the locus of points at which the determinant of the Jacobian matrix of T vanishes, and $LC = T(LC_{-1})$ (see Gumowski and Mira 1980; Mira et al. 1996; Abraham et al. 1997). As LC_{-1} crosses the singular line δ_s out of the focal points, $LC = T (LC_{-1})$ is formed by two unbounded branches asymptotic to the prefocal line δ_Q (see Fig. 6). The knowledge of the curves LC and LC_{-1} is important in the computation of the preimages of the local stable set of S from which $\mathcal{F}$ is obtained according to (47). Indeed, from (48) the rank-1 preimages of $W_{loc}^s (S)$ can be easily computed. The two rank-1 preimages of ω_0, which is entirely included inside Z_2, are one on the same (invariant) W-axis and the other one on the line of equation

$$W = \frac{\mu}{\rho} (z - 1) \tag{51}$$

denoted by ω_{-1} in Fig. 6. This line intersects the line of initial conditions $W = 1$ in the point

$$\bar{z} = 1 + \frac{\rho}{\mu} \tag{52}$$

According to (47), also the line (51) belongs to $\mathcal{F}$. It can be noticed that (51) "crosses" the singular line through the focal point Q_2. The portion of this line located below the critical curve LC belongs to the region Z_2, hence it has two preimages, say ω_{-2}^1 and ω_{-2}^2, whose equation can be obtained from (48) with $W' = \frac{\mu}{\rho} (z' - 1)$. These two preimages are located at opposite sides with respect to the line LC_{-1} and merge in the point H, given by the merging preimages of the point $H_1 = \omega_{-1} \cap LC$ (see Fig. 6). After some algebraic manipulation it is possible to see that such preimages belong to the curve of equation:

$$x = \frac{\mu + \rho W \pm \sqrt{(\mu + \rho W)^2 - 4(1 + \rho W)(\mu + \rho + \rho^2 W)}}{2\mu}. \tag{53}$$

The locus (53) represents an hyperbola if $\rho < \frac{1}{4}$, a parabola if $\rho = \frac{1}{4}$, an ellipse if $\rho > \frac{1}{4}$ (as in Fig. 6, obtained with $\rho = \frac{2}{3}$) and crosses the line with no definition

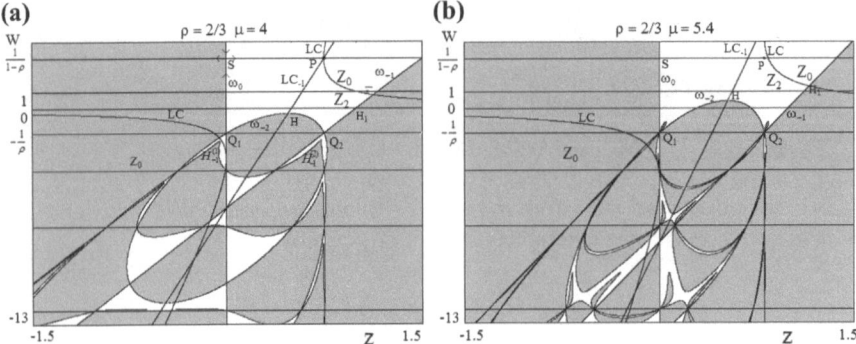

Fig. 6 An extended view of the numerically generated basins of attraction of the two-dimensional map (44): The white region represents the set of points converging to the fixed point P^* and the grey region represents the set of points which generate diverging trajectories. **a** $\rho = \frac{2}{3}$ and $\mu = 4$; **b** $\mu = 5.4$

$W = -1/\rho$ at the focal points. According to (47) also the curve (53) belongs to the frontier $\mathcal{F}$, as well as the preimages of ω_0 of any rank.

The union of all these preimages gives the boundary separating the basin $B(\infty)$ from the basin of the stable fixed point P, represented in Fig. 6, by the grey and the white regions respectively.

With the set of parameters used in Fig. 6a the two merging preimages of the point H_1, represented by the point H in which ω_{-2} intersects LC_{-1}, are below the prefocal line δ_Q. This implies that the two rank-1 preimages of H, denoted by $H_{-1}^{(1)}$ and $H_{-1}^{(2)}$ in Fig. 6a, are below the line δ_s.

As long as the point of intersection H_1 between LC and the line ω_{-1} is below the line $W = W_1 = 1$, the whole curve ω_{-2} lies below the z axis, so that the preimages of ω_{-2} are located below the singular line, as can be easily deduced from the second component of (48).

As μ increases the critical curve LC moves upwards, and when it reaches the line $W = W_1 = 1$ the curve ω_{-2} reaches the z axis, so that its preimages ω_{-3} appear, issuing from the two focal points Q_1 and Q_2. For example, in Fig. 6b the point H_1 is above the line $W = 1$, and consequently its preimage H, which is on the top of the arc ω_{-2}, is above the line $W = 0$. The two preimages of the portion of ω_{-2} above the z axis are the lobes issuing from the focal points Q_1 and Q_2, and the same happens at all the preimages of any rank of the focal points.

However, in order to understand the structure of the basins, we can limit our analysis to the portion of the plane above the line $W = -1/\rho$ (as in Fig. 7a)

For the set of parameters used in Fig. 7a the situation is similar to the one shown in Fig. 6b: the arc ω_{-2} of $\mathcal{F}$ does not intersect the line of initial conditions $W = 1$, thus it does not affect the basin of attraction $D_1(p^*)$ given by the intersection of $\mathcal{B}(\mathcal{P})$ with the line $W = 1$, according to Proposition 1. This is due to the fact that the point $H_1 = \omega_{-1} \cap LC$ is located below the line of equation $W = W_1 = 1 + \rho$, and this implies that its preimage H is located below the line of initial condition

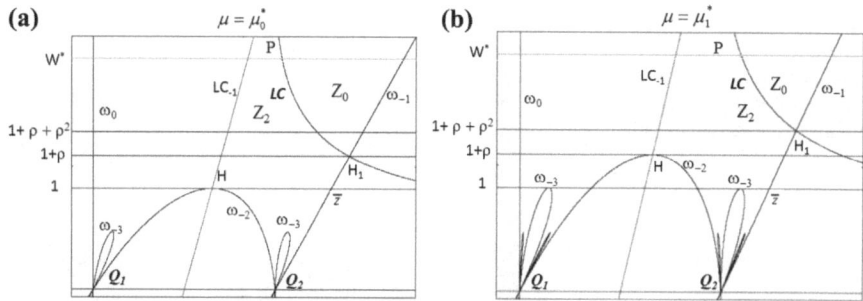

Fig. 7 Numerical computation of the preimages of the segment ω_0 located along the line $z = 0$

$W = W_0 = 1$. In fact, due to the particular structure of the second equation of the map T, the preimages of any point of a line $W = W_t$ are located on the line $W = W_{t-1}$, as can be easily computed by the second equation of (48).

As μ is increased, a value will be reached, say $\mu = \mu_0^*$, at which the point H_1 is on the line $W = W_1 = 1 + \rho$ and, as a consequence, the curve ω_{-2} becomes tangent to the line of initial conditions $W = W_0 = 1$ in the point $H = (z_H, 1)$ (see Fig. 7b) where

$$z_H = \frac{1}{2} + \frac{\rho}{2\mu}. \tag{54}$$

The value $\mu_0^* = 2 + 2\sqrt{1 + \rho}$ (as can be easily computed from the tangency condition between the curve of equation (53) and the line $W = 1$) represents a bifurcation of the basin $D_1(p^*)$ of initial conditions which generate bounded price sequences. In fact for $\mu < \mu_0^*$ the basin $D_1(p^*)$ is the interval $(0, \overline{z})$, with $\overline{z}$ given by (52), whereas for $\mu > \mu_0^*$ a *hole* is created around z_H, whose points belong to $\mathcal{B}(\infty)$, bounded by the two intersections $(h_1, 1)$ and $(h_2, 1)$ between the curve (53) and the line $W = 1$.

The situation becomes even more complex as μ is further increased. The value $\mu = \mu_1^* = 2 + \rho + \sqrt{(1 + \rho)(1 + \rho + \rho^2)}$ is reached at which the point H_1 is on the line $W = W_2 = 1 + \rho + \rho^2$. At this value of μ two lobes of $\mathcal{B}(\infty)$, bounded by ω_{-3}, reach the line of initial conditions, the tangency points being the two preimages H_{-1}^1 and H_{-1}^2 of the point H. This gives a second bifurcation of the basin $D_1(p^*)$, at which two new holes are created around the tangency points, and the basin of the REE p^* is given by the union of 4 disjoint intervals, separated by holes of $\mathcal{B}(\infty)$.

Other similar bifurcations occur at $\mu = \mu_n^*$, where $\mu_n^* = 1 + W_n + 2\sqrt{W_n(1 + \rho W_n)}$ at which ω_{-n-1}, belonging to the set $T^{-n-1}(\omega_0)$, become tangent to the line of initial conditions. This implies that 2^n new holes are created. The result of this sequence of bifurcations is that the basin assumes a structure which is typical of a Cantor set. In fact, at each bifurcation value $\mu = \mu_n^*$, $n \in \mathbb{N}$, the number of lobes of $B(\infty)$ is doubled, and the whole sequence of bifurcations causes a fractalization of the basin boundaries near the focal points (and their preimages) that gives a "finger-shaped" structure of $B(\infty)$. When μ reaches the limiting value

Fig. 8 The case of Bray's learnng $\rho \to 1^-$ with $\mu = 12$. The dots represent a trajectory starting from the initial condition $(z_0, 1)$ with $z_0 = 0.3$

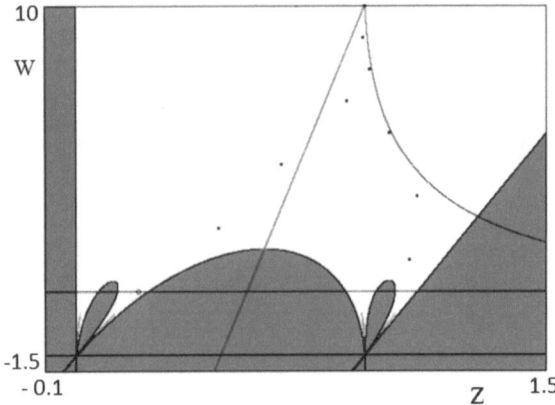

$$\mu_\infty^* = \lim_{n \to \infty} \mu_n^* = \frac{4 - \rho}{1 - \rho},$$

the point H_1, together with all of its infinite preimages located at the top of the lobes, reach the line of the ω-limit sets $W = W^*$. Thus at $\mu = \mu_\infty^*$ infinitely many lobes of $B(\infty)$ have been created, and all have a contact with a chaotic attractor $\mathcal{A}$ located on the line $W = W^*$. This contact between $\partial B(\infty)$ and the chaotic set causes the disappearance of $\mathcal{A}$ and for $\mu > \mu_\infty^*$ only divergent trajectories of the map T can be obtained.

The global analysis of the basin boundaries just described holds for any value of the memory ratio ρ belonging to the interval $(0, 1)$. In particular, it also holds in the limiting case $\rho \to 1^-$. In this case the singular line, where the focal points are located, has equation $W = -1$. The equations of the curves which form the boundary $\mathcal{F}$ are obtained from those given above just substituting $\rho = 1$. So, also in the case of Bray learning (12), the complexity in the structure of the basins is conserved, as shown in Fig. 8, obtained with $\rho = 1$ and $\mu = 12$.

Similar structures of the basins are obtained for other models represented by unimodal maps, like the model proposed in Dimitri (1988), whose global analysis is given in Bischi and Naimzada (1997).

4.2 Bimodal Maps: Coexisting Stable REEs and the Problem Of equilibrium Selection

In the model analyzed above, one of the two "competing" equilibria is a rather unrealistic attractor at infinity. However, similar results hold when two or more bounded coexisting equilibria, or more complex bounded attractors of the limiting map g exist, such as periodic cycles or chaotic sets. An interesting situation arises if a model with expectations is such that two REEs exist which are both stable under a learning rule, i.e. two "competing" rational expectation equilibria whose selection depends on the initial condition. In order to show an example where this happens, we consider a bimodal function F defined as

$$F(x) = -ax^3 + 3ax^2 - 2ax + 1 \tag{55}$$

This map has three fixed points, say $x_1^* < x_2^* < x_3^*$, as shown in Fig. 9a, where the graph of F is represented, for $a = 3$, together with the graph of $g_{0.3} = 0.7x + 0.3F(x)$, the map which governs the one-dimensional dynamics of the expected values when adaptive expectations (4) are introduced with $\alpha = 0.3$, and also represents the limiting map for the two-dimensional dynamics describing the statistical learning (18) with $\rho = 0.7$. From this graph it is evident that x_2^* is always unstable, both for F and for g, whereas for a given value of a the REEs x_1^* and x_3^* are stable for g provided that sufficiently low values of α (or sufficiently high values of ρ) are considered. So situations of two coexisting stable REEs x_1^* and x_3^* are easily obtained. In this case, the problem of equilibrium selection is related to the delimitation of the basins. Such problem is simple as far as adaptive learning (4) is concerned. In fact, the stable set of the unstable REE x_2^*, given by the set of all of its preimages, constitutes the boundary which separates the basin of x_1^* from the basin of x_3^*. These basins are formed by the two immediate basins, which include x_1^* and x_3^* respectively, and infinitely many disjoint portions, preimages of the immediate basins, which accumulate at the two periodic points of a repelling cycle $\{s_1, s_2\}$ which also constitutes the boundary of the basin of infinity, i.e. s_1 and s_2 separate the points which generate trajectories converging to bounded attractors from the ones generating unbounded trajectories.

Instead, if the two-dimensional map equivalent to the model with *statistical learning* (18) is considered, the basins appear to be more complex. In Fig. 9b the two-dimensional basins of $\left(x_1^*, W^*\right)$ and $\left(x_3^*, W^*\right)$, located along the line $W = W^* = 1/(1 - \rho)$, are represented by grey and light-grey regions respectively, whereas the black region represents the basin of infinity. In this case the common boundary of the dark-grey and white regions is given by the stable set of the saddle point $\left(x_2^*, W^*\right)$ and the boundary of the basin of infinity is formed by the stable set of the saddle-cycle $S = \{(s_1, W^*), (s_2, W^*)\}$. As usual, the structure of these boundaries is made up of lobes and crescents originating from the three focal points $Q_i = \left(x_i^*, -1/\rho\right)$, and the complexity of the basins of the model with statistical learning is related to the fact that the boundaries of such lobes and crescents intersect the line of initial conditions $W = 1$ in several points, so that the basins of the two stable REEs are quite different from those observed for the model with adaptive learning. For example, the trajectory starting with the initial condition $x_1^{(e)} = x_0 = 0.3 < x_2^*$, converges to x_3^*, and the one starting from $x_1^{(e)} = x_0 = 1.7 > x_2^*$, converges to x_1^*, whereas with adaptive expectations any trajectory starting with $x_0 < x_2^*$ converges to x_1^*, so that the equilibrium selection results obtained with adaptive expectations are now reversed.

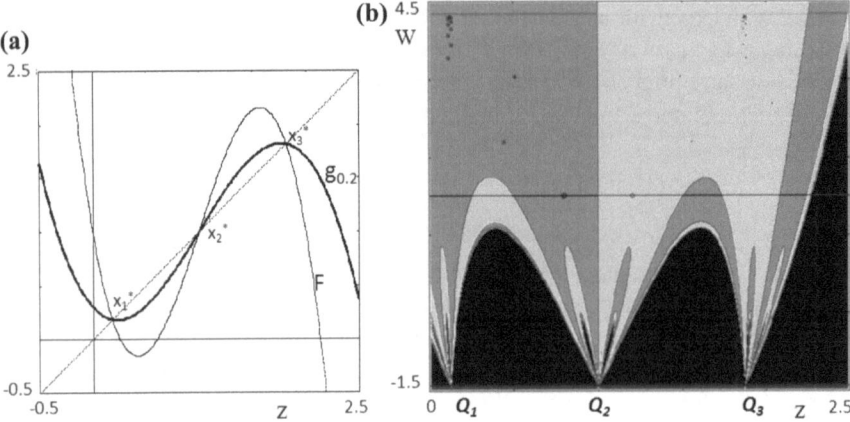

Fig. 9 a The graphs of the map F given in (55) for $a = 3$ together with the graph of $g_{0.3} = 0.7x +$ $0.3F(x)$; **b** Basins of attraction of the map (35) with F in (55): the light-grey region represents the basin of $\left(x_3^*, W^*\right)$, the grey region is the basin of $\left(x_1^*, W^*\right)$, the black region is the basin of infinity

This different equilibrium selection happens when the initial conditions are taken inside the "holes" given by the intersections of lobes and crescents with the line of initial conditions $W = 1$, whereas other initial conditions converge to the same equilibrium as in the model with adaptive expectations. This is true, for example, for the two trajectories represented in the figure, obtained with $x_1^{(e)} = 0.8$ and $x_1^{(e)} = 1.2$.

4.3   An O.G. Forward Looking Model Represented by an Increasing Map

A large class of economic problems are characterized by forward-looking expectations, i.e. are modeled by a discrete time law of motion of the form

$$x_t = F(x_{t+1}^e) \tag{56}$$

Common examples in which such mappings are obtained by Overlapping Generations (O.G.) models, where agents typically living two periods (say young and old) take the consumption and/or saving decision of their whole life in the first period (when young) so that they must guess (i.e. foresee) which will be the status of the economy (e.g. prices) one period ahead, when they will be old. As in the previous sections, x_t represents the *current (or realized) state variable* of the economic system at time t and x_{t+1}^e is the *expected state* for time $t + 1$ according to the agents' forecasting rule and their information set at time t.

Under the assumption of perfect foresight (2) the agents correctly anticipate the future state, i.e. $x_{t+1}^{(e)} = x_{t+1}$ for each t, and this defines the rational expectations equilibrium profiles through the iteration scheme

$$x_t = F(x_{t+1}), \tag{57}$$

called *backward PF dynamics*. Indeed (57) has no dynamical meaning, but it must be simply seen as a recursive scheme which defines an *intertemporal equilibrium with PF* along which expectations are fulfilled, i.e. the equilibrium sequences generated by the recurrence (57) represent the outcomes of the economic system under the strong assumption that the agents are characterized by self-fulfilling RE.

In this case, we can have rational expectation time paths which are more complex than a stationary state. In fact, a recurrence of the form (57) can generate periodic sequences of any period or even aperiodic (i.e. chaotic) bounded sequences.

A fixed point of (57) defines a Rational Expectations Equilibrium (REE), a *k-periodic* cycle $C_k = \{\alpha_1, \ldots, \alpha_k\}$, with $\alpha_i \neq \alpha_j$, $\forall i, j = 1, \ldots, k$, such that $\alpha_i = F(\alpha_{i+1})$, $i = 1, \ldots, k-1$, and $\alpha_k = F(\alpha_1)$, represents a Rational Expectations Cycle (REC) and so on.

In the literature on forward looking models, learning mechanisms are often proposed where the computation of x^e_{t+1} does not involve the current state x_t. Such assumption is usually motivated by saying that in modeling forward looking expectations generally the "subjects are requested to make forecasts at the beginning of period t, when x_t is not in their information set" (from Marimon et al. (1993)).

Under this assumption, the presence of x_t in both sides of the equation (56) is avoided, and it is immediate to realize that the one-dimensional dynamics which describe the time evolution of expected values of the model (56) under adaptive learning as well as the two-dimensional dynamics which describes the time evolution of expected values of the model (56) under the statistical learning, are the same as those described in the previous sections. So, the method and the results described above can be applied to many models with forward looking expectations which have been proposed in the literature. As an example, let us consider an Overlapping Generations model, proposed in Evans and Honkapohja (1995), where a representative consumer is assumed to live for two periods: period t (when young) and period $t + 1$ (when old), and its utility function is $\widetilde{U} = U(c_{t+1}) - V(n_t)$, where c_{t+1} is the consumption when old, n_t the labor supply when young. In Evans and Honkapohja (1995) a concrete illustration is given, with $U(c) = \frac{c^{1-\sigma}}{1-\sigma}, \sigma > 0, V(n) = \frac{n^{1-\varepsilon}}{1-\varepsilon}, \varepsilon > 0$ and a production function $f(n_t, Kn_t)$, where Kn_t is the aggregate labor supply of K consumers, is considered in the separable form $f(n, Kn) = n^\alpha \psi(Kn)$, where $\psi(Kn) = A(I^*)^\beta$. These assumptions allow to obtain, for the consumer optimization problem with budget constraints

$$p^{(e)}_{t+1} c_{t+1} = M_t \quad \text{and} \quad p_t f(n_t, Kn_t) = M_t$$

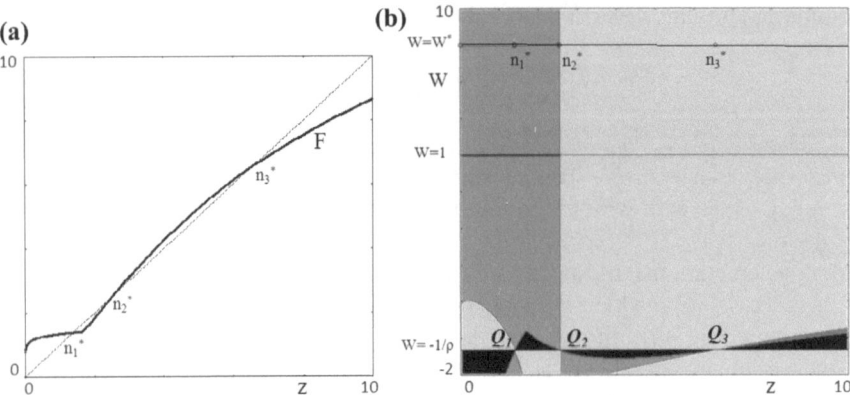

Fig. 10 **a** The graphs of the map F taken from Evans and Honkapohja (1995); **b** Basins of attraction of the map (35) with this map F: the light-grey region represents the basin of (n_3^*, W^*), the grey region is the basin of (n_1^*, W^*), the black region is the basin of infinity

a law of motion in the forward looking form

$$n_t = F(n_{t+1}^{(e)})$$

where F has a graph like the one shown in Fig. 10a, which is obtained with the same set of parameters used in Evans and Honkapohja (1995). The function F has three REEs denoted by n_i^*, $i = 1, 2, 3$, two of which are stable. In Evans and Honkapohja (1995) the following learning scheme is introduced

$$n_{t+1}^{(e)} = (1 - \alpha_t)n_t^{(e)} + \alpha_t n_{t-1} = (1 - \alpha_t)n_t^{(e)} + \alpha_t F(n_t^{(e)})$$

with $\alpha_t = 1/t$, i.e. the Bray learning (12).

If we consider the most general statistical learning (18) we obtain the same basins as the ones obtained for the limiting map $g_{1-\rho}$, i.e. bounded by the unstable REE. This is shown in Fig. 10b, where the basins of the two-dimensional map equivalent to the statistical learning (18) are represented by different grey regions. The intersection of the line of the initial conditions $W = 1$ with the intermediate-grey region represents the basin of the REE n_1^* and the intersection with the light-grey region represents the basin of the REE n_3^*. Since the portions of the stable set of the saddle point (n_2^*, W^*) which are not along the invariant line $z = n_2^*$ (i.e. the arcs originating from the focal points Q_i) are confined below the line of initial conditions $W = 1$, the basins are simply bounded by the unstable REE n_2^*, as in the case of adaptive expectations, as shown in Fig. 10b. Of course, the same holds for $\rho = 1$, i.e. in the case of Bray learning.

5 Conclusions and Further Research

The inclusion of memory of past states in discrete dynamical systems that represent economic models with expectations has been considered in the form of a weighted average with exponentially decreasing weights. This scheme is then compared to adaptive expectations. The two methods to compute expected values share the same attractors but differ for the role played by initial conditions as in general they have different basins of attraction with several coexisting attractors. So, in cases of multistability different equilibrium selections can be obtained. This result has been obtained through the study of the basins of a two-dimensional map equivalent to the statistical learning with fading memory, by using some methods for the study of global bifurcations of plane maps with a denominator that vanishes in a one-dimensional subset of the phase space. The results described in this paper have been illustrated by some simple economic examples, such as cobweb models and an overlapping generations framework. Following the path indicated by some recent works by Matsumoto and Szidarovszky in continuous-time oligopoly models with exponentially fading memory, also the methods described in this paper for discrete-time models may be usefully applied in Cournot or Bertrand oligopoly games in discrete time, see e.g. Deschamps (1975) or Thorlund-Petersen (1990). Such games, endowed with fading memory, will be reduced to an equivalent autonomous maps with denominator of dimension greater than two, a quite challenging mathematical task.

References

Abraham, R., Gardini, L., & Mira, C. (1997). *Chaos in discrete dynamical systems (A visual introduction in two dimensions)*. Springer.

Billings, L., & Curry, J. H. (1996). On noninvertible maps of the plane: Eruptions. *CHAOS, 6*, 108–119.

Billings, L., Curry, J. H., & Phipps, E. (1997). Lyapunov exponents, singularities, and a riddling bifurcation. *Physical Review Letters, 79*(6), 1018–1021.

Bischi, G.I., & Gardini, L. (1996) Mann iterations reducible to plane endomorphisms . In *Quaderni di Economia, Matematica e Statistica*, Facoltà di Economia (Vol. 36). Università di Urbino.

Bischi, G. I., & Naimzada, A. K. (1997). Global analysis of a nonlinear model with learning. *Economic Notes, 26*(3), 143–174.

Bischi, G. I., & Gardini, L. (1997). Basin fractalization due to focal points in a class of triangular maps. *International Journal of Bifurcations & Chaos, 7*(7), 1555–1577.

Bischi, G. I., Gardini, L., & Mira, C. (1999). Maps with denominator. Part 1: some generic properties. *International Journal of Bifurcation & Chaos, 9*(1), 119–153.

Bischi, G. I., Gardini, L., & Mira, C. (2003). Plane maps with denominator. Part II: Noninvertible maps with simple focal points. *International Journal of Bifurcation & Chaos, 13*(8), 2253–2277.

Bischi, G. I., Gardini, L., & Mira, C. (2005). Plane maps with denominator. Part III: Non simple focal points and related bifurcations. *International Journal of Bifurcation & Chaos, 15*(2), 451–496.

Bischi, G. I., Cavalli, F., & Naimzada, A. K. (2015). Mann iteration with power means. *Journal of Difference Equations and Applications, 21*(12), 1212–1233.

Bray, M. (1983) Convergence to rational expectations equilibrium. In R. Friedman & E. S. Phelps (Eds.), *Individual forecasting and aggregate outcomes*. Cambridge University Press.

Cavalli, F., & Naimzada, A. K. (2015). A tâtonnement process with fading memory, stabilization and optimal speed of convergence Chaos. *Solitons & Fractals, 79,* 116–129.

Chiarella, C. (1988). The cobweb model. Its instability and the onset of chaos. *Economic Modelling, 5,* 377–384.

Chiarella, C. (1991). The birth of limit cycles in Cournot oligopoly models with time delays. *Pure Mathematics and Applications, 2,* 81–92.

Cushing, J. M. (1978). *Integrodifferential equations and delay models in population dynamics* (Vol. 20). Lecture notes in biomathematics. Springer.

Deschamps, R. (1975). An algorithm of game theory applied to the duopoly problem. *European Economic Review, 6,* 187–194.

Dimitri, N. (1988) A short remark on learning of rational expectations. *Economic Notes, 3.*

Evans, G. W., & Honkapohja, S. (1995). Increasing social returns, learning and bifurcation phenomena. In A. Kirman & P. Salmon (Eds.), *Learning and rationality in economics* (pp. 216–235). Oxford: Basil Blackwell.

Foroni, I., Gardini, L., & Rosser, B, Jr. (2003). Adaptive and statistical expectations in a renewable resource market. *Mathematics and Computers in Simulation, 63,* 541–567.

Friedman, B. M. (1979). Optimal expectations and the extreme information assumption of rational expectations macromodels. *Journal of Monetary Economics, 5*(1), 23–41.

Fudenberg, D., & Levine, D. K. (1998) *The theory of learning in games.* The MIT Press.

Gardini, L., Bischi, G. I., & Fournier-Prunaret, D. (1999). Basin boundaries and focal points in a map coming from Bairstow's methods. *CHAOS, 9*(2), 367–380.

Gardini, L., Bischi, G. I., & Mira, C. (2007) Maps with vanishing denominators, 16970. www.scholarpedia.org, https://doi.org/10.4249/scholarpedia.3277.

Gu, E. G., & Hao, Y.-D. (2007). On the global analysis of dynamics in a delayed regulation model with an external interference. *Chaos, Solitons & Fractals, 34*(4), 1272–128.

Guesnerie, R., & Woodford, M. (1992) Endogenous fluctuations. In J. J. Laffont (Ed.), *Advances in economic theory* (Vol. II). Cambridge University Press.

Gumowski, I., & Mira, C. (1980). *Dynamique Chaotique.* Toulose: Cepadues editions.

Holmes, J. H., & Manning, R. (1988). Memory and market stability: The case of the cobweb. *Economic Letters, 28,* 1–7.

Hommes, C. (1991). Adaptive learning and roads to chaos. The case of the cobweb. *Economic Letters, 36,* 127–132.

Hommes, C. (1994). Dynamics of the cobweb model with adaptive expectations and nonlinear supply and demand. *Journal of Economic Behavior & Organization, 24,* 315–335.

Hommes, C., Kiseleva, T., Kuznetsov, Y., & Verbic, M. (2012). Is more memory in evolutionary selection (de)stabilizing? *Macroeconomic Dynamics, 16,* 335–357.

Hommes, C. (2013). *(2013) behavioral rationality and heterogeneous agents in complex economic systems.* Cambridge University Press.

Jensen, R. V., & Urban, R. (1984). Chaotic price dynamics in a non-linear cobweb model. *Economic Letters, 15,* 235–240.

Lucas, R. E. (1986). Adaptive behavior and economic theory. *Journal of Business, 59*(4).

MacDonald, N. (1978) *Time lags in biological models.* Lecture notes in biomatemathics (Vol. 27). Springer.

Marimon, R., Spear, S. E., & Sunder, S. (1993). Expectationally driven market volatility: An experimental study. *Journal of Economic Theory, 61,* 74–103.

Marimon, R. (1997) Learning from learning in economics. In D. M. Kreps & K. F. Wallis (Eds.), *Advances in economics and econometrics: Theory and applications,* Vol. I. Cambridge University Press.

Matsumoto, A., & Szidarovszky, F. (2018) *Dynamic oligopolies with time delays.* Springer.

Matsumoto, A., & Szidarovszky, F. (2015). Dynamic monopoly with multiple continuously distributed time delays. *Mathematics and Computers in Simulation, 108,* 99–118.

Matsumoto, A. (2017) Love affairs dynamics with one delay in losing memory or gaining affection. In A. Matsumoto (Ed.), *Optimization and dynamics with their applications.* Springer.

Mira, C., Gardini, L., Barugola, A., & Cathala, J. C. (1996). *Chaotic dynamics in two-dimensional noninvertible maps*. World Scientific.

Naimzada, A. K., & Tramontana, F. (2009). Global analysis and focal points in a model with boundedly rational consumers. *International Journal of Bifurcation & Chaos, 19*(6), 2059–2071.

Nerlove, M. (1958). Adaptive expectations and cobweb phenomena. *Quarterly Journal of Economics, 73*, 227–240.

Pecora, N., & Tramontana, F. (2016) Maps with vanishing denominator and their applications. *Frontiers in Applied Mathematics and Statistics*. https://doi.org/10.3389/fams.2016.00011.

Radner, R. (1983) Comment to Convergence to rational expectations equilibrium by M. Bray. In R. Friedman & E. S. Phelps (Eds.), *Individual forecasting and aggregate outcomes*. Cambridge University Press.

Thorlund-Petersen, L. (1990). Iterative computation of Cournot equilibrium. *Games and Economic Behavior, 2*, 61–95.

Tramontana, F. (2016) Maps with vanishing denominator explained through applications in economics . *Journal of Physics: Conference Series, 692*, Conference 1.

Yee, H. C., & Sweby, P. K. (1994). Global asymptotic behavior of iterative implicit schemes. *International Journal of Bifurcation & Chaos, 4*(6), 1579–1611.

Come Together: The Role of Cognitively Biased Imitators in a Small Scale Agent-Based Financial Market

Giovanni Campisi and Fabio Tramontana

Abstract We analyze the consequences of the presence of imitators in a financial market populated by boundedly rational speculators. We consider imitators that only look at the recent success of the available trading rules. We show that the introduction of this kind of imitators makes the results more complicated but even more realistic. In particular, under some specific circumstances, imitators may stabilize an otherwise unstable market or, at the opposite, make unstable an otherwise stable scenario.

Keywords Cognitive biases · Bounded rationality · Financial markets · Agent-based models

1 Introduction

Over the last few years a lot of attention has been raised on the psychology of financial markets. This is probably a consequence of the failure of the traditional approach to the study of financial markets, which is essentially based on the assumption of perfectly rational agents, cornerstone of the so-called Efficient markets hypothesis (see Fama (1995, 1965)). These theories dramatically failed in anticipating and explaining how financial bubbles like the dot.com bubble or the US real estate bubble originated, grow larger and then burst (see Shiller (2015)). The consequences of the explosions of such bubbles can be huge and nowadays we know that they can also influence real

A earlier version of this paper entitled "The role of cognitively biased imitators in a small scale agent-based financial market" has been published as Working Paper of the Department of Economics and Management of the University of Pavia (Italy).

G. Campisi
Department of Economics Marco Biagi, University of Modena and Reggio Emilia,
Via Jacopo Berengario 51, 41121 Modena, Italy
e-mail: giovanni.campisi@unimore.it

F. Tramontana (✉)
Department of Mathematical Disciplines, Mathematical Finance and Econometrics,
Catholic University of Milan, Milan, Italy
e-mail: fabio.tramontana@unicatt.it

economy, triggering deep recessions. Bubbles and crashes are not the only stylized facts of financial markets that the mainstream approach is unable to explain in a convincing manner. The list of other prominent features of stock markets includes excess volatility, fat tails of returns' distribution together with their virtual unpredictability. We also mention volatility clustering and long memory effects among the facts to be explained. Deviations from the perfectly rational behavior have been founded and analyzed since many years before the current economic crisis.[1]

A systematic study and classification of the irrationalities that plague humans' decision making started with the works of Kahneman and Tversky in the seventies (see Tversky and Kahneman (1974) for a list of the most common biases). They and their scholars proved that people who make decisions follow simple rules of thumb (called *heuristics*) that many times represent an easy way to make a good decision, but may also lead to systematic deviations (called *biases*) from what a perfectly rational agent should do. The good thing is that, given their regularity, these biases can in some sense be foreseen.

Financial traders (professional or not) decide whether to buy or sell an asset following simple heuristics, too. These rules of thumb can be subdivided into two main categories: fundamental and chartists trading rules, followed by fundamentalists and chartists (or technicians), respectively (see Frankel et al. (1986), Menkhoff and Taylor (2007) for empirical validations and Hommes (2011) for a review of laboratory experiments). The former are convinced that, even in the short period, prices will come back to their fundamental values, so they buy undervalued assets and sell overvalued ones. Chartists (or technicians) look at the time series of prices to find a clue for understanding the (near) future price movements.

An interesting strand of research consists in studying small-scale heterogeneous agent-based financial market models (HAMs henceforth) with behavioral assumptions. The pioneering work in this field is due to Day and Huang (1990) and after that the interactions between heterogeneous market participants have been developed in many directions (see Chiarella et al. (2009), Hommes and Wagener (2009), Lux (2009), Westerhoff (2009) for example). The strong point of HAMs lies in the connection between the behavioral assumptions that are supported by empirical and experimental evidence, and the small scale of the dynamical systems that explain asset price movements and the underlying mechanisms that cause them. This permits to analytically study the most of these models, making understandable the endogenous causes of particular price movements. HAMs lead to irregular endogenous price dynamics even in their deterministic version, through the nonlinearities that are introduced in the trading rules and/or in the switching mechanism. The emergence of chaotic dynamics permits to replicate stylized facts like bubbles and crashes and excess volatility. These results reduce the role played by stochastic variables that can still be useful to replicate some quantitative aspects of real time series, but are not necessary for a qualitative explanation of the most of these facts.

HAMs can also be used for taking into consideration the imitative strategies that a (sub)group of traders can adopt to obtain better performances. In this sense the works

---

[1] See for instance the famous Allais paradox (Allais 1953).

of Lux (1995, 1998) and Lux and Marchesi (1998, 1999) deserve to be mentioned. In their works there are fundamentalists and chartists. These latters use a combination of trend following and imitative strategies and can be optimistic or pessimistic. They decide which subgroup to imitate looking at what the majority is doing[2] and at expected and realized excess profits of the available strategies. A similar mechanism is used in Franke and Westerhoff (2016). Further HAMs with imitation are Bischi et al. (2006) and Foroni and Agliari (2008).

Our work is also related to the strand of literature in which financial markets are viewed as evolutionary adaptive systems, populated by boundedly rational interacting agents (see Brock and Hommes (1997, 1998), Chiarella et al. (2000), Chiarella and He (2002), Farmer (2002) among the others). In these models agents are allowed to switch among the different trading strategies, trying to learn the best one. The authors of these papers are interested in the final outcome of this evolutionary competition and remarkable results emerge when the fractions of agents adopting each strategy, continuously vary over time, never reaching a fixed final value.

In our model there are some agents that keep fixed their strategies, no matter what happens in the markets. They are overconfident and can be affected by the so-called *confirmation bias* (see Barber and Odean (2000, 2001)). Roughly speaking, they favor information that support their strategies and give less importance to the others. This bias explains, for instance, why even beliefs that have been heavily discredited survive in the mind of some people (Kunda 1999). On the other hand, we also consider a fraction of traders that are not so self-confident and at each time period reconsider their strategies myopically looking at the performance of the alternatives. We consider the amount of agents adopting each strategy as a consequence of the imitative behavior and not as a cause. In this sense we do not model an *herding behavior* (in which traders follow the crowd). Our approach is even more simple because it does not require for imitators to know the number of traders that use each available strategy. They only look at their neighborhood where they can find examples of traders adopting the various strategies. When they look at the outcome of their strategies, they decide who among them should be imitated. In some context this behavior can be a good one. Especially when a best strategy really exists. Financial markets is not one of these cases because we cannot state, for instance, that the fundamentalists approach is always better than the chartists one or the opposite, and in this case, as we will see, the role of imitators can by quite important and drastically influence price dynamics.

The paper is structured as follows. In Sect. 2 we build a typical HAM with fundamentalists and chartists that keep fixed their strategies. In Sect. 3 we introduce a third group of traders that imitate the behavioral rule of the others by using a simple rule of thumb. The consequences of the introduction of imitators in the market are studied in Sect. 4. Section 5 concludes the paper.

---

[2]In this case we talk about *herding* rather than imitation, that we use in a more generic meaning.

2   The Benchmark Model

In this Section we build a typical HAM describing a financial market where only one asset is exchanged. The market is populated by two kinds of speculators: fundamentalists and chartists.

In the spirit of Day and Huang (1990), a market maker adjusts the log of the asset (P) according to this rule:

$$P_{t+1} = P_t + a\left(D_t^f + D_t^c\right) \tag{1}$$

where D_t^f and D_t^c represent the orders placed at time t by fundamentalists and chartists, respectively. P_t is the current asset price and it is known by all market participants. The positive parameter a measures the intensity of the adjustment.

Fundamentalists are assumed to believe in the reversion of the asset price towards its (exogenously given) fundamental value F. As a consequence, they buy the asset when its price is below the fundamental value and sell it when it is overvalued. Their behavioral rule is the following:

$$D_t^f = f\left(F - P_t\right) \tag{2}$$

where $f > 0$ is a reaction parameter.

At the opposite, chartists optimistically interpret the signal given by a price above the fundamental. So they buy the overvalued asset because they think that the positive trend will go on, at least in the short period. Nevertheless, we introduce a nonlinearity in their trading rule (an arctangent) that permits us to take into consideration a certain degree of prudence when the difference between actual price and fundamental value becomes extremely large. Orders placed by chartists are given by:

$$D_t^c = c * \arctan\left(P_t - F\right) \tag{3}$$

where c is a positive reaction parameter.[3]

The price adjustment rule (1) combined with the trading rules (2) and (3), gives us a dynamic model explaining the movements of the asset price as a function of the previous period price:

$$P_{t+1} = P_t + a\left[f\left(F - P_t\right) + c * \arctan\left(P_t - F\right)\right] \tag{4}$$

---

[3]Note that we would obtain the same qualitative results that are explained in the following by using $(P_t - P_{t-1})$ instead of $(P_t - F)$, so considering some sort of positive feedback investors (see De Long et al. (1990)). We prefer to avoid this characterization of chartists because it would increase the dimensionality of the dynamical system explaining the price dynamics.

We can use the auxiliary variable $x_t = P_t - F$ representing deviation from the fundamental value, to obtain the map:

$$x_{t+1} = f(x_t) = x_t + a[c * \arctan(x_t) - fx_t]. \tag{5}$$

2.1 Steady States and Local Stability

One of the fixed points of the map (5) corresponds to a price equal to the fundamental value: $x^* = 0$ (we call it *fundamental fixed point* in what follows). Nevertheless, the price may not converge to it if it is not locally stable. To check the local stability of x^* we must use the first derivative of $f(x_t)$, i.e.:

$$f'(x_t) = 1 + a\left(\frac{c}{1 + x_t^2} - f\right) \tag{6}$$

and evaluate it at the fundamental fixed point value:

$$f'(0) = 1 + a(c - f) \tag{7}$$

The fixed point is locally stable if the value of the derivative is lower than 1 in absolute value. We can easily obtain the local stability condition in terms of the chartists' reactivity coefficient c:

$$-\frac{2}{a} + f < c < f \tag{8}$$

Condition (8) has a straight interpretation: the asset price converges to the fundamental value provided that fundamentalists are more reactive (or trade more aggressively) than chartists, but not too much.[4]

The fulfillment of condition (8) only ensures the *local* stability of the fundamental fixed point. In other words, we know that starting with an initial value of the price that is close enough to the fundamental value, then the price will converge to it. But how close the price should be to the fundamental value? Note that this is a quite relevant question because another way of formulating the same question is the following: if a shock hits the market, are we sure that the price will come back to the fundamental value? The larger is the interval made up by initial conditions leading to the fundamental value, the more robust is the system that only requires some settlement periods for reabsorbing shocks (see Fig. 1).

---

[4]This second case may appear less easy to interpret but the explanation is straight: when fundamentalists strongly dominate the market the price oscillations are huge and overvalued and undervalued prices alternate. From the mathematical point of view this is a so-called flip (or period-doubling) bifurcation.

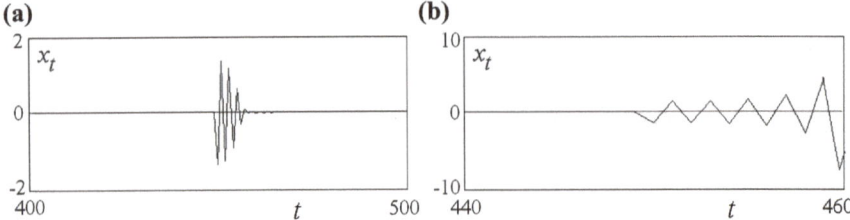

Fig. 1 Timeplots obtained by using the parameters' values: $a = 3$, $c = 0.8$ and $f = 1.2$. At time period $t = 450$ an additive shock is introduced. The shock hitting the system in panel b is larger than the one of panel a

The set of initial conditions leading to a fixed point (x^*) is usually called its *basin of attraction* ($\beta(x^*)$) in the dynamical systems' literature. In our case the basin of attraction of the fundamental value is given by:

$$\beta(x^*) =]\alpha^-, \alpha^+[\tag{9}$$

where α^- and α^+ are the points of an (unstable) cycle of period 2 that we can find by looking at the second iterate $f^2(x)$ (see Fig. 2).

By keeping fixed the value of f and by varying the value of c inside the range of local stability of the fundamental fixed point (8), we find that the larger is c the larger is $\beta(x^*)$. This means that when fundamentalists are much more reactive than chartists, even if condition (8) is not violated, only initial conditions quite close to the fundamental value lead to it.

In other terms, even starting from a scenario in which price is equal to the fundamental value, a small shock may have heavy consequences on the price dynamics.

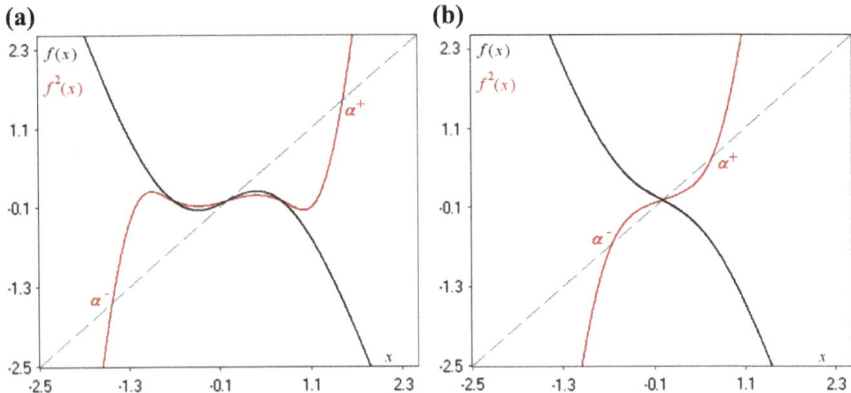

Fig. 2 First and second iterates of the map $f(x)$. In **a** we have that $a = 3$, $f = 1.64$ and $c = 1.5$. The fixed points of the second iterate (the red curve) bound the basin of attraction of the fixed point. This set is smaller when c is reduced to 1.1 as in (**b**)

And what happens to initial conditions outside $\beta(x^*)$? In these cases the strong dominance of fundamentalists combined with a sufficiently large mispricing drives fundamentalists to overreact to the market signal and the price diverges alternating high and low values, each time more distant from the fundamental one (Fig. 1b).

Let us focus on what happens when chartists are more aggressive than fundamentalists (i.e. $c > f$) by looking at the bifurcation diagram of Fig. 3, obtained by keeping fixed $a = 3$ and $f = 1.2$ and varying the values of c between 0.5 and 4.

Until the value of the chartists' reactivity is lower than the fundamentalists' one, the asset price converges to the locally stable fundamental fixed point. At $c = f$ a *pitchfork bifurcation* occurs and two further steady states are created. For values of c not excessively higher than f the fundamental state $x^* = 0$ is locally unstable and price converges to one of the other two steady states, depending on the initial condition. Note that one steady state (x^+) corresponds to an overvalued price while the other one (x^-) to a situation where the price is undervalued. By further increasing the reactivity of chartists, the two coexisting fixed points also become locally unstable via a simultaneous flip (or *period doubling*) bifurcation, giving rise to coexisting cycles of period 2, 4, 8 and so on. After the typical cascade of period-doubling bifurcations, two coexistent chaotic attractors arise. When the system converges to one of them, the price erratically moves in it, keep remaining in the bull or in the bear region (i.e. markets where the prices are overvalued or undervalued, respectively). In the rightmost part of the diagram, for high values of c, the two attractors become one (a so-called homoclinic bifurcation of x^* occurs) and the price oscillates between the bull and the bear regions in an almost unpredictably way, no matter the initial condition. In Fig. 4 we have a typical timeplot obtained by using $c = 4.528$.

This is the most interesting scenario. In fact the dynamics represented in Fig. 4 are hardly distinguishable from those obtained by using a more sophisticated stochastic model. This simple deterministic model is able to qualitatively replicate some important facts of the financial markets like bubbles and crashes and excess volatility. In this scenario, periods in which the price gets closer to the fundamental value, alter-

Fig. 3 Bifurcation diagram. The red diagram is obtained by using an initial condition $x_0 = 0.1$ while the green one is obtained starting from $x_0 = -0.1$. The black portion of the diagram is not influenced by the initial condition chosen

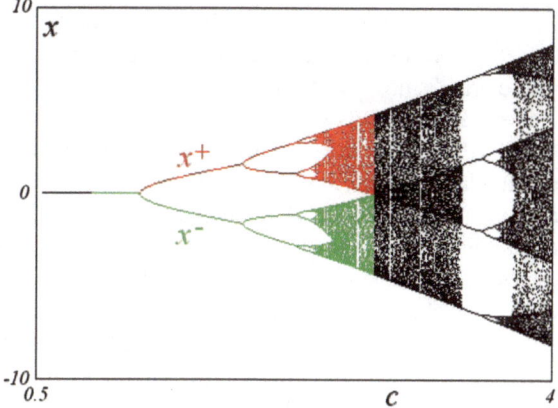

Fig. 4 One hundred
consecutive values of the
price when the attractor is
chaotic and covers both bull
and bear regions

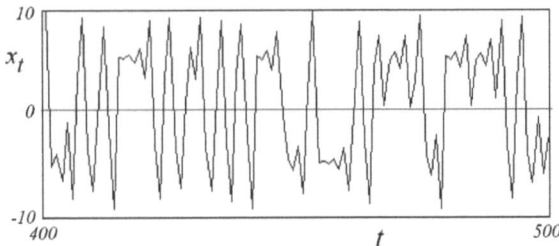

nate with periods in which it moves away from it. This means that in some periods
the fundamentalists trading rule seems to be better than the one used by chartists, but
this situation never persists forever. It is (almost) impossible to identify a rule that
would permit to systematically *beat the market*.

The following table summarizes the main features of the benchmark model:

Parameters' values	Stability of fund. fixed point	Effects of a price shock
$-\frac{2}{a} + f < c < f$	–Locally stable	–Small $\Longrightarrow$ reabsorbed
	–Not globally stable	–Big $\Longrightarrow$ divergence
$c > f$	–Locally unstable	–Reabsorbed
	–Attractors are periodic or chaotic, i.e.	
	price constantly fluctuates possibility of	
	realistic bull and bear dynamics	

To this benchmark model we will add in the next Section a third group of traders,
the imitators, and we will analyze the consequences of their presence.

3   The Model with Imitators

If we want to consider a more realistic asset market, we must take into consideration
that it is not only populated by traders that do not question their beliefs about the
future price's movements. There also exist traders that change their behavioral rules.
Among them there is someone that looks at the beliefs of the other traders, trying
to learn the best strategy. This means to check who has been right between chartists
and fundamentalists, that is if it is meaningful to believe that the price will suddenly
approach the fundamental value or not. We know that a rational behavior consists
in waiting a long time and comparing a long series of daily returns, in order to be
relatively sure that the results are caused by the relative values of the trading strategies
and not by chance. This is a consequence of the so called *law of great numbers* that
permits to the best strategy (if there is one) to asymptotically emerge. Unfortunately,
there exists a huge amount of experimental evidence and data that people believe

instead in the so-called *law of small numbers*.[5] People believing in the "false" law of small numbers, according to definition of Shefrin (2001) "attribute negative serial correlation to an identical and independently distributed stochastic process". In other words, a sample, especially if small is erroneously considered highly representative of the population. For our purposes this means that after some periods in which the asset price has moved closer and closer to the fundamental value, imitators start thinking that the fundamentalists' belief is the best one and decide to behave like them. The opposite is when the price has moved away from the fundamental value.

Besides the fact that this false law may lead to erroneous evaluations even if a best strategy really exists, we will show that it may have heavy consequences when a best strategy does not exists, like in the case of the chaotic motion of price showed in the previous section.

Considering how people are influenced by the actions of the others is not new in the literature. Banerjee (1992) and Bikhchandani et al. (1992) show that in markets with asymmetric information where decisions are taken sequentially, it is possible to create information cascades that sometimes lead to the herding of wrong behaviors. Orléan (1995) removes the hypothesis of sequential decisions in a Bayesian setting. These are all examples of *rational* herding, in the sense that from the point of view of the single decision maker, following the signal given by the actions of the others can be the right thing to do, even if it can lead to undesirable consequences at the aggregate level. In HAMs literature of financial markets, the most important works on imitation are probably those of Lux (1995, 1998) and Lux and Marchesi (1998, 1999) where agents are influenced by the actual price trend and also by the opinion of the majority.

Our model is different from those predecessors in two ways. First of all, we take into consideration an imitative behavior that is not necessarily an herding behavior. In fact, imitators do not want to *follow the crowd*, or the majority, simply because they are unable to survey the opinion of all the traders. Moreover, when we talk about money, it is possible that it is not so relevant to belong to the majority if the majority is going to lose money. We build a model in which there is no majority between the fundamentalists and the chartists behavioral rules before the decision of imitators. The behavior of imitators create a majority that in this sense is a consequence and not a cause of the imitation.[6] The second difference with respect of the existing literature is the simplicity of the decisional mechanism adopted by imitators. As we have seen, a huge amount of evidence exists in support of the idea that people follow simple rule of thumbs for making decisions. We try to keep their behavioral rule as simple as possible, obtaining the same results, from a qualitative point of view, of more complicated models.

---

[5] See Tversky and Kahneman (1971). To be precise, this false belief is a consequence of the so-called *representativeness heuristic* (Tversky and Kahneman 1974) according to which people evaluate the probability of whether A originated from process B by the degree to which they resamble each other.

[6] The pivotal role of imitators in our model resembles the role of undecided voters in elections when the other parties are not able to get majority without them.

3.1 The Complete Model

Let us now introduce a third kind of traders to the benchmark model analyzed in the previous section. As we have seen, we call them *imitators* because they, at each time period, decide which of the other two groups to imitate. In order to take into consideration the relative number of imitators, we also explicitly introduce three parameters, representing the amount of traders belonging to each group: n_c, n_f and n_i denoting the number of chartists, fundamentalists and imitators, respectively.

The market maker rule takes now the following form:

$$P_{t+1} = P_t + a\left(n_f D_t^f + n_c D_t^c + n_i D_t^i\right) \tag{10}$$

where D_t^i denotes the imitators' orders.

Considering that we are interesting in the role played by imitators, we can normalize the values of n_f and n_c to 1. A value of $n_i = 0.5$ implies that imitators are a half with respect to chartists or fundamentalists (and a fifth of the total number of traders), while a value of $n_i = 2$ implies that the number of imitators is twice the number of fundamentalists or chartists (and a half of the total number of traders), and so on. As a consequence, imitators cannot choose looking at what the majority is doing because the number of fundamentalists and chartists is the same. They will create a majority with their decision. This majority can be temporary because they could change their minds in the future.

The behavioral rules of chartists and fundamentalists are the same of the benchmark model, expressed in Eqs. (3) and (2), respectively. We need to specify now the behavioral rule of imitators. We assume that they use a very simple heuristic: at each time t they look at the current price P_t and at its previous value P_{t-1}. If P_t is closer than P_{t-1} to the fundamental value, than they conclude that the fundamentalists' strategy has been successful and they imitate fundamentalists in time $t + 1$. The opposite is true if the distance between the price and the fundamental is grown up in the last period.[7]

Remembering that we have already introduced an auxiliary variable (x_t) that measures the distance between the current price and the fundamental value, the behavioral rule of imitators is the following:

$$D_t^i = \begin{cases} ix_t & \text{if } |x_t| > |x_{t-1}| \\ -ix_t & \text{if } |x_t| < |x_{t-1}| \end{cases} \tag{11}$$

---

[7]This kind of modeling could also represent the net effect of the decisions of subgroups of imitators that look backward and that differ for the number of past periods they consider. The net effect of each new data is that some groups will switch towards the belief they support.

where the positive parameter i measures the reactivity of imitators to the market signal.[8]

By combining behavioral rules (3), (2) and (11) with the market maker rule (10), we obtain the dynamical system regulating how the asset price evolves:

$$x_{t+1} : \begin{cases} x_t + a[c * \arctan(x_t) - (f - n_i i)x_t] & \text{if} \quad |x_t| > |x_{t-1}| \\ x_t + a[c * \arctan(x_t) - (f + n_i i)x_t] & \text{if} \quad |x_t| < |x_{t-1}| \end{cases} \tag{12}$$

that is a system of second-order difference equations. By introducing the auxiliary variable $y_t = x_{t+1}$ we have the equivalent system of first-order difference equations:

$$T : \begin{cases} x_{t+1} = \begin{cases} x_t + a[c * \arctan(x_t) - (f - n_i i)x_t] & \text{if} \quad |x_t| > |y_t| \\ x_t + a[c * \arctan(x_t) - (f + n_i i)x_t] & \text{if} \quad |x_t| < |y_t| \end{cases} \\ y_{t+1} = x_t \end{cases} \tag{13}$$

In Fig. 5 the grey regions (I) of the phase plane represent situations in which the last value of the price is more distant from the fundamental value with respect to the previous price value, so imitators will follow chartists. In the white areas (II), the opposite is true so imitators interpret the market signal as they were fundamentalists, so they are optimistic when price is low (and buy the asset) while they are pessimistic when the price is high (and sell the asset).

We can look at the dynamical system (13) as the combination of two subsystems regulating how the asset price evolves

$$F_{(I)} : \begin{cases} x_{t+1} = x_t + a[c * \arctan(x_t) - (f - n_i i)x_t] \\ y_{t+1} = x_t \end{cases}$$

$$F_{(II)} : \begin{cases} x_{t+1} = x_t + a[c * \arctan(x_t) - (f + n_i i)x_t] \\ y_{t+1} = x_t \end{cases} \tag{14}$$

where system $F_{(I)}$ governs the dynamics in the region (I) of the phase plane while system $F_{(II)}$ is active in the other subregions of the phase plane. Obviously, the dynamics may switch from one region to the other.

In the next Section we analyze what happens to the local stability properties of the fundamental steady state as a consequence of the introduction of imitators.

---

[8]We use in both cases a linear trading rule for imitators in order to avoid introducing a further nonlinearity in the model. By using the inverse tangent function when imitators behave like chartists we would obtain the same results we will show in the rest of the paper.

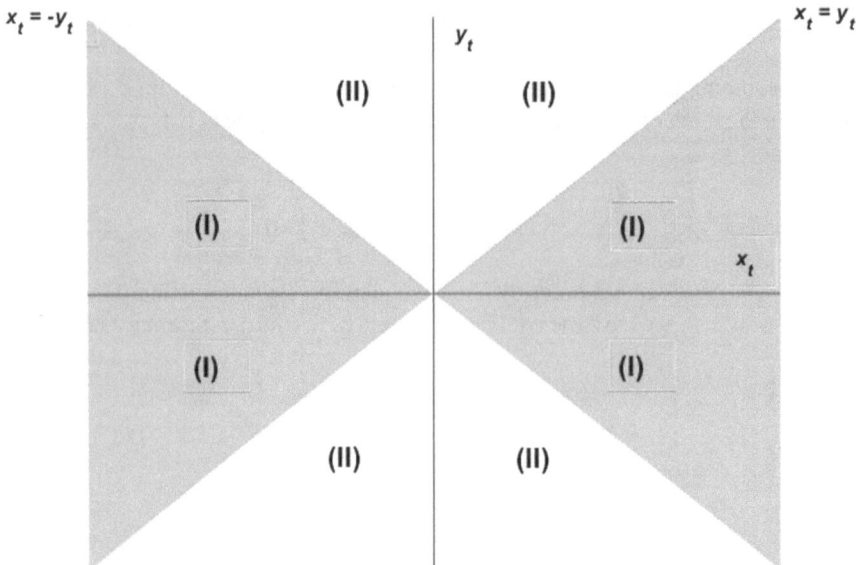

Fig. 5 The regions in grey are those where imitators behave like fundamentalists. In the white regions they interpret the market signals as chartists do

4   Imitators and Local Stability of the Fundamental Steady State

We have seen in the benchmark model that the fundamental fixed point becomes locally unstable via pitchfork bifurcation when chartists are more aggressive than fundamentalists. How does the introduction of imitators affect this result?

In this case the fundamental fixed point belongs to the border that separates Regions I and II. The stability conditions in the two regions become:

$$SC_I : f > c + in_i \quad \text{and} \quad SC_{II} : f + in_i > c \tag{15}$$

respectively.

As a consequence, we can distinguish among four scenarios covering all the possible combinations of parameters:

(a) $f > c + in_i > c$: fundamentalists are much more reactive than chartists and imitators are a few and/or have a low reactivity (*strong fundamentalists dominance scenario*);

(b) $c + in_i > f > c$: fundamentalists are more aggressive than chartists but if imitators behave like chartists they can be predominant (*weak fundamentalists dominance scenario*);

Table 1 Scenarios and local stability of the fundamental fixed point		Region I	Region II
	sfds	Stable	Stable
	wfds	Unstable	Stable
	wcds	Stable	Unstable
	scds	Unstable	Unstable

(c) $f + in_i > c > f$: chartists dominates the market but if imitators and fundamentalists behave similarly they are more reactive than chartists (*weak chartists dominance scenario*);

(d) $c > f + in_i > f$: chartists are much more reactive than fundamentalists and imitators are a few and/or low reactivity (*strong chartists dominance scenario*).

The local stability properties of the fundamental fixed point in each scenario are summarized in Table 1.

4.1 The Strong Fundamentalists Dominance Scenario

The first case we consider is also the easiest to analyze. In fact, in both regions (I) and (II) the fundamental steady state is locally stable. The stability conditions (15) are both fulfilled. As a consequence only the initial condition can belong to the region (I) of the phase plane because fundamentalists are more reactive than chartists and imitators. So the price becomes closer to its fundamental value (that is the system passes to and remains in region (II)) and imitators immediately start behaving like fundamentalists. Only a shock that moves the price in the opposite direction with respect to its fundamental value can make imitators change their minds and make the system come back to region (I), but only for the length of the shock, because the price will start again to approach the fundamental value as soon as the shock is finished.

We must also note that there are two main differences with respect to the corresponding benchmark case (i.e. with $n_i = 0$). First of all, in this case the speed of convergence is faster that the one that we would see without imitators. This is obvious because imitators in this scenario can be considered as additional fundamentalists, so we would obtain the same effect by increasing the value of f in the benchmark model. The second difference lies in the size of the fundamental fixed points' basin of attraction. With imitators behaving as fundamentalists this basin is reduced, and in fact we have seen in the benchmark model that the larger is the value of f the smaller is the set of initial price values leading to the fundamental fixed point. This means that some shocks that were reabsorbed in the benchmark model could not be reabsorbed now.

4.2 The Weak Fundamentalists Dominance Scenario

This case is more interesting than the previous one, because imitators can make the difference in the dynamics of the asset price.

Let us start by considering an initial condition belonging to the subregion (II) of the phase plane. In particular it belongs to the basin of attraction of the fundamental fixed point. This means that the current price is closer to the fundamental value than its previous period value. In this case imitators decide to believe, as fundamentalists do, that the price will converge towards its fundamental value and they behave accordingly. The stability condition for the subregion (II) (see (15)) holds and price moves closer to the fundamental value. In other words the system stays in the subregion (II) of the phase plane, and so on for the next periods. But what would happen as a consequence of a shock that move the asset price away from the fundamental value? The orbit, after the shock, moves from the subregion (II) to the subregion (I) where the stability condition SC_I is violated, in fact $c + in_i > f$. The shock makes the imitators change their mind and they start following the chartists' trading rule, giving them the additional influence that permits to obtain complicated price dynamics. We have seen in the benchmark model that when chartists are more reactive than fundamentalists, periods in which the asset price goes away from the fundamental value alternate with periods in which it moves closer to it. So, we should expect that sooner or later imitators will behave again as fundamentalists, bringing back stability to the system. That is true provided that when fundamentalists prevail, their system is in the basin of attraction of the fundamental fixed point. Otherwise, the overreaction of fundamentalists lead the price to move away from the fundamental value and imitators to switch to the chartists' rule. This is confirmed by the timeplot in Fig. 6 where we introduce a (positive) additive shock to the price after which price movements are characterized by fluctuations.

The same happens starting directly with an initial condition in the subregion (II) of the phase plane but not belonging to the basin of attraction of the fundamental fixed point. In a case like the one represented in Fig. 6 a paradoxical situation seems to occur. In fact, a price value below the fundamental fixed point is followed by an overvalued price, whose deviation from the fundamental fixed point is increased. According to the imitators' behavioral rule, they decide to behave like chartists, even if the fundamentalists' strategy has been successful. Our hypothesis is that imitators

Fig. 6 Timeplot obtained by keeping fixed $a = 3$; $c = 1.1$; $f = 1.2$; $i_m = 0.6$ and $n_i = 0.9$. The shock is additive and introduced at iteration 425

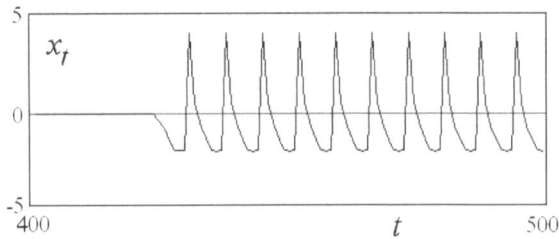

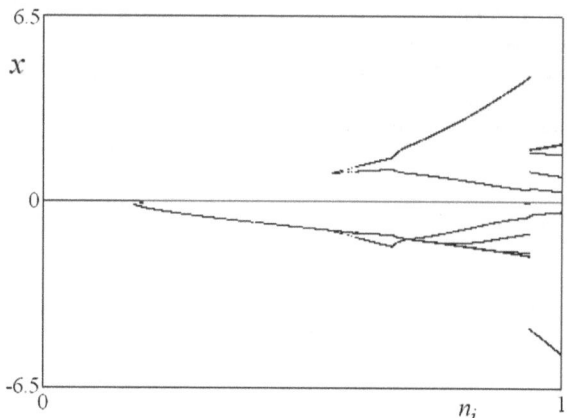

Fig. 7 Bifurcation diagram obtained keeping fixed $a = 3; c = 1.1; f = 1.2$ and $i = 0.6$. The relative number of imitators varies between 0 and 1

look at the beliefs of the other groups and not at their gains. Fundamentalists take their decisions thinking that the price will converge to the fundamental fixed point, so their success, in this case, is not considered a signal of the accuracy of their belief. The overreaction of fundamentalists is more probable when the number of imitators is not negligible. This is confirmed by the bifurcation diagram in Fig. 7.

We can see that if the number of imitators is high enough, the fundamental steady state could not be reached and the complexity of the dynamics increases with the number of imitators.

So imitators play a key role in this scenario and if we think at more frequent shocks we can easily imagine how complicated the dynamics may appear.

4.3 The Weak Chartists' Dominance Scenario

This case is specular to the previous one and share with it the importance of the initial condition. Starting from Region II, that is from a value of the asset price closer to its fundamental value than the former period price (or immediately after a shock that moved the price towards its fundamental value), the system may start to converge to the fundamental value, despite the fact that without imitators we would see convergence to an attractor different from the fundamental steady state. The bifurcation diagram in Fig. 8, obtained by using an initial condition in Region II can clarify this situation.

As can be seen, until imitators are too few to compete with chartists, they contribute in originating dynamics that do not converge to the fundamental value. The initial condition (or the shock) becomes even more relevant when imitators are numerous enough to make the price change direction when they opt for imitating the fundamentalists' trading rule. Stabilizing forces are now dominating with respect to the destabilizing role played by chartists and price starts moving towards its fundamen-

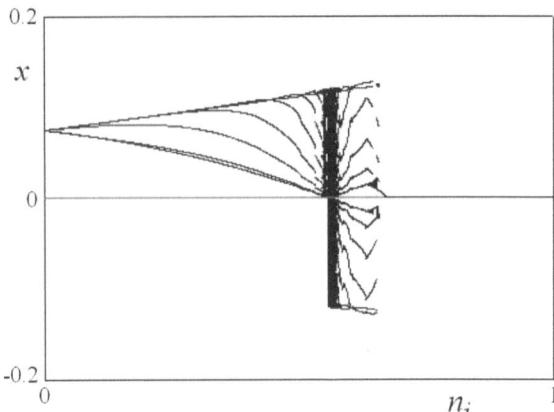

Fig. 8 Bifurcation diagram obtained keeping fixed $a = 3; f = 1.2; c = 1.4; i = 0.3$. The share of imitators n_i varies between 0 and 1

tal value, as can be seen in the right part of Fig. 8, the one representing this third scenario. This dominance of stabilizing forces is strong because even if a new price shock happens, moving away the price from the fundamental value, we know that sooner or later the price will move again towards the fundamental value and imitators definitely behave like fundamentalists. Unless the system would not be outside the basin of attraction of the fundamental fixed point (as we have seen in the previous subcase).

The role of imitators is extremely important here because their role is pivotal: when they believe that price will go close to the fundamental value, it actually does.

The left part of Fig. 8 is also representative of what would happen starting from an initial condition in Region I. Imitators follow the trading rule of chartists and the more they are the more complicated the price dynamics becomes. We move into this point analyzing the last scenario.

4.4 The Strong Chartists' Dominance Scenario

In this case, the asset price never converge to its fundamental value. Stability conditions (15) are both violated. Chartists dominate the market so strongly that even a shock that moves the price towards the fundamental value does not have long period consequence. After the first period in which imitators follow the fundamentalists' trading rule, the price starts again moving towards an attractor that is periodic or chaotic. We know that in this case imitators often change their minds and alternate behavioral rules. Let us consider the case in which, even without imitators, the attractor is chaotic and bull and bear dynamics (in the sense explained before) occur, i.e. when chartists strongly dominate the market (as in the case represented in Fig. 4). Under these circumstances, values of the asset price close the fundamental value

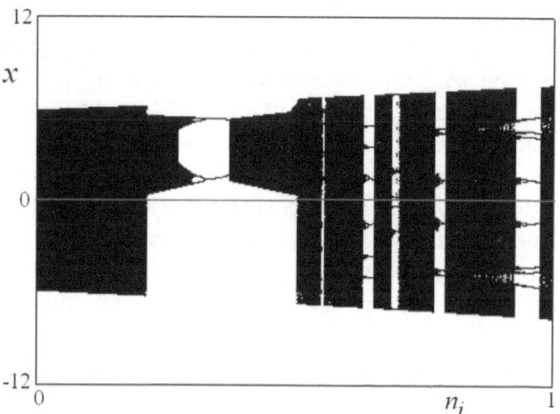

Fig. 9 Bifurcation diagram obtained by keeping fixed $a = 3; f = 1.2; c = 3.3; i = 0.3$. The share of imitators varies between 0 and 1

alternate with values far away from it in an almost unpredictable way. In such a case it can be natural to expect that by introducing imitators the price's variance will increase. That is fluctuations should be amplified. This is only partially confirmed by simulations. Let us look at the bifurcation diagram represented in Fig. 9

We can see that for certain values of the relative number of imitators n_i the chaotic attractor reduces its size, sometimes even becoming periodic. This result is quite interesting and can be seen as a possible base model for reproducing some stylized facts of financial markets like volatility clustering. In our simple model, in fact, we consider the model's parameters as exogenously given. In real markets we expect that they can vary over time. It is not strange to consider the number of imitators one of them. If this is so, the changing number of imitators can drastically modify the boundaries of the price movements as we have just seen. So, a general tendency for an increase of the price variations as imitators increases can be observed but cannot be considered a general rule. This interesting scenario deserves, in our opinion, to be deepen investigated. We plan to do that in future research.

5 Conclusions

The discovery of deviations from the rational behavior described in the microeconomics textbooks, has a long history. In the heuristics-and-biases program, Kahneman and his collaborators find a lot of deviations that are systematic and in some sense predictable. Among the various strands of the theoretical economic literature that try to incorporate such behavioral assumptions, there is one in which the mathematics of dynamical systems meets behavioral economics. The aim of the researchers working in this field consists in building small scale models representing markets populated by heterogeneous and boundedly rational agents. The small scale of the models usually permits their analytical study that, together with the use of numerical simulations, makes possible to understand the causes of the emergence of some phenomena. These models are usually not so sophisticated to be immediately calibrated. Hardly the simulated time series could quantitatively be compared with real ones.[9] The main aim of these models consists in qualitatively replicating some of the most puzzling stylized facts. At a later stage, these models can be the base for the building of more complicated and complete models that also quantitatively replicate some market phenomena. One of the main results of this strand of literature is that in order to qualitatively reproduce a lot of stylized facts, the introduction of some well thought behavioral assumptions in a completely deterministic setting may be enough.

The most of these models tries to explain some features of financial markets. In this model we have analyzed the role of cognitively biased imitators in a single asset market. Imitators trade together with fundamentalists and chartists and at each time period they decide which of the two available trading rules is the best for the next trading session. This decision is based on the last performance of the available strategies. Our results permit to enlight the role of imitators in destabilizing or in stabilizing the market. In particular, some of the scenarios we obtain seems to be suitable for being used as a skeleton for replicating important facts of financial markets like long memory effects and volatility clustering.

The model can be extended in several directions. Different kinds of trading strategies can be added (a first attempt in this direction is Brianzoni and Campisi (2020)) and some stochastic elements can be inserted in order to replicate more quantitative features of financial markets and to better understand the role of cognitive biases.

References

Allais, M. (1953). Le comportement de l'homme rationnel devant le risque: Critique des postulats et axiomes de l'école américaine. *Econometrica: Journal of the Econometric Society*, 503–546.

Banerjee, A. V. (1992). A simple model of herd behavior. *The Quarterly Journal of Economics*, *107*(3), 797–817.

Barber, B. M., & Odean, T. (2000). Trading is hazardous to your wealth: The common stock investment performance of individual investors. *The Journal of Finance*, *55*(2), 773–806.

Barber, B. M., & Odean, T. (2001). Boys will be boys: Gender, overconfidence, and common stock investment. *The Quarterly Journal of Economics*, *116*(1), 261–292.

Bikhchandani, S., Hirshleifer, D., & Welch, I. (1992). A theory of fads, fashion, custom, and cultural change as informational cascades. *Journal of Political Economy*, *100*(5), 992–1026.

Bischi, G.-I., Gallegati, M., Gardini, L., Leombruni, R., & Palestrini, A. (2006). Herd behavior and nonfundamental asset price fluctuations in financial markets. *Macroeconomic Dynamics*, *10*(4), 502–528.

Brianzoni, S., & Campisi, G. (2020). Dynamical analysis of a financial market with fundamentalists, chartists, and imitators. *Chaos, Solitons & Fractals*, *130*, 109434.

---

[9]Even if there are also some recent and successful attempts to do that (see Franke and Westerhoff (2012), Tramontana and Westerhoff (2013)).

Brock, W. A., & Hommes, C. H. (1997). A rational route to randomness. *Econometrica: Journal of the Econometric Society*, 1059–1095.

Brock, W. A., & Hommes, C. H. (1998). Heterogeneous beliefs and routes to chaos in a simple asset pricing model. *Journal of Economic Dynamics and Control*, 22(8), 1235–1274.

Chiarella, C., & He, X.-Z. (2002). Heterogeneous beliefs, risk and learning in a simple asset pricing model. *Computational Economics*, 19(1), 95–132.

Chiarella, C., He, X.-Z., et al. (2000). *Stability of competitive equilibria with heterogeneous beliefs and learning*. Technical report, Quantitative Finance Research Centre, University of Technology, Sydney.

Chiarella, C., Dieci, R., & He, X.-Z. (2009). Heterogeneity, market mechanisms, and asset price dynamics. In *Handbook of financial markets: Dynamics and evolution* (pp. 277–344). Elsevier.

Day, R. H., & Huang, W. (1990). Bulls, bears and market sheep. *Journal of Economic Behavior & Organization*, 14(3), 299–329.

De Long, J. B., Shleifer, A., Summers, L. H., & Waldmann, R. J. (1990). Positive feedback investment strategies and destabilizing rational speculation. *The Journal of Finance*, 45(2), 379–395.

Fama, E. F. (1965). The behavior of stock-market prices. *The Journal of Business*, 38(1), 34–105.

Fama, E. F. (1995). Random walks in stock market prices. *Financial Analysts Journal*, 51(1), 75–80.

Farmer, J. D. (2002). Market force, ecology and evolution. *Industrial and Corporate Change*, 11(5), 895–953.

Foroni, I., & Agliari, A. (2008). Complex price dynamics in a financial market with imitation. *Computational Economics*, 32(1–2), 21–36.

Franke, R., & Westerhoff, F. (2012). Structural stochastic volatility in asset pricing dynamics: Estimation and model contest. *Journal of Economic Dynamics and Control*, 36(8), 1193–1211.

Franke, R., & Westerhoff, F. (2016). Why a simple herding model may generate the stylized facts of daily returns: Explanation and estimation. *Journal of Economic Interaction and Coordination*, 11(1), 1–34.

Frankel, J. A., Froot, K. A., et al. (1986). Understanding the us dollar in the eighties: The expectations of chartists and fundamentalists. *Economic Record*, 62(1), 24–38.

Hommes, C. (2011). The heterogeneous expectations hypothesis: Some evidence from the lab. *Journal of Economic Dynamics and Control*, 35(1), 1–24.

Hommes, C., & Wagener, F. (2009). Complex evolutionary systems in behavioral finance. In *Handbook of financial markets: Dynamics and evolution* (pp. 217–276). Elsevier.

Kunda, Z. (1999). *Social cognition: Making sense of people*. MIT press.

Lux, T. (1995). Herd behaviour, bubbles and crashes. *The Economic Journal*, 105(431), 881–896.

Lux, T. (1998). The socio-economic dynamics of speculative markets: Interacting agents, chaos, and the fat tails of return distributions. *Journal of Economic Behavior & Organization*, 33(2), 143–165.

Lux, T. (2009). Stochastic behavioral asset-pricing models and the stylized facts. In *Handbook of financial markets: Dynamics and evolution* (pp. 161–215). Elsevier.

Lux, T., & Marchesi, M. (1998). Volatility clustering in financial markets: A micro-simulation of interacting agents. *IFAC Proceedings Volumes*, 31(16), 7–10.

Lux, T., & Marchesi, M. (1999). Scaling and criticality in a stochastic multi-agent model of a financial market. *Nature*, 397(6719), 498.

Menkhoff, L., & Taylor, M. P. (2007). The obstinate passion of foreign exchange professionals: Technical analysis. *Journal of Economic Literature*, 45(4), 936–972.

Orléan, A. (1995). Bayesian interactions and collective dynamics of opinion: Herd behavior and mimetic contagion. *Journal of Economic Behavior & Organization*, 28(2), 257–274.

Shefrin, H. (2001). *Behavioral finance*. Edward Elgar Publishing.

Shiller, R. J. (2015). *Irrational exuberance: Revised and expanded* (3rd ed.). Princeton university press.

Tramontana, F., & Westerhoff, F. (2013). One-dimensional discontinuous piecewise-linear maps and the dynamics of financial markets. In *Global analysis of dynamic models in economics and finance* (pp. 205–227). Springer.

Tversky, A., & Kahneman, D. (1971). Belief in the law of small numbers. *Psychological Bulletin*, *76*(2), 105.

Tversky, A., & Kahneman, D. (1974). Judgment under uncertainty: Heuristics and biases. *Science*, *185*(4157), 1124–1131.

Westerhoff, F. H. (2009). Exchange rate dynamics: A nonlinear survey. In B. Rosser Jr. (Ed.), *Handbook of research on complexity*. Citeseer.

Evolutionary Tax Evasion, Prospect Theory and Heterogeneous Taxpayers

Domenico De Giovanni, Fabio Lamantia and Mario Pezzino

Abstract This work studies the dynamics of compliance and optimal auditing in a population of boundedly rational agents who decide whether to engage in tax evasion depending on an evolutionary adaptation process. If they decide to evade taxes, taxpayers can choose different ways to engage in tax evasion and face different auditing probabilities. Moreover, taxpayers make decisions according to the (realistic) principles of Prospect Theory. The analysis studies the intertemporal optimal auditing of a tax authority that targets tax revenues maximization and strategically selects audit probabilities to manage the trade-off created by controlling different modes of evasion with a resource constraint.

Keywords Tax evasion · Prospect theory · Optimal control · Auditing · Controlled evolutionary dynamics

JEL Codes: D8 · C61 · C73 · H26

1 Introduction

The way economic literature traditionally has defined tax evasion as a form of decision under risk[1] has faced criticism by those studies that have not found sufficient empirical nor experimental support to the key results of the theory.[2]

---

[1] See for example Allingham and Sandmo (1972), Yitzhaki (1974), Slemrod and Yitzhaki (2002), Slemrod and Weber (2012).

[2] See Alm et al. (1992), Alm (1999), Torgler (2002), Frey and Feld (2002), Alm (2018).

D. De Giovanni · F. Lamantia (✉)
Department of Economics, Statistics and Finance, University of Calabria, Rende, Italy
e-mail: fabio.lamantia@unical.it

D. De Giovanni
e-mail: ddegiovanni@unical.it

F. Lamantia · M. Pezzino
School of Social Sciences, University of Manchester, Manchester, UK
e-mail: mario.pezzino@manchester.ac.uk

© Springer Nature Singapore Pte Ltd. 2020
F. Szidarovszky and G. I. Bischi (eds.), *Games and Dynamics in Economics*,
https://doi.org/10.1007/978-981-15-3623-6_5

Recently a growing body of economic literature has reconciled the results of the theoretical models and the empirical evidence introducing aspects of bounded rationality and social interactions in the analysis.[3] It seems increasingly apparent in particular that individuals may tend to overestimate detection probabilities[4] and their behavior may be best described by the type of value functions considered in prospect theory (PT) rather than traditional expected utility theory. Examples of works that have applied the cumulative prospect theory framework, first introduced in Kahneman and Tversky (1979), to study the risky decision of boundedly rational taxpayers are provided by Bernasconi and Zanardi (2004), Dhami and Al-Nowaihi (2007), Dhami and Al-Nowaihi (2010), De Giovanni et al. (2019).[5]

Our work is most closely related to De Giovanni et al. (2019), cited as DLP in further discussions. DLP extends the framework of tax evasion under PT to a *dynamic evolutionary setting*.[6] Evolutionary dynamics consider individuals to be boundedly rational, i.e. assumed to be "programmed" to behave in a certain manner. Social interactions play an important role: through social interactions agents can learn of the payoffs obtained by other individuals and, over time, change their conduct (i.e. *evolve*). Specifically, DLP describes the dynamic effects of tax reforms (e.g. changes in tax rates or auditing approaches) and the effects of the bounded rationality of taxpayers under PT on the long-run level of compliance. Assuming that the tax authority and the agents have different degrees of rationality (taxpayers assumed to be following PT principles and the tax authority to be rational),[7] DLP studies the intertemporal optimization of tax revenues by a regulator who can choose auditing effort. The authors show that the long-run evolution of the controlled dynamic system depends on how taxpayers react to auditing policies and, in particular, on the way they may distort the probability of auditing. In addition, the authors show the possibility of the existence of a discontinuity in the regulator's optimal control created by a threshold level of tax evasion. Auditing costs are increasing and convex in the level of tax evasion and, for higher levels of tax evasion, there may be a threshold level of evasion that would make the regulator suddenly decide to drastically reduce auditing effort.

This work extends the analysis proposed in DLP to consider the possibility that taxpayers may evade taxes in different ways and that, depending on the particular form of evasion chosen, they will face different, endogenous audit probabilities. Assuming that individuals can evade taxes in different ways is a rather realistic exten-

---

[3]Recent contributions have studied the way social norms and forms of intrinsic motivation (often referred as *tax morale*) may affect individuals' behavior and, ultimately, compliance rates. See Andreoni et al. (1998), Luttmer and Singhal (2014), Lamantia and Pezzino (2018), Alm (2018).

[4]See Chetty (2009).

[5]See also Trotin (2012), Piolatto and Trotin (2016), Piolatto and Rablen (2017).

[6]See also Antoci et al. (2014), Lamantia and Pezzino (2018) where evolutionary dynamics are applied to the study of tax compliance. Pickhardt and Prinz (2014) provide a review of the works that study the behavioral dynamics of tax evasion, with a particular focus on the way interaction among individuals playing different roles (e.g. taxpayers, tax practitioners, tax authorities, etc.) can affect the level of compliance in a population.

[7]See also Dhami and Al-Nowaihi (2010), Petrohilos-Andrianos and Xepapadeas (2016).

sion. In addition, it is often the case where tax authorities worldwide tend to classify taxpayers into different groups that may differ on the type of income and expected incidence of tax evasion.[8] Naturally, it is expected that different categories of taxpayers may engage differently with tax evasion and, consequently, may require different levels of auditing effort. Specifically, in what follows we assume that all taxpayers in the population earn the same level of income and may be "programmed".[9] They may be honest and fully report their income. They may otherwise decide to engage in tax evasion. Those who decide to do so can choose whether to consider more or less aggressive modes of evasion. Similarly, we could think of individuals with the same level but different types of income (e.g. business and non-business) who, consequently, can entertain different modes of evasion.

Extending the analysis of DLP to include more than one form of tax evasion is not trivial. Tax evasion can now take multiple forms and the tax authority can attempt to optimally control the behavior of tax payers strategically choosing two audit probabilities. Auditing, however, comes at a cost that, realistically, we assume to be increasing and convex in the auditing effort of the tax authority. The convexity of the cost of auditing indirectly expresses the effects of a resource constraint that the regulator faces and implies that the optimal control, even if the regulator has two probabilities at her disposal, can only produce a second best solution. In other words, the tax authority faces a trade-off between the benefit of reducing more aggressive (and more difficult to detect) forms of tax evasion and saving resources allowing the existence of an non-empty set of individuals who engage in less aggressive evasion. This type of trade off could not be identified in a framework with only one mode of evasion.

The way the tax authority deals with the trade off described above will depend on the level and distribution of tax evasion modes that she will initially face and on the way individuals may deform (in line with the boundedly rational behavior proposed in PT) auditing probabilities. If individuals overestimate audit probabilities, then it will get easier for the auditor to enforce auditing for less aggressive evaders; the optimal audit probability for this mode of evasion displays non-monotonic patterns due to the convexity of the auditing costs. Interesting, and in line with the results in DLP, now that individuals may be deforming audit probabilities, we may observe a discontinuity on the optimal auditing probability of high evaders. This happens when a sufficiently large number of individuals engages with aggressive forms of tax evasion; the tax authority identifies a threshold level of evasion for which enforcement becomes too expensive and a drastic reduction in effort is required. The long rung equilibria (the *good*, in which all individual in the population tend to act hon-

---

[8]For example, tax authorities tend to distinguish between taxpayers who are employed or self-employed; those whose income is generated by business or non-business activities; those who have filled a tax assessment with or without the support of a tax advisor.

[9]Frey (1999) shows that in a population there may be taxpayers who simply do not look for opportunities to evade taxes. On similar lines, Long and Swingen (1991) (p. 130) argue that some individuals are not naturally predisposed to evade taxes. This is in line with experimental evidence that shows that some individuals never choose to evade taxes (see Feld and Tyran 2002), even in the absence of enforcement.

estly, and the *bad*, in which tax evasion is the prevailing behavior in the population) continue to exist, but the probability deformation increases the basin of attraction of the *good* equilibrium. Perhaps not surprisingly, for a sufficiently strong probability deformation, auditing becomes easier enough and both optimal auditing probabilities become increasing functions of their respective levels of evasion modes and, in terms of dynamic evolution of the population of taxpayers, the population will always end up with all taxpayer being honest.

The work is organized as follows. Section 2 describes agents' preferences, the evolutionary setup and intertemporal optimization problem of tax auditing. Section 3 performs a series of numerical experiments highlighting the main insights of our analysis. Section 4 concludes the paper with further research directions.

2 Model

We model the evolution of tax evasion by means of a continuous-time infinite-horizon population game. The population consists of boundely rational taxpayers (agents), all with the same income W and subject to tax rate r. To model situations in which tax evasion may take different forms, we make available to each taxpayer at each time three strategies. First, taxpayers might adopt strategy H meaning that they decide to pay the full amount of taxes. We name as *honest* the taxpayers adopting strategy II. Second, taxpayers might choose strategy M. This strategy refers to situation in which taxpayers decide to evade taxes, but the tax evasion is mild, in the sense that agents adopting strategy M choose a profile corresponding to a low amount of evasion. More formally, adopting strategy M implies declaring to the tax authority an income $D_M < W$, so that the amount of tax evasion is $E_M = W - D_M$. We refer to agents adopting strategy M as *mild evaders*. Third, taxpayers might choose strategy A, which refers to situation in which the evasion is aggressive. More formally, adopting strategy A entails declaring to the tax authority a low income, $D_A < D_M$, which implies a high evasion level $E_A = W - D_A > E_M$. We call *aggressive evaders* taxpayers choosing strategy A.

With probability p_h, $h = M, A$, an evader is audited and sanctioned. Sanction is proportional to evasion: if detected, the sanctioned agent pays $\lambda r E_h$, where $\lambda \geq 1$ measures the additional fine if found guilty. Summing up, if the evader playing strategy h ($h = M, A$) is not found guilty, then his net income is:

$$Y_h^N = W(1 - r) + r E_h;$$

On the other hand, if audited, then agent's net income is:

$$Y_h^G = W(1 - r) - \lambda r E_h.$$

Honest taxpayers' net income is:

$$Y^H = W(1 - r).$$

2.1 Agents Preferences

In this section we introduce the framework of Prospect Theory (PT) to our analysis. We standardize agents' net income using the after-tax income $(1 - r)W$ as reference point. This implies that agents are interested in the utility coming from their net income relative to the reference points. Using this change of variable, honest agents have a relative income equal to zero while E_h evaders gets a relative income equal to $X_h^G = Y_h^G - (1 - r)W = -\lambda r E_h$ if detected and to $X_h^N = Y_h^N - (1 - r)W = r E_h$ if not detected.

Another relevant element of PT is represented by the *Probability Weighting function*, which models the empirical evidence that people tend to underweight "high" probabilities and overweight "low" probabilities. The Probability Weighting function $w(p) : [0, 1] \to [0, 1]$ is increasing in the probability of being found guilty p and satisfies the usual properties in PT. Here we employ the *Prelec* probability weighting function $w(p) = e^{-[(-\log p)^\alpha]}$ proposed in Prelec (1998).

For the value function (utility) $v(x)$ associated to outcome x (i.e. $X_h^q, q = N, G$), here we borrow the well-known one suggested in Tversky and Kahneman (1992) (as also done in Dhami and Al-Nowaihi 2007) and assume

$$v(x) = \begin{cases} x^\beta & \text{if } x \geq 0 \\ -\theta (-x)^\beta & \text{if } x < 0 \end{cases}$$

where $\theta > 1$ measures loss aversion and $\beta \in [0, 1]$ is a preference parameter (notice that Tversky and Kahneman (1992) suggest to use $\beta = 0.88$ and $\theta = 2.25$).

Summing up, for $h = M, A$, the expected value of evaders playing h is

$$V^h = w(p_h)v(-\lambda r E_h) + w(1 - p_h)v(r E_h),$$

while honest taxpayers have utility $V^H = w(0)v(0) = 0$.

2.2 Evolutionary Setup

This section describes a dynamic model of tax evasion based on the evolution of agent types in a population. The population's state at time t is represented by the couple $x(t) = (x_M(t), x_A(t))$, where $(x_M(t), x_A(t))$ denotes the share of mild (aggressive) evaders at time t. The remaining fraction of agents in the population, $1 - x_M(t) - x_A(t)$, is the share of honest taxpayers. We also assume that the auditing probabilities depend on the current state of the system, that is $p_h(t) = p_h(x_M(t), x_A(t))$, $h = M, A$. This gives the regulator the ability to adjust the auditing probabilities according to the current state of the population.

According to the static model described above, the expected value of evading E_h, $h = M, A$, at time t is given by:

$$V^h(t) := V^h(x_M(t), x_A(t)) = \tag{1}$$
$$w(p_h(t)) v(-\lambda r E_h) + w(1 - p_h(t)) v(r E_h).$$

In continuous time, replicator dynamics for the shares $(x_M(t), x_A(t))$ is expressed by the following system of ordinary differential equations (ODE), see Weibull (1997) for details:

$$\begin{cases} \dot{x}_M(t) = x_M(t) \left(V^M(t) (1 - x_M(t)) - x_A(t) V^A(t) \right) \\ \dot{x}_A(t) = x_A(t) \left(V^A(t) (1 - x_A(t)) - x_M(t) V^M(t) \right) \end{cases}. \tag{2}$$

According to dynamical system (2), the proportion of agents playing a given strategy increases whenever its fitness is above the average fitness in the population (recall that the utility of honest taxpayers is set to zero).

2.3 Optimal Enforcement

Next, we turn on the enforcement side of the model, as done in Petrohilos-Andrianos and Xepapadeas (2016). We assume that the regulator selects the effort put into auditing each type of evaders in order to control the dynamical system (2), with the long-term target of maximizing the present value of the future stream (net auditing costs) of cash-flows deriving from tax income. Without loss of generality, we assume that there is a one-to-one correspondence between regulator effort and auditing probability for each type of evaders. While innocuous, this assumption allows us to get rid of the efforts and treat the auditing probabilities, $p_M(t)$ and $p_A(t)$, as the control variables of the optimization problem. Also, we assume that the cost selecting an auditing probability p_h is quadratic, that is $c_h(p_h) = \gamma_h p_h^2$, $h = M, A$. Parameters $\gamma_h > 0, h = M, A$ describe the inefficiency in detecting tax evasion. Here we are also assuming $\gamma_A > \gamma_M$: more aggressive forms of tax evasion imply also more sophisticated strategies employed by dishonest agents and, consequently, higher auditing costs for the regulator.

The tax authority collects tax and fines as indicated in Table 1, and is subject to auditing cost. The net tax revenue, is defined as $NTR(t) = TR_A(t) + TR_N(t) - c(t)$, where:

Table 1 Summary of taxes and fines collected by the regulator

	Compliance	Mild evasion	Aggressive evasion
Audit	rW	$r(W - E_M) + \lambda r E_M$	$r(W - E_A) + \lambda r E_A$
No audit	rW	$r(W - E_M)$	$r(W - E_A)$

1. $TR_A(t)$ is the expected time-t gross tax revenues coming from honest and audited agents, that is:

$$TR_A(t) = (1 - x_M(t) - x_A(t))r\,W +$$
$$\sum_{h=M,A} p_h(t)x_h(t)\,(\lambda r E_h + r\,(W - E_h));$$

2. $TR_N(t)$ is the expected time-t gross revenue coming from non-audited agents, that is:

$$TR_N(t) = r \sum_{h=M,A} (1 - p_h(t))x_h(t)(W - E_h);$$

3. $c(t)$ is the time-t cost of auditing, that is:

$$c(t) = \sum_{h=M,A} c_h\,(p_h(t)).$$

We are assuming that the regulator is a rational forward looking agent. Thus, the regulator's dynamic problem includes selection of the feedback rules, $p_h(t) = p_h(x_M(t), x_A(t)) \in [0, 1]$, $h = M, A$, such that the following objective function

$$\int_0^{+\infty} e^{-\delta t} NTR(t)dt \tag{3}$$

is maximized subject to the replicator state equations (2) and the additional constraints $x_h(t) \in [0, 1]$.

3 Dynamic Analysis

In this section, we present the main insights of our analysis when the forward looking auditor implements the auditing policy that maximizes the discounted sum of future Net Tax Revenues (3). Here, our interest concerns two specific issues, namely: *1.* the optimal auditing policy and its long-run implications on the shares of both types of

Table 2 Parameter values used in the analysis

W	E_M	E_A	λ	γ_M	γ_A	β	θ	δ
5	1	2.5	1.5	3	4	0.88	2.55	0.05

evaders; 2. the impact of key parameters on the optimal solution. In what follows, we use the parametrization displayed in Table 2.

Since the optimal reinforcement model does not admit a closed-form solution, we rely on numerical approximation to compute value functions, optimal auditing rules and optimal vector field. We have performed a series of experiments with different parametric setup, all highlighting the tax rate and the parameter of probability deformation as the main drivers of the optimal auditor behavior and, consequently, the long-run evolution of the system. In what follows, we present the main insights of our analysis, again making use of the parametric setup in Table 2.

A common feature shared by the optimal auditing policies is that, for a given type of evaders, they appears to depend only on the percentage of evaders of the same type. In different terms, the current fraction of evaders of type M (A) affects only the auditing probability for type M (A). This seems reasonable as the contribution of type M to the NTR, both in terms of auditor's expected reward and auditing costs, is independent on the fraction of evaders of type A and vice-versa. Also, observe that selecting the optimal auditing rule is a matter of balance between two forces. First, the auditor has an incentive to choose a high auditing probability to discourage evasion. Second, as evasion increases, convex auditing costs make the auditor willing to choose a low auditing probability, which force dominates the other depending on the current state of the system and the situation at hand. To elaborate, we discuss the results by highlighting the effect of probability deformation, which is the factor affecting the long-run dynamics of the system.

We solve the optimal regulation problem by means of a semi-Lagrangian method that discretizes the corresponding Hamilton-Jacobi-Bellman equation. We first apply the Euler scheme to replace the continuous-time problem with a discrete-time version. We then apply the finite element method to the infinite-dimensional discrete-time problem. A detailed explanation of the method is far beyond the scope of this paper. We refer to Grüne and Semmler (2004), De Giovanni and Lamantia (2018) for more details.

Figure 1 shows the solution of the dynamic optimal auditing problem when tax-payers do not deform probabilities.[10]

The optimal auditing policies for evaders of type M and A display quite different qualitative (and quantitative) patterns. The left panel displays a non-monotonic pattern. The optimal auditing probability the auditor uses for mild evaders, p_M, is strictly

---

[10]Here we adopt the usual representation for a three-strategy game (a game with one population game and three pure strategies): instead of depicting the shares of the population employing the different pure strategies in the simplex of $\mathbb{R}^3$, these shares are shown in a two-dimensional equilateral triangle where the vertices correspond to population distributions where all agents employ the same pure strategy and the sides to population distributions where only two pure strategies are adopted.

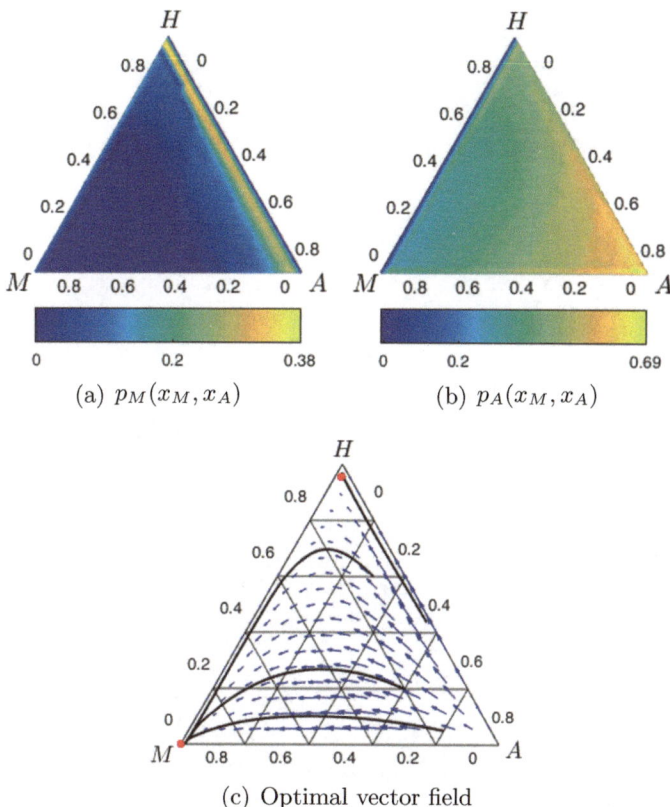

(a) $p_M(x_M, x_A)$ (b) $p_A(x_M, x_A)$

(c) Optimal vector field

Fig. 1 Solution of the optimal auditing problem with no probability deformation. **Panel a**: Optimal auditing probability rule for mild evaders; **Panel b**: Optimal auditing probability rule for aggressive evaders. **Panel c**: System dynamics. The tax rate is set to $r = 0.25$ and the remaining parameter values are those shown in Table 2

positive even when low evaders are absent (that is, $x_M = 0$). Here, the far-sighted auditor is caring not only for the current profit, but also for the future loss of profits due to possible increase in the fraction of low evaders. The positive optimal probability despite the current absence of mild evaders then reflects the auditor's willingness to keep mild evaders' prospect at a low level, so as to discourage future incentives to become mild evader. As the share of mild evaders increases, we observe a natural increase in the auditing probability, reflecting the auditor's urgency to reduce both the loss of current profits and mild evaders' prospect at the same time. However, the incentive to increase the auditing probability as x_M increases is persistent only as far as the share of mild evaders is sufficiently low. After reaching a certain level, convex auditing costs make economically convenient for the auditor to decrease the auditing probability. On the other hand, the auditing probability for aggressive evaders turns out to be always increasing with x_A. In fact, with aggressive evaders the auditor's

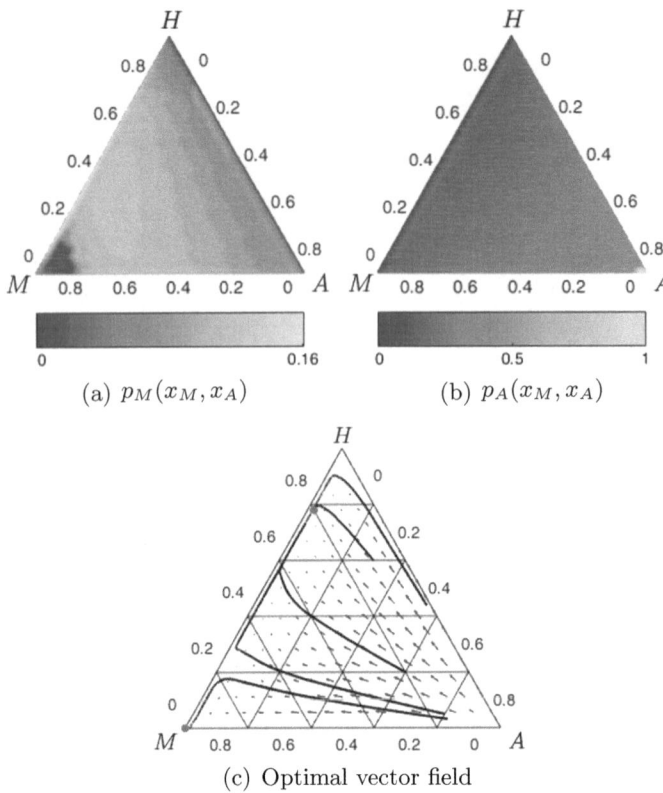

(a) $p_M(x_M, x_A)$

(b) $p_A(x_M, x_A)$

(c) Optimal vector field

Fig. 2 Solution of the optimal auditing problem with a probability deformation level equal to $\alpha = 0.5$. **Panel a**: Optimal auditing probability rule for mild evaders; **Panel b**: Optimal auditing probability rule for aggressive evaders. **Panel c**: System dynamics. The tax rate is set to $r = 0.25$ and the remaining parameter values are those shown in Table 2

willingness to keep the level of x_A as low as possible always prevails. The effects of the optimal auditing policy on the long-run evolution of the shares of evaders is shown in the optimal vector field displayed in panel (c) of Fig. 1.

The system exhibits the coexistence of two stable long-run equilibria, each with its own basin of attraction. In the first equilibrium, all tax payers will end up to be mild evaders. In the second equilibrium, the population is characterized by a polymorphic population with a small, positive, fraction of both type of evaders and the majority of taxpayers that choose to be honest. The boundary of the basin of attraction of the two final outcomes, which determines the long-run population configuration, is only driven by the initial level of mild evaders. Suppose the initial population configuration lies in the region in which the optimal auditing probability p_M is increasing. Then, the system will end up to the *good* equilibrium where the large part of the population behaves honestly. In this region, indeed, the auditor is able to keep the prospect of both types of evaders low enough, so as to discourage evasion of both types. In

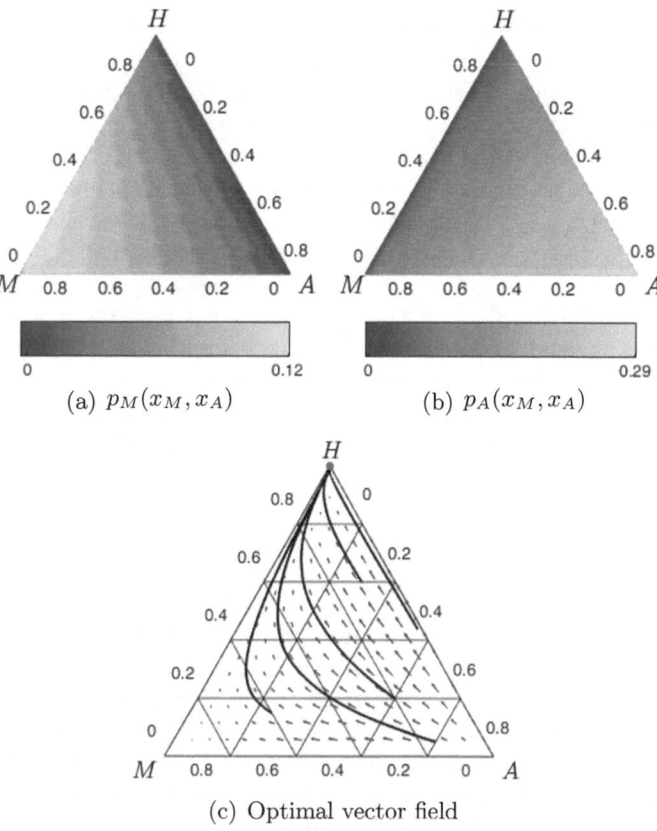

(a) $p_M(x_M, x_A)$

(b) $p_A(x_M, x_A)$

(c) Optimal vector field

Fig. 3 Solution of the optimal auditing problem with a probability deformation level equal to $\alpha = 0.25$. **Panel a**: Optimal auditing probability rule for mild evaders; **Panel b**: Optimal auditing probability rule for aggressive evaders. **Panel c**: System dynamics. The tax rate is set to $r = 0.25$ and the remaining parameter values are those shown in Table 2

the complementary region, the increasing auditor's incentive to rise the prospect of low evaders in order to balance the increasingly auditing cost, makes the system converge to the *bad* configuration where all taxpayers are mild evaders. At this point it is interesting to observe a peculiar difference between the one-state model proposed in DLP and the two-state model of the present work. In DLP, the absence of probability deformation always leads to the *bad* scenario in which all taxpayers end up being evaders. Here, instead, we observe long-run population heterogeneity even without probability deformation.

Continuing the discussion, Fig. 2 shows the optimal auditing rule when a strong probability deformation is introduced. The probability deformation makes easier for the auditor to enforce auditing for mild evaders. The optimal auditing rule again displays a non-monotonic patterns. However, it reaches its maximum level at a higher share of aggressive evaders. On the other hand, we observe a discontinuity on the optimal auditing probability of high evaders, at very high levels of x_A. The effects of the probability deformation in terms of the dynamic evolution of the system are shown in panel c of Fig. 2. Again the system admits the coexistence of two different equilibria, where the *bad* one consists in the monomorphic configuration where all taxpayers are mild evaders and the *good* one represents a polymorphic configuration. Compared with the previous case, we observe that the basin of attraction of the *good* equilibrium in this case is larger than the case where no deformation is assumed. However, the *good* equilibrium observed is characterized by a higher share of mild evaders.

Figure 3 looks at the effects of a even stronger probability deformation. Auditing becomes easier enough to make the incentive to increase auditing dominate the cost effect. As a result, both optimal auditing probabilities become increasing functions of their respective target. In terms of dynamic evolution of the population of taxpayers, the population will always end up with all taxpayer being honest.

4 Concluding Remarks

In this work we have extended the analysis proposed in DLP to allow for heterogeneous taxpayers. Our analysis reveals two interesting insights. First, in contrast with the one-state model of DLP, we observe long-run heterogeneity even without the probability deformation of prospect theory. Second, taxpayers that behave according to prospect theory are more easily enforced, the degree of probability deformation impacting positively. This increases the chances of observing a long-run distribution of evaders in which all agents are mild evaders.

Given the relevance of the subject under investigation and the promising results presented here, we believe that there is room for further exploration in two directions. First, a calibration of taxpayers' preferences based on real world data is an important step in understanding the factors that drive tax evasion in different countries. Second, an analysis of the uncontrolled dynamical system would, on both the policy and theoretical side, sheds light on the role of tax auditor on the long run distribution of tax evasion. We are currently working on such extensions and will present the results in a separate paper.

References

Allingham, M. G., & Sandmo, A. (1972). Income tax evasion: A theoretical analysis. *Journal of Public Economics*, *1*(3–4), 323–338.

Alm, J. (1999). Tax compliance and administration. In W. Bartley Hildreth and James A. Richardson (Eds.), *Handbook on taxation* (pp. 741–768). New York: Marcel Dekker.

Alm, J. (2018). What motivates tax compliance? *Journal of Economic Surveys*, *33*, 353–388.

Alm, J., McClelland, G. H., & Schulze, W. D. (1992). Why do people pay taxes? *Journal of Public Economics*, *48*(1), 21–38.

Andreoni, J., Erard, B., & Feinstein, J. (1998). Tax compliance. *Journal of Economic Literature, 36*(2), 818–860.

Antoci, A., Russu, P., & Zarri, L. (2014). Tax evasion in a behaviorally heterogeneous society: An evolutionary analysis. *Economic Modelling, 42*, 106–115.

Bernasconi, M., & Zanardi, A. (2004). Tax evasion, tax rates, and reference dependence. In *FinanzArchiv/public finance analysis* (pp. 422–445).

Chetty, R. (2009). Is the taxable income elasticity sufficient to calculate deadweight loss? The implications of evasion and avoidance. *American Economic Journal: Economic Policy, 1*(2), 31–52.

De Giovanni, D., & Lamantia, F. (2018). Dynamic harvesting under imperfect catch control. *Journal of Optimization Theory and Applications, 176*, 252:267.

De Giovanni, D., Lamantia, F., & Pezzino, M. (2019). A behavioral model of evolutionary dynamics and optimal regulation of tax evasion. *Structural Change and Economic Dynamics, 50*, 79–89.

Dhami, S., & Al-Nowaihi, A. (2007). Why do people pay taxes? Prospect theory versus expected utility theory. *Journal of Economic Behavior & Organization, 64*(1), 171–192.

Dhami, S., & Al-Nowaihi, A. (2010). Optimal taxation in the presence of tax evasion: Expected utility versus prospect theory. *Journal of Economic Behavior & Organization, 75*(2), 313–337.

Feld, L. P., & Tyran, J. R. (2002). Tax evasion and voting: An experimental analysis. *Kyklos, 55*(2), 197–221.

Frey, B. S. (1999). *Economics as a science of human behaviour: Towards a new social science paradigm.* Springer Science & Business Media.

Frey, B. S., & Feld, L. P. (2002). Deterrence and morale in taxation: An empirical analysis. CESifo Working Paper Series 760, CESifo Group Munich.

Grüne, L., & Semmler, W. (2004). Using dynamic programming with adaptive grid scheme for optimal control problems in economics. *Journal of Economic Dynamics and Control, 28*, 2427–2456.

Kahneman, D., & Tversky, A. (1979). Prospect theory: An analysis of decision under risk. *Econometrica, 47*, 263–291.

Lamantia, F., & Pezzino, M. (2018). The dynamic effects of fiscal reforms and tax competition on tax compliance and migration. *Review of International Economics, 26*, 672–690.

Long, S. B., & Swingen, J. A. (1991). The conduct of tax-evasion experiments: Validation, analytical methods, and experimental realism. In P. Webley, H. Robben, H. Elffers, & D. Hessing (Eds.), *Tax evasion: An experimental approach* (pp. 128–138). Cambridge: Cambridge University Press.

Luttmer, E. F. P., & Singhal, M. (2014). Tax morale. *Journal of Economic Perspectives, 28*(4), 149–68.

Petrohilos-Andrianos, Y., & Xepapadeas, A. (2016). On the evolution of compliance and regulation with tax evading agents. *Journal of Dynamics & Games, 3*(3), 231–260.

Pickhardt, M., & Prinz, A. (2014). Behavioral dynamics of tax evasion-a survey. *Journal of Economic Psychology, 40*, 1–19.

Piolatto, A., & Rablen, M. D. (2017). Prospect theory and tax evasion: A reconsideration of the Yitzhaki puzzle. *Theory and Decision, 82*(4), 543–565.

Piolatto, A., & Trotin, G. (2016). Optimal income tax enforcement under prospect theory. *Journal of Public Economic Theory, 18*(1), 29–41.

Prelec, D. (1998). The probability weighting function. *Econometrica, 66*(3), 497–527.

Slemrod, J., & Weber, C. (2012). Evidence of the invisible: Toward a credibility revolution in the empirical analysis of tax evasion and the informal economy. *International Tax and Public Finance, 19*(1), 25–53.

Slemrod, J., & Yitzhaki, S. (2002). Tax avoidance, evasion, and administration. In A. J. Auerbach, & M. Feldstein (Eds.), *Handbook of public economics,* (1st ed., vol. 3, pp. 1423–1470). Elsevier.

Torgler, B. (2002). Speaking to theorists and searching for facts: Tax morale and tax compliance in experiments. *Journal of Economic Surveys, 16*(5), 657–683.

Trotin, G. (2012). Solving the Yitzhaki paradox: Income tax evasion and reference dependence under prospect theory. Technical Report HAL Id: halshs-00793664.

Tversky, A., & Kahneman, D. (1992). Advances in prospect theory: Cumulative representation of uncertainty. *Journal of Risk and Uncertainty*, *5*(4), 297–323.

Weibull, J. W. (1997). *Evolutionary game theory*. MIT Press.

Yitzhaki, S. (1974). Income tax evasion: A theoretical analysis. *Journal of Public Economics*, *3*(2), 201–202.

On Dynamics of a Three-Country Kaldorian Model of Business Cycles with Fixed Exchange Rates

Toshio Inaba and Toichiro Asada

Abstract In this paper, we consider a three-country Kaldorian nonlinear macrodynamic model of business cycles with fixed exchange rates. The term 'Kaldorian' means that our model is a three-country extension of Kaldor's nonlinear business cycle model that is characterized by the dynamic interaction of the real income and the real capital stock. It is supposed that three countries are connected through international trade and international capital movement with imperfect capital mobility under fixed exchange rates. This paper is a sequel to our previous study of the two-country Kaldorian business cycle model under fixed exchange rates. We find by means of numerical simulations that the addition of one country to the previous analytical framework makes the dynamic behavior of the model much more complex compared with the two-country version.

Keywords Kaldorian model of business cycles · Three country model · Fixed exchange rates · Chaotic dynamics

1 Introduction

Akio Matsumoto devoted his quite prolific and brilliant research career to the study of nonlinear economic dynamics. His recent contributions contain a series of the applications of nonlinear continuous time delay differential equations to economic dynamics (see, for example, Matsumoto and Szidarovszky (2011, 2019), Matsumoto et al. (2018)). However, his early works include the economic applications of the relatively small dimensional (one or two dimensional) nonlinear discrete time dynamic systems, which often produces the complex dynamics (see, for example, Matsumoto (1994, 1996, 1997, 1999). If the dimension of the system is relatively small, analytical treatment of the system as well as the numerical simulations becomes relatively

T. Inaba
School of Education, Waseda University, Tokyo, Japan
e-mail: inaba@waseda.jp

T. Asada (✉)
Faculty of Economics, Chuo University, Tokyo, Japan
e-mail: asada@tamacc.chuo-u.ac.jp

© Springer Nature Singapore Pte Ltd. 2020
F. Szidarovszky and G. I. Bischi (eds.), *Games and Dynamics in Economics*,
https://doi.org/10.1007/978-981-15-3623-6_6

easy. But, the analytical treatment becomes almost impossible so that we must exclusively resort to the numerical simulations if the dimension of the nonlinear discrete time dynamic system is large.

Our present study is somewhat related to Matsumoto's early works, that is, the study of a nonlinear discrete time dynamic economic model. In this paper, however, we study an unusually large dimensional nonlinear economic system, that is, three country Kaldorian business cycle model with fixed exchange rates, which is reduced to a discrete time eight dimensional nonlinear dynamic system. Analytical treatment of such a complicated system is quite difficult, so that we are obliged to derive some tentative conclusions through some numerical simulations. Analytical treatment of the model in this paper is restricted to only a special case that is reduced to a series of the unconnected two dimensional systems. Next, we shall describe the motivation of our research.

Recent development of the dynamic approaches to international and regional economics is quite remarkable.[1] In line with this development, open economy extension of Kaldor's (1940) nonlinear business cycle model, which is based on Keynes' (1936) theory of effective demand with underemployment, was studied by several authors. For example, Asada (1995), Asada et al. (2000a, b, 2012), and Medved'ová (2011) tried to extend the Kaldorian business cycle theory to the models of small open economy under fixed and flexible exchange rates. On the other hand, Asada (2004), Asada et al. (2001, 2010, 2011) and Malička and Zimka (2010, 2012) studied two-country or two-regional model of Kaldorian business cycles under fixed and flexible exchange rates. Incidentally, original Kaldorian business cycle model consists of a two-dimensional system of nonlinear dynamic equations that can interpret the dynamic interaction of real national income and real capital stock in a closed economy with underemployment.[2] In this paper, we try to extend Kaldorian business cycle theory to three-country model under fixed exchange rates.

It is worth noting that Lorenz (1987, 1993) already studied such a three country Kaldorian model under fixed exchange rates. But, in the model in Lorenz (1987, 1993) only the international trade is introduced and the international capital movement is neglected. In the three-country Kaldorian model in this paper, the international capital movement as well as international trade is explicitly introduced. We suppose that the money capital rather than real capital moves between countries according to the differences of nominal interest rates of the countries with imperfect capital mobility, which means that the degree of capital mobility (β) is positive and finite. In our model, the total money stock as a whole is fixed by the trans-national central bank such as European Central Bank under currency integration that is a kind of fixed exchange rate system, and the money moves between countries through international trade and international capital movement. Therefore, our three-country Kaldorian model under fixed exchange rates can be applicable to the theoretical analysis of the economic system under currency integration such as European Union.

---

[1] See, for example, Asada et al. (2003), Krugman (1996), Nijkamp and Reggiani (1992), Puu (1997) and Rosser (1991).

[2] See Dohtani et al. (1996) as well as Kaldor's (1940) original paper.

Our model consists of an eight-dimensional system of nonlinear difference equations. By means of numerical simulations, we find that the addition of one country makes the dynamic behavior of the model much more complex compared with the two-country version. In Sect. 2, we present the full system of equations in our model, and in Sect. 3 we study the dynamic properties of the model mathematically under some special conditions. In Sect. 4 we report some results of our numerical simulations of the general case, which contain several dynamic properties of the system such as stability, instability, cyclical and chaotic movements of the main variables. Section 5 is devoted to the concluding remarks.

2 Formulation of the Model

Our three-country model with fixed exchange rates consists of the following system of equations, where t denotes the time period and the subscript i ($i = 1, 2, 3$) is the index number of a country. The exchange rates of the currencies of countries 2 and 3 that are measured in terms of the currency of country 1 (E_2, E_3) are fixed at the levels $E_2 = E_3 = 1$ without loss of generality.

Disequilibrium adjustment process of the goods market in country i is

$$Y_i(t + 1) - Y_i(t) = \alpha_i[C_i(t) + I_i(t) + G_i + J_i(t) - Y_i(t)]; \alpha_i > 0. \quad (1)$$

Capital accumulation equation in country i is

$$K_i(t + 1) - K_i(t) = I_i(t). \quad (2)$$

Consumption function in country i is

$$C_i(t) = c_i\{Y_i(t) - T_i(t)\} + C_{0i}; 0 < c_i < 1, C_{0i} \geq 0. \quad (3)$$

Investment function in country i is

$$I_i(t) = I_i(Y_i(t), K_i(t), r_i(t)); \frac{\partial I_i}{\partial Y_i} > 0, \frac{\partial I_i}{\partial K_i} < 0, \frac{\partial I_i}{\partial r_i} < 0. \quad (4)$$

Income tax function in country i is

$$T_i(t) = \tau_i Y_i(t) - T_{0i}; 0 < \tau_i < 1, T_{0i} \geq 0. \quad (5)$$

Equilibrium condition of the money market in country i is

$$\frac{M_i}{p_i} = L_i(Y_i(t), r_i(t)); \frac{\partial L_i}{\partial Y_i} > 0, \frac{\partial L_i}{\partial r_i} < 0. \quad (6)$$

Net export (current account) functions in three countries are

$$J_1(t) = \delta H_1(Y_1(t), Y_2(t), Y_3(t)); \quad \frac{\partial H_1}{\partial Y_1} < 0, \quad \frac{\partial H_1}{\partial Y_2} > 0, \quad \frac{\partial H_1}{\partial Y_3} > 0, \quad (7)$$

$$J_2(t) = \delta H_2(Y_1(t), Y_2(t), Y_3(t)); \quad \frac{\partial H_2}{\partial Y_1} > 0, \quad \frac{\partial H_2}{\partial Y_2} < 0, \quad \frac{\partial H_2}{\partial Y_3} > 0, \quad (8)$$

$$J_3(t) = -J_1(t) - J_2(t) \equiv \delta H_3(Y_1(t), Y_2(t), Y_3(t)); \quad \frac{\partial H_3}{\partial Y_1} = -\frac{\partial H_1}{\partial Y_1} - \frac{\partial H_2}{\partial Y_1},$$

$$\frac{\partial H_3}{\partial Y_2} = -\frac{\partial H_1}{\partial Y_2} - \frac{\partial H_2}{\partial Y_2}, \quad \frac{\partial H_3}{\partial Y_3} = -\frac{\partial H_1}{\partial Y_3} - \frac{\partial H_2}{\partial Y_3} < 0, 0 \leq \delta \leq 1. \quad (9)$$

Capital account functions in three countries are

$$Q_1(t) = \beta\{r_1(t) - r_2(t)\} + \beta\{r_1(t) - r_3(t)\} \equiv \beta\{2r_1(t) - r_2(t) - r_3(t)\}, \quad (10)$$

$$Q_2(t) = \beta\{r_2(t) - r_1(t)\} + \beta\{r_2(t) - r_3(t)\} \equiv \beta\{-r_1(t) + 2r_2(t) - r_3(t)\}, \quad (11)$$

$$Q_3(t) = -Q_1(t) - Q_2(t) \equiv \beta\{-r_1(t) - r_2(t) + 2r_3(t)\}; \beta \geq 0. \quad (12)$$

The definition of total balance of payments in country i is

$$A_i(t) = J_i(t) + Q_i(t). \quad (13)$$

Specification of the monetary policy of the trans-national central bank is

$$M_1(t) + M_2(t) + M_3(t) = \overline{M} > 0, M_i(t) > 0. \quad (14)$$

Equations that describe the international movement of money stock between three countries are

$$M_1(t+1) - M_1(t) = A_1(t), \quad (15)$$

$$M_2(t+1) - M_2(t) = A_2(t), \quad (16)$$

$$M_3(t+1) - M_3(t) = -A_1(t) - A_2(t) = A_3(t). \quad (17)$$

The meanings of the symbols are as follows. $Y_i =$ real net national income. $C_i =$ real consumption expenditure. $I_i =$ real net private expenditure on physical capital. $G_i =$ real government expenditure (fixed). $J_i =$ real net export $(J_1 + J_2 + J_3 = 0)$. $K_i =$ real physical capital stock. $T_i =$ real income tax. $r_i =$ nominal rate of interest. $M_i =$ nominal money stock. $p_i =$ price level. $Q_i =$ real capital account $(Q_1 + Q_2 + Q_3 = 0)$. $A_i =$ real total balance of payments $(A_1 + A_2 + A_3 = 0)$.

Furthermore, we have the following three kinds of important parameters. $\alpha_i =$ adjustment speed in the goods market in a country i. $\delta =$ degree of international trade. $\beta =$ degree of capital mobility.

Equation (1) describes the Keynesian/Kaldorian quantity adjustment process of the disequilibrium in the goods market in country i, which says that the real output changes according as the excess demand in the goods market is positive or negative.[3] Equation (2) is the capital accumulation equation in country i, which says that the net investment contributes to the changes of the physical capital stock. Equation (3) is the standard Keynesian consumption function in country i, which says that the real consumption is an increasing function of the real disposable income. Equation (4) is the standard Kaldorian/Keynesian investment function in country i, which says that the firms' real net expenditure on the investment goods is an increasing function of real national income, a decreasing function of the real capital stock, and a decreasing function of the nominal interest rate.[4] Equation (5) is a standard income tax function, and the right hand side of Eq. (6) is a standard Keynesian money demand function. Equations (7), (8) and (9) are the standard Keynesian net export functions of three countries. It is worth noting that by definition, we always have $J_1(t) + J_2(t) + J_3(t) = 0$. Equations (10), (11) and (12) are the standard capital account equations of three countries under imperfect capital movement. In this model, the 'capital movement' does not mean the movement of the physical capital stocks K_i between countries but it is supposed that only money capitals (bonds) move between countries. We also have $Q_1(t) + Q_2(t) + Q_3(t) = 0$ by definition. Equation (13) is nothing but the definition of the total balance of payments in country i. Needless to say, we always have $A_1(t) + A_2(t) + A_3(t) = 0$. Equation (14) is a specification of the monetary policy, which means that the total money stock as a whole is fixed by the trans-national central bank of the currency union such as European Central Bank. Equations (15), (16) and (17) mean that the money stock of a country i changes endogenously according as the total balance of payments of this country is positive or negative. Incidentally, by adding Eqs. (15), (16) and (17) we have

$$\sum_{i=1}^{3} M_i(t+1) - \sum_{i=1}^{3} M_i(t) = \sum_{i=1}^{3} A_i(t) = 0, \tag{18}$$

which implies Eq. (14).

For simplicity, we assume that price levels of three countries are fixed at the level such that $p_1 = p_2 = p_3 = 1$.

By solving Eq. (6) with respect to $r_i(t)$, we obtain

$$r_i(t) = r_i(Y_i(t), M_i(t)); \quad \frac{\partial r_i}{\partial Y_i} > 0, \quad \frac{\partial r_i}{\partial M_i} < 0. \tag{19}$$

---

[3] See Kaldor (1940) and Keynes (1936).
[4] See also Kaldor (1940) and Keynes (1936).

Then, we can reduce the above system to the following eight dimensional system of difference equations.

$$Y_1(t+1) = Y_1(t) + \alpha_1[c_1(1-\tau_1)Y_1(t) + C_{01} + c_1 T_{01} + G_1$$
$$+ I_1(Y_1(t), K_1(t), r_1(Y_1(t), M_1(t))) + \delta H_1(Y_1(t), Y_2(t), Y_3(t)) - Y_1(t)]$$
$$\equiv F_1(Y_1(t), K_1(t), Y_2(t), Y_3(t), M_1(t); \alpha_1, \delta), \tag{20a}$$

$$K_1(t+1) = K_1(t) + I_1(Y_1(t), K_1(t), r_1(Y_1(t), M_1(t)))$$
$$\equiv F_2(Y_1(t), K_1(t), M_1(t)), \tag{20b}$$

$$Y_2(t+1) = Y_2(t) + \alpha_2[c_2(1-\tau_2)Y_2(t) + C_{02} + c_2 T_{02} + G_2$$
$$+ I_2(Y_2(t), K_2(t), r_2(Y_2(t), M_2(t))) + \delta H_2(Y_1(t), Y_2(t), Y_3(t)) - Y_2(t)]$$
$$\equiv F_3(Y_1(t), Y_2(t), K_2(t), Y_3(t), M_2(t); \alpha_2, \delta), \tag{20c}$$

$$K_2(t+1) = K_2(t) + I_2(Y_2(t), K_2(t), r_2(Y_2(t), M_2(t)))$$
$$\equiv F_4(Y_2(t), K_2(t), M_2(t)), \tag{20d}$$

$$Y_3(t+1) = Y_1(t) + \alpha_3[c_3(1-\tau_3)Y_3(t) + C_{03} + c_3 T_{03} + G_3$$
$$+ I_3(Y_3(t), K_3(t), r_3(Y_3(t), \bar{M} - M_1(t) - M_2(t)))$$
$$+ \delta H_3(Y_1(t), Y_2(t), Y_3(t)) - Y_3(t)]$$
$$\equiv F_5(Y_1(t), Y_2(t), Y_3(t), K_3(t), M_1(t), M_2(t); \alpha_3, \delta), \tag{20e}$$

$$K_3(t+1) = K_3(t) + I_3(Y_3(t), K_3(t), r_3(Y_3(t), \bar{M} - M_1(t) - M_2(t)))$$
$$\equiv F_6(Y_3(t), K_3(t), M_1(t), M_2(t)), \tag{20f}$$

$$M_1(t+1) = M_1(t) + \delta H_1(Y_1(t), Y_2(t), Y_3(t))$$
$$+ \beta\{2r_1(Y_1(t), M_1(t)) - r_2(Y_2(t), M_2(t)) - r_3(Y_3(t), \bar{M} - M_1(t) - M_2(t))\}$$
$$\equiv F_7(Y_1(t), Y_2(t), Y_3(t), M_1(t), M_2(t); \delta, \beta), \tag{20g}$$

$$M_2(t+1) = M_2(t) + \delta H_2(Y_1(t), Y_2(t), Y_3(t))$$
$$+ \beta\{-r_1(Y_1(t), M_1(t)) + 2r_2(Y_2(t), M_2(t)) - r_3(Y_3(t), \bar{M} - M_1(t) - M_2(t))\}$$
$$\equiv F_8(Y_1(t), Y_2(t), Y_3(t), M_1(t), M_2(t); \delta, \beta). \tag{20h}$$

In this model, we have three kinds of important parameters, $\alpha_i (i = 1, 2, 3)$, δ and β. The larger α_i, more quick is the adjustment in the goods market in response to the excess demand in the goods market in country i. The larger δ, more active is the international trade between three countries. The larger β, higher is the international mobility of money capital between three countries. It will be interesting to study how the changes of these parameter values affect the dynamic properties of the model.

3 Mathematical Analysis of a Special Case

In the special case of $\delta = \beta = 0$, a system of Eqs. (20a–20h) is reduced to the following system of equations ($i = 1, 2, 3$).

$$Y_i(t + 1) = Y_i(t) + \alpha_i[c_i(1 - \tau_i)Y_i(t) + C_{0i} + c_iT_{0i} + G_i$$
$$+ I_i(Y_i(t), K_i(t), r_i(Y_i(t), \bar{M}_i)) - Y_i(t)] \equiv F_1^i(Y_i(t), K_i(t); \bar{M}_i, \alpha_i),$$
$$\tag{21a}$$

$$K_i(t + 1) = K_i(t) + I_i(Y_i(t), K_i(t), r_i(Y_i(t), \overline{M}_i)) \equiv F_2^i(Y_i(t), K_i(t); \overline{M}_i),$$
$$\tag{21b}$$

$$M_i(t + 1) = M_i(t) = \overline{M}_i. \tag{21c}$$

In case of $\delta = \beta = 0$, there is no international trade and no international capital movement between countries. In other words, three countries are completely isolated economically. In mathematical term, this is a decomposable system, in which there are three isolated systems of two dimensional nonlinear difference equations.

The equilibrium values (Y_i^*, K_i^*) of Eqs. (21a–21c) such that $Y_i(t + 1) = Y_i(t) = Y_i^*$, $K_i(t + 1) = K_i(t) = K_i^*$ can be expressed as follows.

$$Y_i^* = \frac{1}{1 - c_i(1 - \tau_i)}(C_{0i} + c_iT_{0i} + G_i) > 0, \tag{22a}$$

$$I_i(Y_i^*, K_i^*, r_i(Y_i^*, \overline{M}_i)) = 0. \tag{22b}$$

We can see that the equilibrium value K_i^* is uniquely determined because of the assumption $\frac{\partial I_i}{\partial K_i} < 0$. In this section, we assume that $K_i^* > 0$.

The Jacobian matrix of the dynamic system (21a–21c) at the equilibrium point can be written as

$$J_i = \begin{bmatrix} F_{11}^i & F_{12}^i \\ F_{21}^i & F_{22}^i \end{bmatrix} \tag{23}$$

where $F_{11}^i = 1 + \alpha_i\left[\underset{(+)}{I_{Y_i}^i} + \underset{(-)(+)}{I_{r_i}^i r_{Y_i}^i} - \left\{\underset{(+)}{1 - c_i(1 - \tau_i)}\right\}\right]$, $F_{12}^i = \underset{(-)}{\alpha_i I_{K_i}^i} < 0$, $F_{21}^i = \underset{(+)}{I_{Y_i}^i} + \underset{(-)(+)}{I_{r_i}^i r_{Y_i}^i}$, $F_{22}^i = 1 + \underset{(-)}{I_{K_i}^i}$, $I_{Y_i}^i = \partial I_i/\partial Y_i > 0$, $I_{r_i}^i = \partial I_i/\partial r_i < 0$, $r_{Y_i}^i = \partial r_i/\partial Y_i > 0$, $I_{K_i}^i = \partial I_i/\partial K_i < 0$, and all partial derivatives are evaluated *at the equilibrium point*. Furthermore, we assume as follows.

Assumption 1 $A_i \equiv \underset{(+)}{I_{Y_i}^i} + \underset{(-)(+)}{I_{r_i}^i r_{Y_i}^i} - \underset{(+)}{\{1 - c_i(1 - \tau_i)\}} > 0.$

Assumption 2 $-I^i_{K_i} < 1.$
$(-)$

Assumption 1 means that the sensitivity of investment activity with respect to the changes of the national income of country i ($I^i_{Y_i}$) is sufficiently large *at the equilibrium point*, which is the standard hypothesis of Kaldorian business cycle model.[5] Assumption 2 means that the sensitivity of investment activity with respect to the changes of the capital stock in country i ($\left|I^i_{K_i}\right|$) is not extremely large.

In this case, the characteristic equation of the dynamic system (21a–21c) *at the equilibrium point* becomes as follows:

$$\Delta_i(\lambda) \equiv |\lambda I - J_i| = \begin{vmatrix} \lambda - F^i_{11} & -F^i_{12} \\ -F^i_{21} & \lambda - F^i_{22} \end{vmatrix} = \lambda^2 + b_{1i}\lambda + b_{2i} = 0 \qquad (24)$$

where

$$b_{1i} = -trace J_i = -F^i_{11} - F^i_{22} = \underset{(+)}{-2 - \alpha_i A_i} - \underset{(-)}{I^i_{K_i}}, \qquad (25)$$

$$b_{2i} = \det J_i = F^i_{11} F^i_{22} - F^i_{12} F^i_{21} = 1 + \underset{(-)}{I^i_{K_i}} + \alpha_i \left[\underset{(+)}{A_i} - \underbrace{\{1 - c_i(1 - \tau_i)\}}_{(+)} \underset{(-)}{I^i_{K_i}} \right]. \qquad (26)$$

The characteristic Eq. (24) has the following two roots:

$$\lambda_{1i}(\alpha_i) = \left(-b_{1i} + \sqrt{b^2_{1i} - 4b_{2i}} \right)/2, \; \lambda_{2i}(\alpha_i) = \left(-b_{1i} - \sqrt{b^2_{1i} - 4b_{2i}} \right)/2 \quad (27)$$

Then, we have the following proposition.

Proposition 1 Let us define the value α_{i0} as follows:

$$\alpha_{i0} \equiv \underset{(-)}{-I^i_{K_i}} \bigg/ \left[\underset{(+)}{A_i} - \underbrace{\{1 - c_i(1 - \tau_i)\}}_{(+)} \underset{(-)}{I^i_{K_i}} \right] > 0 \qquad (28)$$

Then, the characteristic roots that are expressed by Eq. (27) become a set of conjugate complex roots with $|\lambda_i(\alpha_i)| = 1$ at $\alpha_i = \alpha_{i0}$, where $|\lambda_i(\alpha_i)|$ is the modulus of the characteristic roots. Furthermore, we have $d|\lambda_i(\alpha_i)|/d\alpha_i > 0$.

Proof It follows from Eq. (27) that the characteristic roots become a set of conjugate complex roots if and only if the inequality

---

[5]See Kaldor (1940).

$$b_{1i}^2 - 4b_{2i} < 0 \tag{29}$$

is satisfied, and in this case we have

$$|\lambda_i(\alpha_i)| = \sqrt{(b_{1i}/2)^2 + \left(\sqrt{(-b_{1i}^2 + 4b_{2i})/2}\right)^2} = \sqrt{b_{2i}}. \tag{30}$$

We can easily see that at $\alpha_i = \alpha_{i0}$ we have $b_{2i} = 1$, and in this case we obtain

$$b_{1i}^2 - 4b_{2i} = (-2 - \alpha_{i0}A_i - I_{K_i}^i)^2 - 4$$

$$= \left\{-2 + \underbrace{\frac{-\{1 - c_i(1 - \tau_i)\}(I_{K_i}^i)}{A_i - \{1 - c_i(1 - \tau_i)\} I_{K_i}^i}}_{(-)}(-I_{K_i}^i)\right\}^2 - 4 < 0 \tag{31}$$

from Eq. (28) under Assumption 2. Furthermore, we obtain

$$db_{2i}/d\alpha_i = A_i > 0 \tag{32}$$

under Assumption 1. This completes the proof. □

Proposition 1 implies that the cyclical fluctuations of the real national income and the real capital stock of country i occur because of the existence of a set of complex roots if α_i is close to α_{i0}, and the equilibrium point is locally stable (unstable) if $\alpha_i < \alpha_{i0}(\alpha_i > \alpha_{i0})$ but α_i is close to α_{i0}. Furthermore, the point $\alpha_i = \alpha_{i0}$ becomes a Hopf Bifurcation point of the two dimensional discrete time system under the additional assumptions $\lambda_{ij}^n(\alpha_{i0}) \neq \pm 1$ $(n = 1, 2, 3, 4)$ $(j = 1, 2)$.[6] In this case, there exist some non-constant periodic solutions at some parameter values α_i that are sufficiently close to α_{i0}.

In this section, we studied the mathematical analysis of the system under the special assumption $\delta = \beta = 0$, which means that there is no international trade and no international capital movement. However, the system becomes the fully indecomposable eight dimensional nonlinear system of difference equations and the dynamic behavior of the system may become much complex if we introduce the international trade and international capital movement, that is, $\delta > 0$ and $\beta > 0$. It is quite difficult to analyze such a complex system mathematically, so that we shall provide some results of tentative numerical simulations to study the dynamic behavior of the system.

---

[6]See Gandolfo (2009) Chap. 24.

4 Numerical Simulations

In this section, we adopt the following specifications of the functional forms of a system of dynamic Eqs. (20a–20h), and report some results of our numerical simulations. Some parameter values are based on empirical studies. For example, 0.64 in Eq. (33a) is derived from marginal propensity 0.7 to consume.

$$Y_1(t+1) = Y_1(t) + \alpha_1[0.64Y_1(t) + 75 + f(Y_1(t)) - 0.3K_1(t) - 10Y_1(t)^{0.5} + M_1(t)$$
$$+ \delta\{-0.2Y_1(t) + 0.1Y_2(t) + 0.1Y_3(t)^{0.5} - Y_1(t)\}], \tag{33a}$$

$$K_1(t+1) = K_1(t) + f(Y_1(t)) - 0.3K_1(t) - 10Y_1(t) + M_1(t), \tag{33b}$$

$$Y_2(t+1) = Y_2(t) + \alpha_2[0.64Y_2(t) + 75 + f(Y_2(t)) - 0.3K_2(t) - 10Y_2(t)^{0.45} + M_2(t)$$
$$+ \delta\{0.1Y_1(t) - 0.2Y_2(t) + 0.1Y_3(t)\} - Y_2(t)], \tag{33c}$$

$$K_2(t+1) = K_2(t) + f(Y_2(t)) - 0.3K_2(t) - 10Y_2(t)^{0.45} + M_2(t), \tag{33d}$$

$$Y_3(t+1) = Y_3(t) + \alpha_3[0.64Y_3(t) + 75 + f(Y_3(t)) - 0.3K_3(t) - 10Y_3(t)^{0.4} + M_3(t)$$
$$+ \delta\{0.1Y_1(t) + 0.1Y_2(t) - 0.2Y_3(t)\} - Y_3(t)], \tag{33e}$$

$$K_3(t+1) = K_3(t) + f(Y_3(t)) - 0.3K_3(t) - 10Y_3(t)^{0.4} + M_3(t), \tag{33f}$$

$$M_1(t+1) = M_1(t) + \delta\{-0.2Y_1(t) + 0.1Y_2(t) + 0.1Y_3(t)\}$$
$$+ \beta\{20Y_1(t)^{0.5} - 2M_1(t) - 10Y_2(t)^{0.45} + M_2(t) - 10Y_3(t)^{0.4} + M_3(t)\}, \tag{33g}$$

$$M_2(t+1) = M_2(t) + \delta\{0.1Y_1(t) - 0.2Y_2(t) + 0.1Y_3(t)\}$$
$$+ \beta\{-10Y_1(t)^{0.5} + M_1(t) + 20Y_2(t)^{0.45} - 2M_2(t) - 10Y_3(t)^{0.4} + M_3(t)\}, \tag{33h}$$

$$M_3(t) = 1200 - M_1(t) - M_2(t), \tag{33i}$$

where $f(Y_i(t))$ is given by

$$f(Y_i(t)) = \frac{80}{\pi}\text{Arctan}\left\{\frac{2.25\pi}{20}(Y_i(t) - 200)\right\} + 35 \ (i = 1, 2, 3), \tag{34}$$

which is an example of Kaldor's (1940) S-shaped nonlinear investment function.[7]

Kaldor (1940) rationalizes the hypothesis of S-shaped investment function as follows. The sensitivity of investment expenditure with respect to national income $(\partial I_i/\partial Y_i)$ will become relatively small for both of the extremely small and extremely large national incomes levels "because when there is a great deal of surplus capacity,

[7]See, for example, Asada et al. (2000a, b, 2001, 2010, 2011), and Dohtani et al. (1996).

an increase in activity will not induce entrepreneurs to undertake additional construction: the rise of profit will not stimulate investment. ... But it will also be small for unusually high levels of activity, because rising costs of construction, increasing costs and increasing difficulty of borrowing will dissuade entrepreneurs from expanding still faster – at a time when they already have large commitments." Kaldor (1940).

We consider the following five cases to study the numerical simulations of this eight-dimensional system of nonlinear dynamic equations:

Case 1: Isolated economy ($\delta = \beta = 0$) with $\alpha_2 = \alpha_3 = 0.6$.
Case 2: Not isolated but strictly restricted in both international trade and international capital mobility ($\delta = \beta = 0.1$) with $\alpha_2 = \alpha_3 = 0.6$.
Case 3: Only international trade restriction is slightly relaxed ($\delta = 0.3$, $\beta = 0.1$) with $\alpha_2 = \alpha_3 = 0.6$.
Case 4: Both of the restrictions of international trade and international capital mobility are relaxed ($\delta = 1$, $\beta = 0.6$) with $\alpha_2 = \alpha_3 = 0.6$.
Case 5: The high adjustment speed in goods market ($\alpha_2 = \alpha_3 = 0.9$) with $\delta = 1$, $\beta = 0.6$.

In all cases, we adopt the parameter α_1 as a bifurcation parameter to construct bifurcation diagram of Y_1.

Case 1: Isolated economy
Figures 1, 2 and 3 illustrate the case of $\delta = \beta = 0$, which means that there is no international trade and no international capital movement, in other words, three countries are economically isolated.

Figure 1, which is a bifurcation diagram of Y_1 with respect to the parameter α_1, shows that the equilibrium point is stable for sufficiently small adjustment speed

Fig. 1 Bifurcation diagram of Y_1 with $\delta = \beta = 0$, $\alpha_2 = \alpha_3 = 0.6$

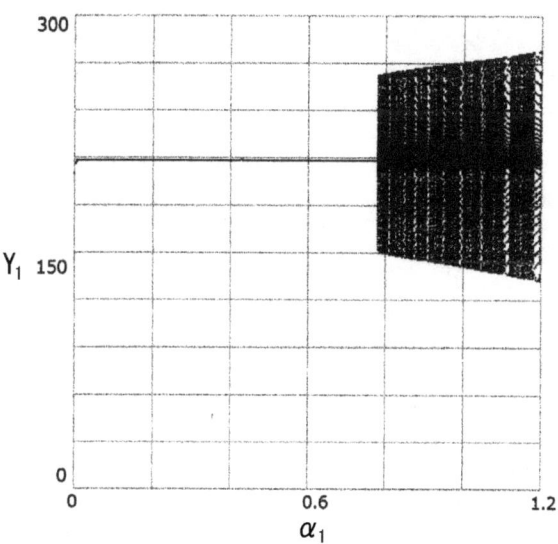

Fig. 2 Attractor of Y_1
versus Y_2 with $\delta = \beta = 0$, α_1
$= \alpha_2 = \alpha_3 = 0.6$

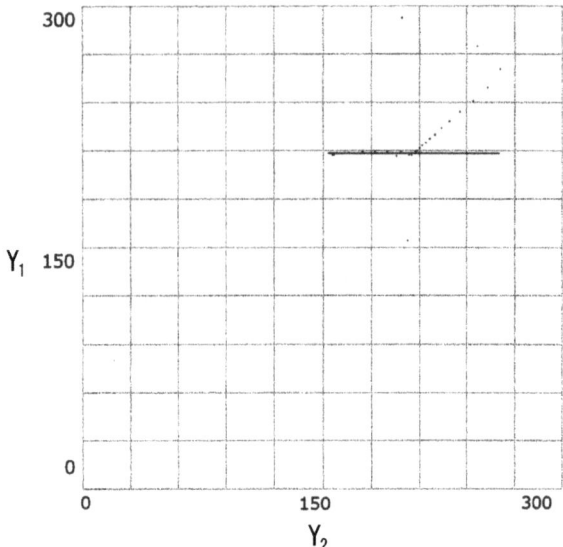

Fig. 3 Lyapunov exponent
(λ) with $\delta = \beta = 0$, $\alpha_2 = \alpha_3$
$= 0.6$

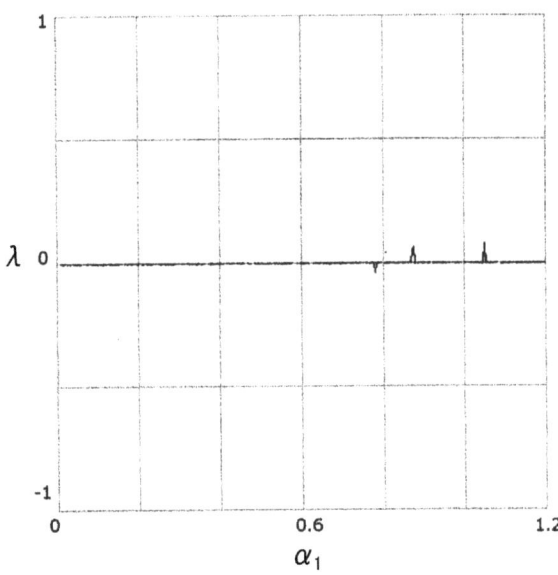

of the goods market, but it becomes unstable and persistent fluctuations emerge for
sufficient large adjustment speed even if the countries are economically isolated.
Figure 2, which is an attractor of the variables Y_1 and Y_2, shows that Y_1 is fixed but
Y_2 fluctuates under the parameter set $\delta = \beta = 0$, $\alpha_1 = \alpha_2 = \alpha_3 = 0.6$. Figure 3
shows that for relevant range of adjustment speed in country 1, the chaotic movement
scarcely occurs, because the Lyapunov exponent for that range is almost zero.

Case 2: Not isolated but strictly restricted in both international trade and international capital mobility

Figures 4, 5 and 6 illustrate the case of $\delta = \beta = 0.1$, which means that three countries are not economically isolated, but international trade and international capital mobility are strictly restricted.

Comparing Figs. 1 and 4, we can see that the introduction of international trade and international capital movement may contribute to increase the dynamic instability of

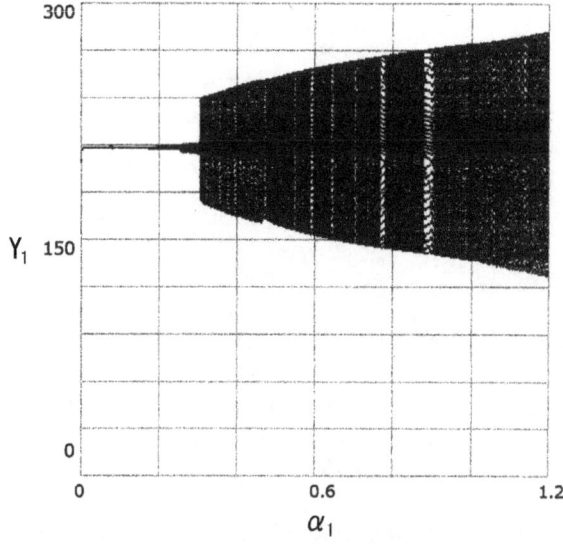

Fig. 4 Bifurcation diagram of Y_1 with $\delta = \beta = 0.1$, $\alpha_2 = \alpha_3 = 0.6$

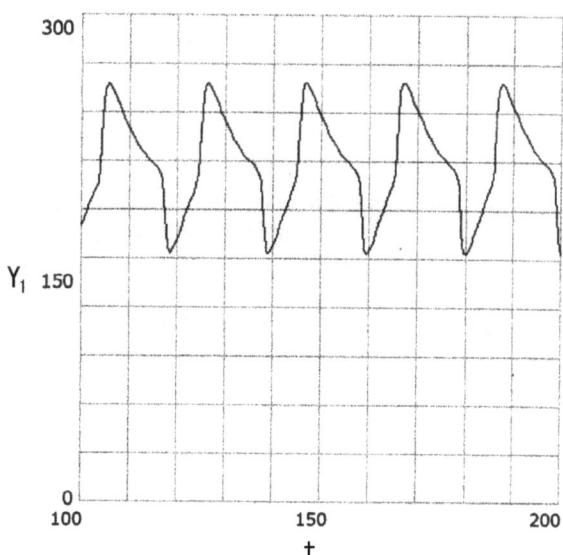

Fig. 5 Time tragectory of Y_1 with $\delta = \beta = 0.1$, $\alpha_1 = \alpha_2 = \alpha_3 = 0.6$

Fig. 6 Attractor of Y_1
versus Y_2 with $\delta = \beta = 0.1$,
$\alpha_1 = \alpha_2 = \alpha_3 = 0.6$

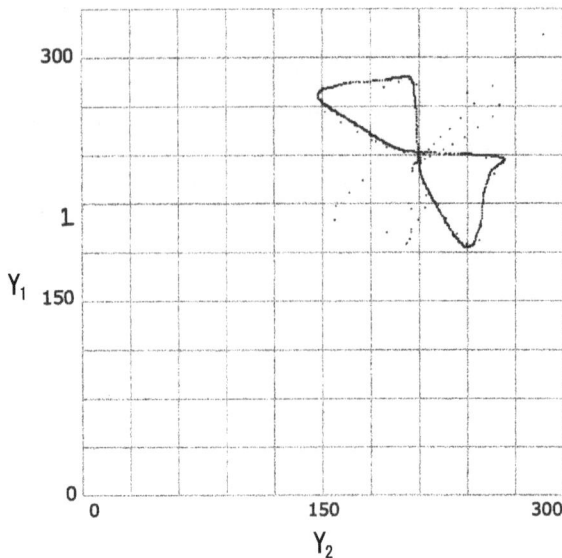

the system and enhance the cyclical fluctuations of the economic variables. Figure 5 is an example of the time trajectory of periodic fluctuations. Figure 6 is a butterfly-like attractor of the real national incomes of countries 1 and 2, which produces the partly synchronized and partly counter-synchronized fluctuations of the real national incomes of two countries.

Case 3: Only international trade restriction is slightly relaxed
Figures 7, 8 and 9 illustrate the case of $\delta = 0.3$ and $\beta = 0.1$, which means that only international trade restriction is slightly relaxed compared with the case 2.

Comparing Figs. 4 and 7, we can see that the slight relaxation of the restriction of international trade alone with the same degree of capital mobility may contribute to stabilize the system, in the sense that it may enlarge the stability region of the adjustment speed parameter. Figure 8 is an example of the time trajectory of the periodic fluctuations. Comparison of Figs. 5 and 8 suggests that the increase of the degree of international trade (δ) may enlarge the period of the cycle. Figure 9 is an attractor of the real national incomes of countries 1 and 2, which produces almost synchronized movements of the real national incomes of two countries.

Case 4: Both of the restrictions of international trade and international capital mobility are relaxed
Figures 10, 11, 12 and 13 illustrate the case of $\delta = 1$ and $\beta = 0.6$, which means that both of international trade and international capital mobility are relaxed.

Comparison of Figs. 7 and 10 suggests that the increases of both of the degree of international trade and the degree of international capital mobility will not destabilize the system if the adjustment speed of the goods market in a country is sufficiently

Fig. 7 Bifurcation diagram of Y_1 with $\delta = 0.3$, $\beta = 0.1$, $\alpha_2 = \alpha_3 = 0.6$

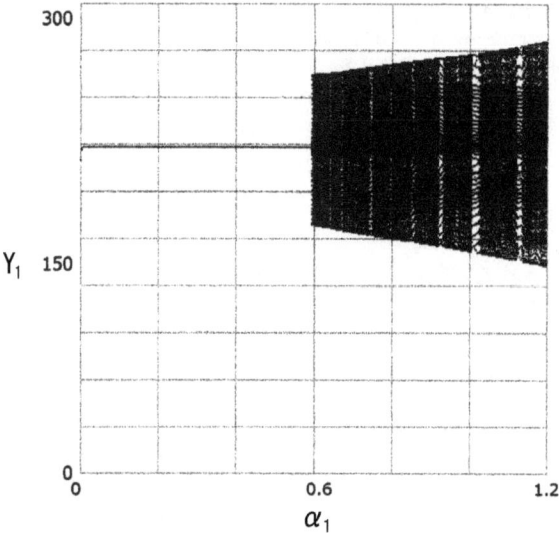

Fig. 8 Time trajectory of Y_1 with $\delta = 0.3$, $\beta = 0.1$, $\alpha_1 = \alpha_2 = \alpha_3 = 0.6$

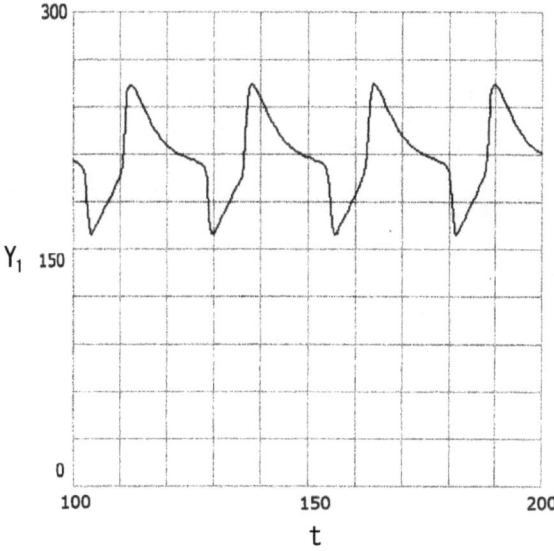

small, but they will destabilize the system for sufficiently large adjustment speed of the goods market in that country. Figure 11 is a ribbon-like attractor of the real national incomes of countries 1 and 2, which produces the partly synchronized and partly counter-synchronized cyclical movements of the real national incomes of two countries. Figure 12 is a totally synchronized attractor of the real national incomes of countries 1 and 3 under the same parameter values as those of Fig. 11. Figure 13 shows that the chaotic movements occur because of positive values of Lyapunov

Fig. 9 Attractor of Y_1
versus Y_2 with $\delta = 0.3$, $\beta = 0.1$, $\alpha_1 = \alpha_2 = \alpha_3 = 0.6$

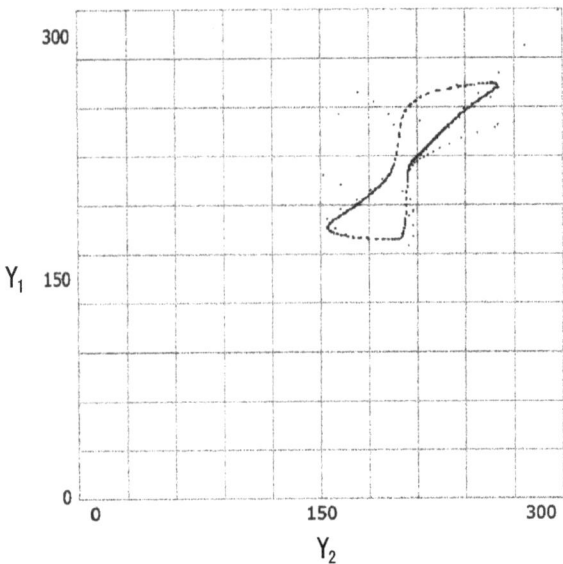

Fig. 10 Bifurcation diagram
of Y_1 with $\delta = 1$, $\beta = 0.6$, $\alpha_2 = \alpha_3 = 0.6$

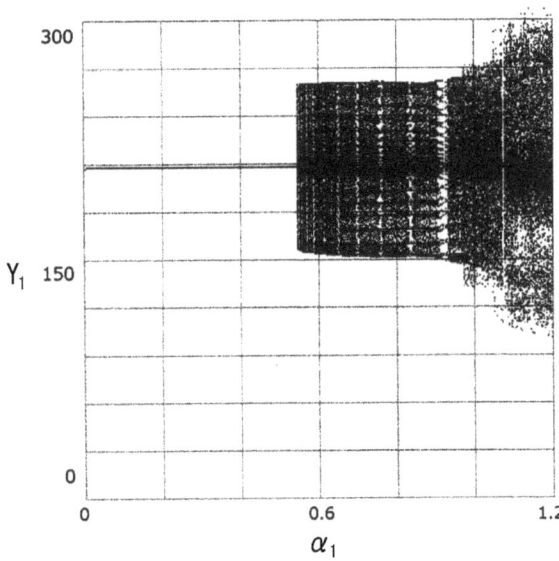

exponent under sufficiently high adjustment speed of the goods market in a country
when both of the degree of international trade and the degree of the international
capital mobility are sufficiently high.

Case 5: The high adjustment speeds in goods market
Figures 14, 15, 16 and 17 illustrate the case of $\delta = 1$, $\beta = 0.6$, and $\alpha_2 = \alpha_3 = 0.9$,

Fig. 11 Attractor of Y_1 versus Y_2 with $\delta = 1$, $\beta = 0.6$, $\alpha_1 = \alpha_2 = \alpha_3 = 0.6$

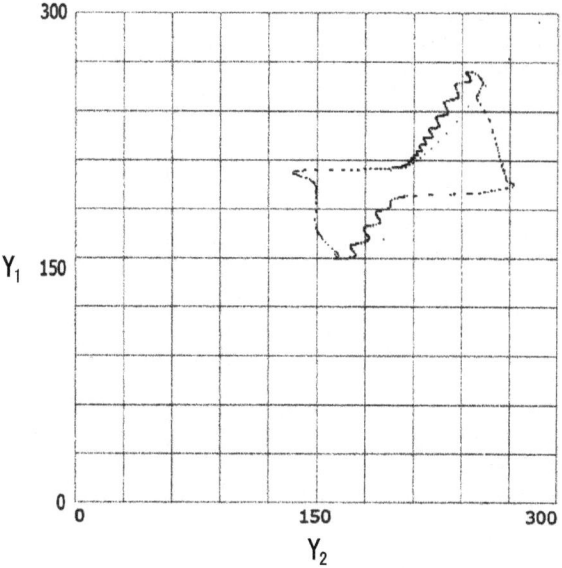

Fig. 12 Attractor of Y_1 versus Y_3 with $\delta = 1$, $\beta = 0.6$, $\alpha_1 = \alpha_2 = \alpha_3 = 0.6$

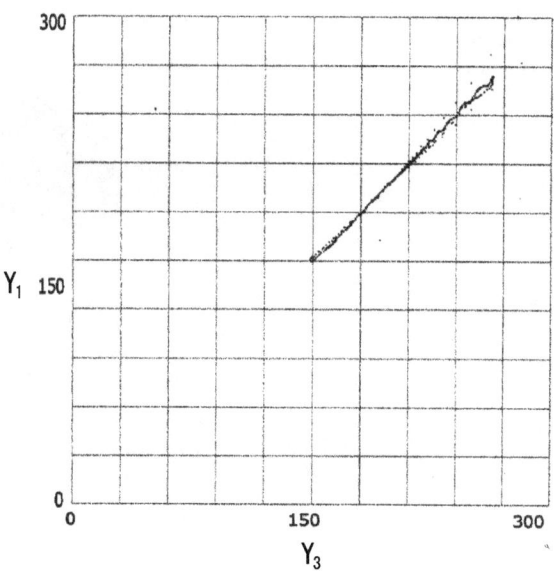

which means that the adjustment speeds of the goods market in countries 2 and 3 as well as the degrees of international trade and international capital mobility are sufficiently high.

Comparison of Figs. 10 and 14 suggests that the increases of the adjustment speed of the goods market in each country is a destabilizer of the system under high degrees of international trade and international capital mobility. Figures 15 and 16 are

Fig. 13 Lyapunov exponent
(λ) with $\delta = 1$, $\beta = 0.6$, $\alpha_2 =$
$\alpha_3 = 0.6$

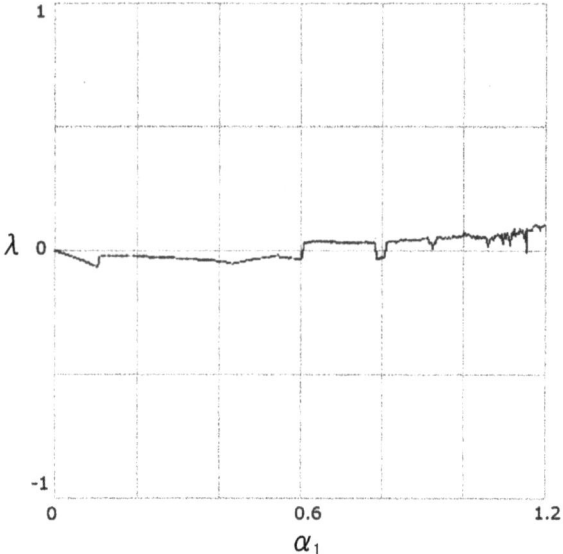

Fig. 14 Bifurcation diagram
of Y_1 with $\delta = 1$, $\beta = 0.6$, α_2
$= \alpha_3 = 0.9$

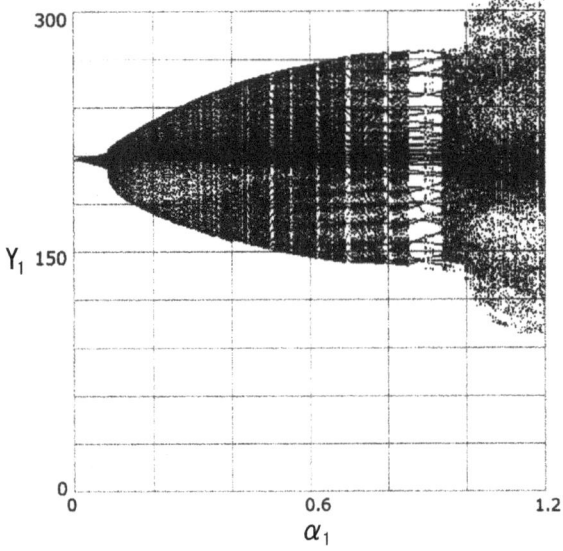

attractors of the real national incomes of three countries, in which case the movements
of the national incomes of all countries are totally synchronized. Figure 17 shows
that in this case the chaotic movements occur under wide range of the adjustment
speed of the goods market in country 1 because of the positive values of Lyapunov
exponent.

Fig. 15 Attractor of Y_1 versus Y_2 with $\delta = 1$, $\beta = 0.6$, $\alpha_1 = \alpha_2 = \alpha_3 = 0.9$

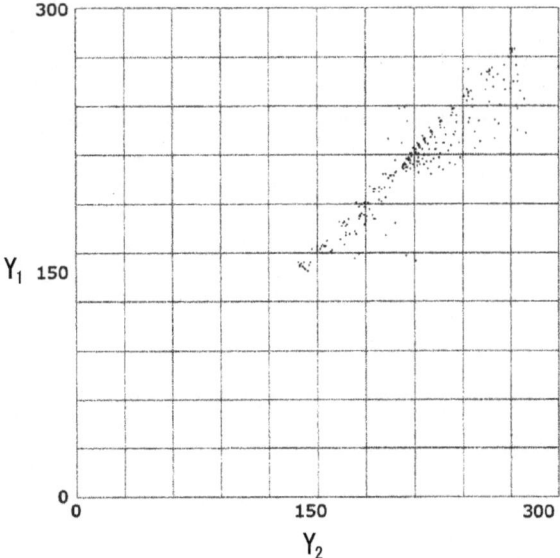

Fig. 16 Attractor of Y_1 versus Y_3 with $\delta = 1$, $\beta = 0.6$, $\alpha_1 = \alpha_2 = \alpha_3 = 0.9$

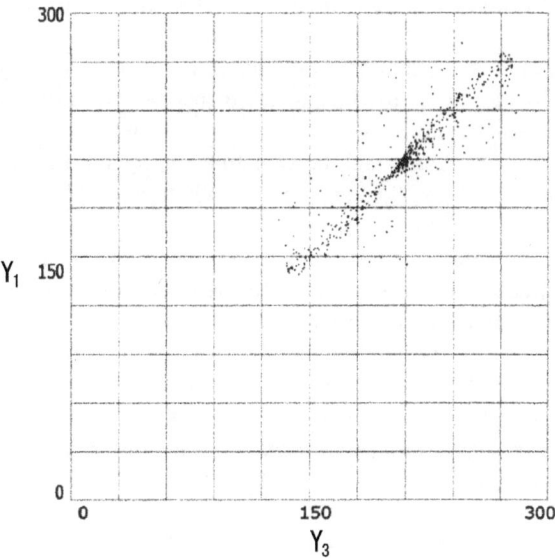

5 Concluding Remarks

In this paper, we extended two country Kaldorian business cycle model with fixed exchange rates that was developed by Asada et al. (2011, 2001) to a three country model, and studied its dynamics numerically. As a result, we found that some

Fig. 17 Lyapunov exponent (λ) with $\delta = 1$, $\beta = 0.6$, $\alpha_2 = \alpha_3 = 0.9$

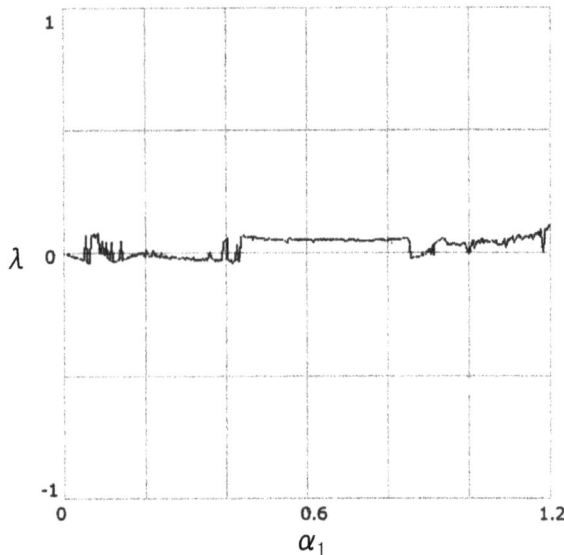

interesting dynamic phenomena can occur in such an extended model. For example, the partially synchronized and partially counter-synchronized movements of the real national incomes of countries 1 and 2 can coexist with the totally synchronized movements of the real national incomes of countries 1 and 3 under some set of parameter values such as the degree of international trade, degree of international capital mobility, and the adjustment speeds of the goods market in three countries. We also found that chaotic movements as well as cyclical fluctuations can occur for some set of the parameter values.

Incidentally, in our model the real government expenditures and the tax rates of three countries are fixed, and the total money stock of the currency union that is determined by the trans-national central bank are also fixed. It will be interesting to study what kinds of fiscal and monetary policies can stabilize the economy and reduce the amplitude of the fluctuations in this three country model under fixed exchange rates. It is also interesting to study the dynamic properties of three-country model with flexible exchange rates. Investigation of these topics is an important research program that is left for studies in future.

Acknowledgements The authors are grateful to the valuable comment by an anonymous referee. This research was financially supported by Chuo University. Needless to say, however, only the authors are responsible for possible remaining errors.

References

Asada, T. (1995). Kaldorian dynamics in an open economy. *Journal of Economics, 62,* 239–269.

Asada, T. (2004). A two-regional model of business cycles with fixed exchange rates: A Kaldorin approach. *Studies in Regional Sciences, 34,* 19–38.

Asada, T., Chiarella, C., Flaschel, P., & Franke, R. (2003). *Open economy macrodynamics: An integrated disequilibrium approach.* Berlin: Springer.

Asada, T., Douskos, C., Kalantonis, V., & Markellos, P. (2010). Numerical exploration of Kaldorian interregional macrodynamics: Enhanced stability and predominance of period doubling under flexible exchange rates. *Discrete Dynamics in Nature and Society,* Article ID 263041, 1–29.

Asada, T., Douskos, C., & Markellos, P. (2011). Numerical exprolation of Kaldorian interregional macrodynamics: Stability and the trade threshold for business cycles under fixed exchange rates. *Nonlinear Dynamics, Psychology, and Life Sciences, 15,* 105–128.

Asada, T., Inaba, T., & Misawa, T. (2000a). A nonlinear macrodynamic model with fixed exchange rates: Its dynamics and noise effects. *Discrete Dynamics in Nature and Society, 4,* 319–331.

Asada, T., Misawa, T., & Inaba, T. (2000b). Chaotic dynamics in a flexible exchange rate system: A study of noise effects. *Discrete Dynamics in Nature and Society, 4,* 309–317.

Asada, T., Inaba, T., & Misawa, T. (2001). An interregional dynamic model: The case of fixed exchange rates. *Studies in Regional Sciences, 31,* 29–41.

Asada, T., Kalantonis, T., Markellos, M., & Markellos, P. (2012). Analytical expressions of periodic disequilibrium fluctuations generated by Hopf bifurcations in economic dynamics. *Applied Mathematics and Computation, 218,* 7066–7077.

Dohtani, A., Misawa, T., Inaba, T., Yokoo, M., & Owase, T. (1996). Chaos, complex transients and noise: Illustration with a Kaldor model. *Chaos, Solitons & Fractals, 7,* 2157–2174.

Gandolfo, G. (2009). *Economic dynamics* (4th ed.). Berlin: Springer.

Kaldor, N. (1940). A model of the trade cycle. *Economic Journal, 50,* 69–86.

Keynes, J. M. (1936). *The general theory of employment, interest and money.* London: Macmillan.

Krugman, P. (1996). *The self-organizing economy.* Oxford: Blackwell.

Lorenz, H. W. (1987). International trade and the possible occurrence of chaos. *Economics Letters, 23,* 135–138.

Lorenz, H. W. (1993). *Nonlinear dynamical economics and chaotic motion* (2nd ed.). Berlin: Springer.

Maličky, P., & Zimka, R. (2010). On the existence of business cycles in Asada's two-regional model. *Nonlinear Analysis: Real World Applications, 11,* 2787–2795.

Maličky, P., & Zimka, R. (2012). On the existence of Tori in Asada's two-regional model. *Nonlinear Analysis: Real World Applications, 13,* 710–724.

Matsumoto, A. (1994). Complex dynamics in a simple macro disequilibrium model. *Journal of Economic Behavior & Organization, 24,* 297–313.

Matsumoto, A. (1996). Ergodic chaos in inventory oscillations: An example. *Chaos, Solitons & Fractals, 7,* 2175–2188.

Matsumoto, A. (1997). Endogenous oscillations in a discrete dynamic model with inventory. *Discrete Dynamics in Nature and Society, 1,* 203–223.

Matsumoto, A. (1999). Preferable disequilibrium in a nonlinear cobweb economy. *Annals of Operations Research, 89,* 101–123.

Matsumoto, A., & Szidarovszky, F. (2011). Delay differential neoclassical growth model. *Journal of Economic Behavior & Organization, 78,* 272–289.

Matsumoto, A., Merlone, U., & Szidarovszky, F. (2018). Goodwin accelerator model revisited with fixed time delays. *Communications in Nonlinear Science and Numerical Simulation, 58,* 233–248.

Matsumoto, A., & Szidarovszky, F. (2019). *Dynamic oligopolies with time delays.* Tokyo: Springer.

Medved'ová, P. (2011). A dynamic model of a small open economy under flexible exchange rates. *Acta Polytechnica Hungarica, 8,* 13–26.

Nijkamp, P., & Reggiani, A. (1992). *Interaction, evolution and chaos in space.* Berlin: Springer.

Puu, T. (1997). *Nonlinear economic dynamics* (4th ed.). Berlin: Springer.

Rosser, J. B., Jr. (1991). *From catastrophe to chaos: A general theory of economic discontinuities.* Boston: Kluwer Academic Publishers.

Eductive Stability, Heterogeneous Information Costs and Period-Two Cycle Multiplicity

Ahmad Naimzada and Marina Pireddu

Abstract Starting from a Muthian cobweb model, we here extend the profit-based evolutionary settings in Hommes and Wagener (2010) and in Naimzada and Pireddu (2020a) by assuming that unbiased fundamentalists and several groups of biased fundamentalists face information costs that are inversely proportional to their bias. Like in those works, we deal with the case in which the model is globally eductively stable, being globally stable under naive expectations. Similarly to Naimzada and Pireddu (2020a), we find that the stability of the unique steady state, which coincides with the fundamental, holds either for every value of the bias, like in Hommes and Wagener (2010), or just for suitably small and large values of the bias. On the other hand, introducing into the economy new couples of symmetrically biased groups of fundamentalists with a sufficiently high bias, multiple coexisting locally stable period-two cycles emerge. While in Hommes and Wagener (2010) such phenomenon occurs only when the steady state is locally stable, we observe the coexistence of multiple locally stable period-two cycles also when the equilibrium is unstable, thanks to information costs. Moreover, we show that the relative position of the newly arisen period-two cycles may not coincide with and without information costs.

Keywords Muthian cobweb model · Heterogeneous agents · Evolutionary learning · Information costs · Period-two cycle multiplicity

A. Naimzada
Department of Economics, Management and Statistics, University of Milano-Bicocca,
U6 Building, Piazza dell'Ateneo Nuovo 1, 20126 Milan, Italy
e-mail: ahmad.naimzada@unimib.it

M. Pireddu (✉)
Department of Mathematics and its Applications, University of Milano-Bicocca,
U5 Building, Via R. Cozzi 55, 20125 Milan, Italy
e-mail: marina.pireddu@unimib.it

© Springer Nature Singapore Pte Ltd. 2020
F. Szidarovszky and G. I. Bischi (eds.), *Games and Dynamics in Economics*,
https://doi.org/10.1007/978-981-15-3623-6_7

1 Introduction

In the present contribution, following the approach adopted in Naimzada and Pireddu (2020a), we extend the Muthian cobweb model with heterogeneous agents by Hommes and Wagener (2010)—in which a profit-based evolutionary mechanism operates—by assuming that producers face differentiated information costs, directly proportional to their rationality degree. In particular, while in Naimzada and Pireddu (2020a), in addition to unbiased fundamentalists, a unique couple of groups of symmetrically biased agents is considered, we here deal with the case of $N \geq 2$ couples of groups of symmetrically biased optimists and pessimists, which differ in the strength of the bias. We recall that, despite the heterogeneity in the forecasting rules, in Hommes and Wagener (2010) all agents face a common zero information cost.

The starting point of this research strand is given by Brock and Hommes (1997), where a Muthian cobweb type demand-supply model was presented, with producers which choose between rational and naive expectations about prices, selecting the strategy on the basis of the recent profits the two forecasting rules allowed to realize. In particular, an information cost is associated to the use of the more sophisticated forecasting rule. Dealing with the same share updating mechanism adopted in Brock and Hommes (1997) for the case without memory, Hommes and Wagener (2010) consider a Muthian cobweb model framework in which producers can choose among three different forecasting rules: fundamentalists predict that prices will always be at their fundamental value, optimists predict that the price of the good will always be above the fundamental price, whereas pessimists predict prices below the fundamental price. Hommes and Wagener (2010) focus on the case in which the Muthian model is globally eductively stable in the sense of Guesnerie (2002), that is, on the case in which the model is stable under naive expectations, as the slopes of demand and supply satisfy the familiar "cobweb theorem" by Ezekiel (1938). They show that under evolutionary learning the steady state, which is always (locally or globally) stable, may coexist with a locally stable period-two cycle, along which prices fluctuate around the rational expectations price and most agents switch between optimistic and pessimistic strategies. This means that, although the model in Hommes and Wagener (2010) is globally eductively stable, the evolutionary system therein admits both the steady state and the period-two cycle as possible long-run outcomes, and thus, contrarily to the setting in Brock and Hommes (1997), it may be not globally evolutionary stable.

Extending the model in Hommes and Wagener (2010) by assuming that producers face heterogeneous information costs, inversely proportional to the degree of their bias, for the case of one couple of groups of symmetrically biased agents in Naimzada and Pireddu (2020a) we found that the equilibrium, when globally eductively stable, may be unstable under evolutionary learning. Hence, the introduction of differentiated information costs, in addition to making the characterization of agents' heterogeneity more complete than in Hommes and Wagener (2010), allows us to give a cleaner negative answer to the question *does eductive stability always imply evolutive stability?* addressed in that paper, which was in turn inspired by the claim in Guesnerie (2002) that *"reasonable" adaptive learning processes are asymptotically stable*.

In more detail, the setting in Naimzada and Pireddu (2020a) was mainly analyzed by measuring the influence of agents' heterogeneity through the parameter describing the degree of optimism and pessimism, that was also introduced, in financial markets contexts, e.g. in Cavalli et al. (2017), in De Grauwe and Rovira Kaltwasser (2012), in Naimzada and Pireddu (2015), and in Naimzada and Ricchiuti (2008, 2009). We found that the unique steady state, which coincides with the fundamental, may be stable either for all values of the bias or just for suitably small and for suitably large values of the bias. The possible destabilization of the steady state occurs via a flip bifurcation, at which a stable period-two cycle emerges, which persists even after the pitchfork bifurcation through which the steady state recovers its local stability. On the other hand, since the map governing the dynamics in Naimzada and Pireddu (2020a) is monotonically decreasing, like it happened in Hommes and Wagener (2010) in the absence of information costs, no richer dynamics could arise.

As mentioned above, in the present work we extend the setting considered in Naimzada and Pireddu (2020a) by dealing with $N \geq 2$ couples of groups of symmetrically biased fundamentalists, which differ in the degree of the bias, and which face heterogeneous information costs, that are inversely proportional to their bias. We find that, although the map governing the dynamics is decreasing even in such more general context, multiple coexisting locally stable period-two cycles can emerge, one for each added couple of groups, when their bias is sufficiently high. Similarly to what happened in Hommes and Wagener (2010), such locally stable period-two cycles emerge, together with unstable period-two cycles, through double fold bifurcations of the second iterate of the map governing the dynamics, even if they may arise in a different position with respect to the already existing cycles, when compared to the case without information costs. In either case, the stability of the preexisting attractors in not affected by the emergence of the new period-two cycles. Indeed, the main difference between our framework and that considered in Hommes and Wagener (2010) lies in the fact that, while in that setting the emergence of new period-two cycles can occur only when the steady state is locally stable—since in the absence of information costs the steady state can not lose stability—dealing with at least two couples of groups of biased fundamentalists we can observe the coexistence of multiple locally stable period-two cycles also when the equilibrium is unstable, thanks to the introduction of information costs. Thus, we can say that eductive stability may not imply evolutionary stability when information costs are taken into account, also in the presence of several couples of groups of biased fundamentalists.

In addition to the just described results, along the paper we investigate the effect produced by the main model parameters on the stability of the steady state when several coupled groups of biased agents populate the economy. We stress that there exists a unique equilibrium and its expression is influenced neither by the introduction of information costs, nor by the number of groups of biased agents. Namely, as in Hommes and Wagener (2010), the steady state always coincides with the fundamental. Moreover, like in the simplified setting considered in Naimzada and Pireddu

(2020a), we still find that increasing the information cost of unbiased fundamentalists may have a destabilizing effect on the equilibrium. Indeed, the choices of fundamentalists lead prices towards the fundamental value and raising their information cost makes their share decrease, due to their resulting lower fitness in terms of profits, not only for prices far from the equilibrium, but also in a neighborhood of it, and this may lead to a destabilization of the steady state. On the other hand, increasing the information cost of a group of biased agents may have a stabilizing effect, when their bias is excessively large. Namely, in this case, raising the information cost of those optimists and pessimists makes their fitness fall, not only for prices close to the equilibrium, but also far from it, making the share of agents opting for such strategies decrease, so that prices more likely converge towards the fundamental value. As concerns the effect produced on the system stability by the introduction of a further couple of groups of symmetrically biased agents, comparing the stability conditions derived in Naimzada and Pireddu (2020a) for the case of one group of optimists and one group of pessimists with the stability conditions for the case with several groups of biased fundamentalists, we find that further coupled groups of biased fundamentalists may have either a destabilizing or a stabilizing effect on the equilibrium according to the considered parameter configuration.

The remainder of the paper is organized as follows. In Sect. 2 we recall the model in Naimzada and Pireddu (2020a) with differentiated information costs and two types of biased fundamentalists, together with the corresponding main findings. In Sect. 3 we present and analyze the setting with several types of biased fundamentalists. In Sect. 4 we describe some possible extensions of the model.

2   The Model with Two Types of Biased Fundamentalists

At first we recall the discrete-time evolutionary cobweb setting in Hommes and Wagener (2010) with two types of biased fundamentalists, to which in Naimzada and Pireddu (2020a) we added information costs in the profits (see (2.5) below).

The economy is populated by unbiased fundamentalists, that we will just call fundamentalists, and by two types of biased fundamentalists, i.e., one group of optimists and one group of pessimists, which are symmetrically biased. In the Muthian farmer model, agents have to choose the quantity q of a certain good to produce in the next period and are expected profit maximizers. Assuming a quadratic cost function

$$\gamma(q) = \frac{q^2}{2s}, \tag{2.1}$$

with $s > 0$, the supply curve is given by

$$S(p^e) = sp^e, \tag{2.2}$$

where p^e is the expected price and s describes its slope. The demand function is supposed to be linearly decreasing in the market price, i.e.,

$$D(p) = A - dp,$$

with A and d positive parameters, representing respectively the market size and the slope of the demand function. We stress that the demand is positive for sufficiently large values of A.

At the fundamental price $p = p^*$ demand equals supply, i.e.,

$$p^* = \frac{A}{d + s}. \tag{2.3}$$

This is also the expression of the unique model steady state in Hommes and Wagener (2010) and in Naimzada and Pireddu (2020a). Like in those works, along the paper we will deal with the case in which the Muthian model is globally eductively stable in the sense of Guesnerie (2002), that is, with the case in which the model is stable under naive expectations, as the slopes of demand and supply satisfy the familiar "cobweb theorem" by Ezekiel (1938) and thus it holds that $s/d < 1$.

Agents have heterogeneous expectations about the price of the good they have to produce. In particular, fundamentalists predict that prices will always be at their fundamental value, while optimists (pessimists) predict that the price of the good will always be above (below) the fundamental price. Hence, assuming a symmetric disposition of the beliefs and characterizing the fundamentalists, pessimists and optimists by subscripts 0, 1, 2, respectively, in symbols we have that their expectations at time t are given by

$$p_{i,t}^e = p^* + b_i, \ i \in \{0, 1, 2\}, \text{ with } b_0 = 0, \ b_1 = -b, \ b_2 = b, \tag{2.4}$$

where $b > 0$ describes the bias degree of pessimists and optimists. In order to avoid a negative expectation for pessimists, we will restrict our attention to the bias values $b \in (0, p^*)$, with p^* as in (2.3).

Denoting by $\omega_{i,t}$ the share of agents choosing the forecasting rule $i \in \{0, 1, 2\}$ at time t, the total supply is given by $\sum_{i=0}^{2} \omega_{i,t} S(p_{i,t}^e)$ and thus the market equilibrium condition reads as

$$A - dp_t = \sum_{i=0}^{2} \omega_{i,t} S(p_{i,t}^e).$$

As concerns the share updating mechanism, Hommes and Wagener (2010) deal with the discrete choice model in Brock and Hommes (1997) for the case without memory, in which only the most recently realized net profits $\pi_{j,t-1}$, $j \in \{0, 1, 2\}$, are taken into account. In symbols

$$\omega_{i,t} = \frac{\exp(\beta\pi_{i,t-1})}{\sum_{j=0}^{2}\exp(\beta\pi_{j,t-1})}, \; i \in \{0, 1, 2\},$$

where $\beta > 0$ is the intensity of choice parameter.

When considering information costs, net profits $\pi_{j,t}$, $j \in \{0, 1, 2\}$, at time t are defined as

$$\pi_{j,t} = p_t S(p_{j,t}^e) - \gamma(S(p_{j,t}^e)) - c_j, \tag{2.5}$$

with γ and S as in (2.1) and (2.2), respectively, and with the nonnegative parameter c_j representing the information cost deriving by the adoption of forecasting rule j. Since optimists and pessimists do not perfectly know the economic fundamentals and make symmetric errors in estimating them, we may affirm that those agents display the same degree of rationality, and thus we will assume that $c_1 = c_2 = c$, for a certain $c \geq 0$. On the other hand, unbiased fundamentalists exactly know the formulations of demand and supply functions and they are able to correctly compute the fundamental value. Due to their higher degree of rationality with respect to optimists and pessimists, we will suppose that the information cost c_0 of fundamentalists satisfies $0 \leq c \leq c_0$, i.e., that the information costs are inversely proportional to the bias degree.[1] We stress that for $c = c_0 = 0$ we are led back to the framework in Hommes and Wagener (2010), while setting $c_0 = C$ and $c = 0$ we obtain the same information costs as in Anufriev et al. (2013) where, in a DSGE model with heterogeneous expectations, the equilibrium predictor for the inflation rate is available at cost $C \geq 0$, while biased agents do not face any information cost.

Introducing, like in Hommes and Wagener (2010), the variable $x_t = p_t - p^*$, we can write the model dynamic equation in deviation from fundamental as

$$x_t = -\frac{s}{d}\sum_{i=0}^{2}\omega_{i,t}\,b_i$$

with

$$\omega_{i,t} = \frac{\exp\left(-\frac{\beta s}{2}(x_{t-1} - b_i)^2 - \beta c_i\right)}{\sum_{j=0}^{2}\exp\left(-\frac{\beta s}{2}(x_{t-1} - b_j)^2 - \beta c_j\right)},$$

or, more explicitly, recalling (2.4), as

$$x_t = \frac{sb}{d}(\omega_{1,t} - \omega_{2,t})$$
$$= \frac{sb}{d}\frac{\exp\left(-\frac{\beta s}{2}(x_{t-1}+b)^2\right) - \exp\left(-\frac{\beta s}{2}(x_{t-1}-b)^2\right)}{\exp\left(-\frac{\beta s}{2}(x_{t-1}+b)^2\right) + \exp\left(-\frac{\beta s}{2}(x_{t-1}-b)^2\right) + \exp\left(-\frac{\beta s}{2}x_{t-1}^2 - \beta(c_0-c)\right)}. \tag{2.6}$$

---

[1]Indeed, according to Hommes (2013), p. 150, *A fundamentalists strategy, however, requires structural knowledge of the economy and information about "economic fundamentals", and therefore we assume positive information-gathering costs for fundamentalists. In the cobweb model the fundamental forecast requires structural knowledge of demand and supply curves in order to compute the fundamental steady state price p^*.*

For the model formulation in terms of x_t, the unique steady state is given by $x^* = 0$ and we will report the corresponding stability condition in terms of the intensity of choice parameter in Proposition 2.1. We can rewrite (2.6) as

$$x_t = f(x_{t-1}), \tag{2.7}$$

where the one-dimensional map $f : (-p^*, +\infty) \to \mathbb{R}$ is defined as

$$f(x) = \frac{sb}{d} \frac{\exp\left(-\frac{\beta s}{2}(x+b)^2\right) - \exp\left(-\frac{\beta s}{2}(x-b)^2\right)}{\exp\left(-\frac{\beta s}{2}(x+b)^2\right) + \exp\left(-\frac{\beta s}{2}(x-b)^2\right) + \exp\left(-\frac{\beta s}{2}x^2 - \beta(c_0 - c)\right)}. \tag{2.8}$$

We stress that f is differentiable and that, recalling the expression of p^* in (2.3), its domain is enlarged by considering increasing values of A. Moreover, when extending its domain to $\mathbb{R}$, the map is odd, and it is possible to prove that f is decreasing also for nonzero information costs, using the same argument based on the nonnegativity of the variance in relation to a suitable stochastic process concerning the biases, employed in Theorem A in Hommes and Wagener (2010) for the case $c = c_0 = 0$. The monotonicity of f excludes the possibility of complex dynamics in the presence of information costs, too, and indeed in Naimzada and Pireddu (2020a) we observed at most a period-two cycle, either coexisting with the locally stable steady state, or being the unique attractor. In fact, differently from what obtained in Theorem A in Hommes and Wagener (2010)—where in the absence of information costs the fundamental steady state is always (locally or globally) stable for $s/d < 1$— in Proposition 2.1 below (which coincides with Proposition 3.1 in Naimzada and Pireddu 2020a) we find that the intensity of choice parameter β may also be either destabilizing or it may play an ambiguous role on the equilibrium stability when information costs are taken into account.

Along the present section, we call a scenario destabilizing (stabilizing) with respect to a parameter when the steady state is stable (unstable) below a certain threshold of that parameter and unstable (stable) above it. We say that a scenario is mixed if the steady state is unstable inside an interval of intermediate parameter values and stable for lower and higher values of the parameter. We say that a scenario is unconditionally unstable (unconditionally stable) when the steady state is unstable (stable) for all the parameter values.

Proposition 2.1 *Equation* (2.7) *admits* $x = 0$ *as unique steady state. The equilibrium* $x = 0$ *is locally asymptotically stable for* (2.6) *if*

$$\beta < \frac{d\left(2 + \exp\left(\frac{\beta b^2 s}{2} - \beta(c_0 - c)\right)\right)}{2b^2 s^2}.$$

Hence, if $s/d < 1$, *depending on the considered parameter configuration, on increasing* β *we may have an* (i) *unconditionally stable,* (ii) *mixed or* (iii) *destabilizing scenario.*

We stress that when setting $c = c_0 = 0$, i.e., in the absence of information costs, cases (ii) and (iii) in Proposition 2.1 can not occur since, as shown in Theorem A in Hommes and Wagener (2010) and as we explained above, without information costs the steady state is always (locally or globally) asymptotically stable. Hence, if we neglected information costs we would not observe, in particular, the most classical effect produced by the intensity of choice parameter, that is, the destabilizing scenario.

We recall that with the introduction of information costs, when the steady state loses stability, a flip bifurcation occurs at which a globally stable period-two cycle emerges. The period-two cycle becomes locally stable only in case the steady state recovers its stability for increasing values of β, otherwise the period-two cycle remains globally stable. When instead the steady state is always stable like in Hommes and Wagener (2010), the locally stable period-two cycle emerges, together with an unstable period-two cycle, for sufficiently large values of β through a double fold bifurcation of the second iterate of f, and both the steady state and one of the two period-two cycles remain locally stable when raising β.

Rewriting the stability conditions in Proposition 2.1 in terms of the bias, the following result follows:

Corollary 2.1 *The equilibrium $x = 0$ is locally asymptotically stable for* (2.6) *if*

$$b^2 < \frac{d\left(2 + \exp\left(\frac{\beta b^2 s}{2} - \beta(c_0 - c)\right)\right)}{2\beta s^2}. \tag{2.9}$$

Hence, depending on the considered parameter configuration, on increasing b we may have an unconditionally stable or mixed scenario.

Thus, according to Proposition 2.1 and Corollary 2.1, there are up to two possible stability thresholds for $x = 0$ with respect to β and b, respectively, and $x = 0$ may be locally stable just for sufficiently low and for sufficiently high values of the intensity of choice parameter and of the bias. In particular, this means that the introduction of information costs may not only produce a destabilization of the system for intermediate values of the bias of optimistic and pessimistic agents, but that a sufficiently strong beliefs' heterogeneity can be stabilizing in the presence of information costs.

We now report in Figs. 1 and 2 the scenarios compatible with Corollary 2.1 for increasing values of b, while fixing the other parameters as follows: $A = 8$, $s = 0.95$, $d = 1$, $\beta = 10$, $c = 0.1$. In such configuration, setting in Fig. 1 $c_0 = 0.11$, so that the information costs for fundamentalists and for biased agents have almost coinciding values, we observe just the two frameworks which can occur in the absence of information costs. Namely, in Fig. 1a for $b = 0.4$ we find that the steady state $x = 0$ is globally stable and in Fig. 1c for $b = 0.8$ we observe, in addition to the locally stable steady state, denoted by a black dot, a stable and an unstable period-two cycles, denoted respectively by black and empty squares, which are born for $b \approx 0.690$ through a double fold bifurcation of the second iterate of f, that we illustrate in Fig. 1b. On the other hand, in Fig. 2 for $c_0 = 0.2$ we observe that the

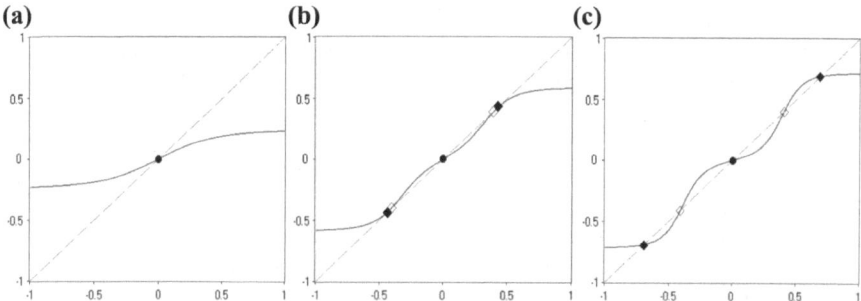

Fig. 1 The graph of the second iterate of f for $c_0 = 0.11$, and $b = 0.4$ in **a**, $b = 0.690$ in **b** and $b = 0.8$ in **c**

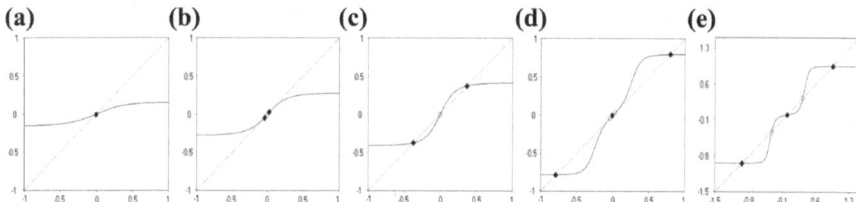

Fig. 2 The graph of the second iterate of f for $c_0 = 0.2$, and $b = 0.3$ in **a**, $b = 0.393$ in **b**, $b = 0.5$ in **c**, $b = 0.850$ in **d** and $b = 1$ in **e**

steady state $x = 0$ is globally stable in Fig. 2a for $b = 0.3$, but in Fig. 2c the steady state has become unstable for $b = 0.5$, and it is denoted by a small circle, being surrounded by a globally stable period-two cycle, born for $b \approx 0.393$ through a pitchfork bifurcation of the second iterate of f, that we illustrate in Fig. 2b and which corresponds to a flip bifurcation of f. In Fig. 2e for $b = 1$ the steady state $x = 0$ is again locally stable thanks to a further pitchfork bifurcation of the second iterate of f that has occurred for $b \approx 0.850$ at $x = 0$ (see Fig. 2d). The basin of attraction of $x = 0$ is separated by that of the locally stable period-two cycle by an unstable period-two cycle, born through that pitchfork bifurcation. We stress that, for suitable parameter configurations, $x = 0$ may lose and recover stability for increasing values of β through a flip bifurcation of f and a pitchfork bifurcation of the second iterate of f, respectively, as it happens in Fig. 2 when raising b.

Rewriting the stability conditions in Proposition 2.1 in terms of the information costs, the following result follows:

Corollary 2.2 *The equilibrium $x = 0$ is locally asymptotically stable for (2.6) for every value of the information costs $0 \leq c \leq c_0$ if $b \leq \sqrt{d/(\beta s^2)}$. If instead $b > \sqrt{d/(\beta s^2)}$, then $x = 0$ is locally asymptotically stable for (2.6) if*

$$c_0 - c < \log \left(\frac{d \, \exp\left(\frac{\beta b^2 s}{2}\right)}{2(b^2 \beta s^2 - d)} \right)^{\frac{1}{\beta}} .$$

Hence, if $b > \sqrt{d/(\beta s^2)}$, depending on the considered parameter configuration, on increasing c we may have an unconditionally stable, stabilizing or unconditionally unstable scenario, while on increasing c_0 we may have an unconditionally unstable or a destabilizing scenario.

Thus, according to Corollary 2.2, increasing the information cost of biased agents has either no effect on the equilibrium stability or it is stabilizing, when their bias is excessively large. On the other hand, Corollary 2.2 tells us that raising the information cost of fundamentalists is destabilizing, when the equilibrium is not always unstable and if the bias of optimists and pessimists is large enough.

3   The Model with Several Types of Biased Fundamentalists

We now extend the set of expectation rules in (2.4) by assuming that the economy, in addition to encompassing unbiased fundamentalists facing an information cost $c_0 \geq 0$, is populated by $N \geq 1$ couples of groups of symmetrically biased fundamentalists, which face information costs that are inversely proportional to their bias degree. Hence, for $i \in \{0, 1, 2, 3, \ldots, 2N\}$ and p^* as in (2.3) we have

$$p^e_{i,t} = p^* + b_i, \text{ with } b_0 = 0, \ b_{2n} = -b_{2n-1} = \mathfrak{b}_n > 0, \text{ for } 1 \leq n \leq N, \quad (3.1)$$

with b_0 which describes the bias of fundamentalists (denoted by index 0) and $\mathfrak{b}_n$ describing the bias of the n-th coupled groups, composed respectively by pessimists (denoted by index $2n - 1$) and by optimists (denoted by index $2n$). Since for the bias of fundamentalists we suppose that $b_0 = 0$, by (2.4) we find $p^e_{0,t} = p^*$. In regard to the remaining biases, in order to avoid negative expectations for pessimists, we will restrict our attention to the bias values $\mathfrak{b}_n < p^*$, for $1 \leq n \leq N$. Moreover, without loss of generality we can assume that $0 = b_0 < \mathfrak{b}_1 < \mathfrak{b}_2 < \cdots < \mathfrak{b}_N$, so that for the information costs it is reasonable to suppose that $c_0 \geq \mathfrak{c}_1 \geq \mathfrak{c}_2 \geq \cdots \geq \mathfrak{c}_N \geq 0$, where $\mathfrak{c}_n$, for $1 \leq n \leq N$, describes the information cost faced by the n-th coupled groups of pessimists and optimists, and, as explained above, c_0 is the information cost faced by unbiased fundamentalists. Since pessimists (denoted by index $2n - 1$) and optimists (denoted by index $2n$) belonging to the n-th coupled groups face the same information cost $\mathfrak{c}_n$, it holds that $c_{2n} = c_{2n-1} = \mathfrak{c}_n > 0$ for $1 \leq n \leq N$.

In the case $N = 1$, the model reduces to that studied in Naimzada and Pireddu (2020a) and recalled in Sect. 2, when identifying $\mathfrak{b}_1$ with b and $\mathfrak{c}_1$ with c. For $N > 1$ the net profits $\pi_{j,t}$ realized at time t by unbiased fundamentalists and by the $2N$ coupled groups of pessimists and optimists are described by the expression in (2.5) for $j \in \{0, 1, 2, 3, \ldots, 2N\}$, respectively, while the share $\omega_{i,t}$ of agents choosing the forecasting rule $i \in \{0, 1, 2, 3, \ldots, 2N\}$ at time t is given by

$$\omega_{i,t} = \frac{\exp(\beta \pi_{i,t-1})}{\sum_{j=0}^{2N} \exp(\beta \pi_{j,t-1})}.$$

Our model dynamic equation can be written in terms of $x_t = p_t - p^*$ as

$$x_t = -\frac{s}{d} \sum_{i=0}^{2N} w_{i,t} b_i$$

with

$$w_{i,t} = \frac{\exp\left(-\frac{\beta s}{2}(x_{t-1} - b_i)^2 - \beta c_i\right)}{\sum_{j=0}^{2N} \exp\left(-\frac{\beta s}{2}(x_{t-1} - b_j)^2 - \beta c_j\right)}$$

for $i \in \{0, 1, 2, 3, \ldots, 2N\}$. More explicitly, by (3.1) and recalling the relation between c_i, for $i \in \{1, 2, 3, \ldots, 2N\}$, and c_n, for $n \in \{1, \ldots, N\}$, we obtain

$$\begin{aligned} x_t &= \frac{s}{d}\sum_{n=1}^{N} b_n (w_{2n-1,t} - w_{2n,t}) \\ &= \frac{s\left(\sum_{n=1}^{N} b_n \left(\exp\left(-\frac{\beta s}{2}(x_{t-1}+b_n)^2 - \beta c_n\right) - \exp\left(-\frac{\beta s}{2}(x_{t-1}-b_n)^2 - \beta c_n\right)\right)\right)}{d\left(\exp\left(-\frac{\beta s}{2}x_{t-1}^2 - \beta c_0\right) + \sum_{n=1}^{N}\left(\exp\left(-\frac{\beta s}{2}(x_{t-1}+b_n)^2 - \beta c_n\right) + \exp\left(-\frac{\beta s}{2}(x_{t-1}-b_n)^2 - \beta c_n\right)\right)\right)} . \end{aligned} \qquad (3.2)$$

The unique steady state for the extended system is still given by $x^* = 0$ (see Proposition 3.1 below), and its expression is influenced neither by the introduction of information costs, nor by the number of groups of biased agents.

In view of the next analysis, we rewrite (3.2) as

$$x_t = f_N(x_{t-1}), \qquad (3.3)$$

where the one-dimensional map $f_N : (-p^*, +\infty) \to \mathbb{R}$ is defined as

$$f_N(x) = \frac{s\left(\sum_{n=1}^{N} b_n \left(\exp\left(-\frac{\beta s}{2}(x + b_n)^2 - \beta c_n\right) - \exp\left(-\frac{\beta s}{2}(x - b_n)^2 - \beta c_n\right)\right)\right)}{d\left(\exp\left(-\frac{\beta s}{2}x^2 - \beta c_0\right) + \sum_{n=1}^{N}\left(\exp\left(-\frac{\beta s}{2}(x + b_n)^2 - \beta c_n\right) + \exp\left(-\frac{\beta s}{2}(x - b_n)^2 - \beta c_n\right)\right)\right)} . \qquad (3.4)$$

We stress that f_N reduces to f in (2.8) for $N = 1$, when identifying b_1 and c_1 with b and c, respectively. Like for f, it holds that f_N is differentiable and that its domain is enlarged by considering increasing values of A. Moreover, when extending its domain to $\mathbb{R}$, the map is odd. Namely, replacing x with $-x$ leaves the denominator unchanged, while the numerator of f_N changes sign, so that $f_N(-x) = -f_N(x)$ for every $x \in \mathbb{R}$. Moreover, still employing the argument in Theorem A in Hommes and Wagener (2010), based on the nonnegativity of the variance in relation to a suitable stochastic process, it can be proven that f_N is decreasing for every $N \geq 1$, so that no complex dynamics can occur. Indeed, at most we will observe multiple locally stable coexisting period-two cycles, both when the steady state is locally stable and when it is unstable (cf. Figs. 3 and 4, respectively).

In the next result we derive the expression of the market stationary equilibrium for (3.3) and we describe the corresponding stability condition.

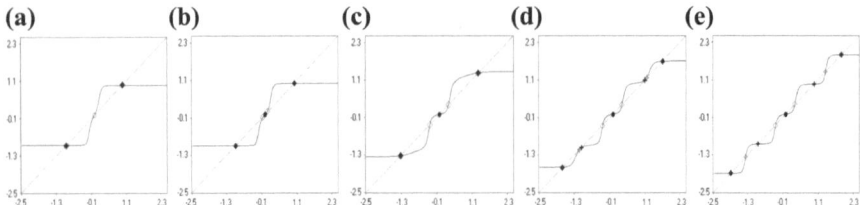

Fig. 3 The graph of the second iterate of f_2 for $c_0 = 0.3$, and $b_2 = 1.01$ in **a**, $b_2 = 1.06$ in **b**, $b_2 = 1.5$ in **c**, $b_2 = 1.82$ in **d** and $b_2 = 2$ in **e**

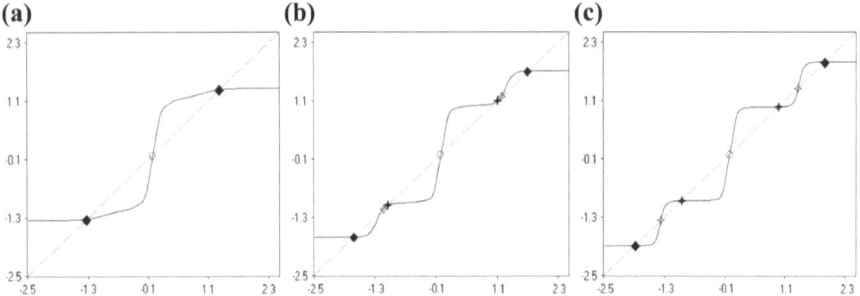

Fig. 4 The graph of the second iterate of f_2 for $c_0 = 0.5$, and $b_2 = 1.5$ in **a**, $b_2 = 1.83$ in **b** and $b_2 = 2$ in **c**

Proposition 3.1 *Equation* (3.3) *admits* $x = 0$ *as unique steady state. The equilibrium* $x = 0$ *is locally asymptotically stable for map* f_N *in* (3.4) *if*

$$\sum_{n=1}^{N} \left(\beta s^2 b_n^2 - d \right) \exp\left(\frac{-\beta s b_n^2}{2} - \beta c_n \right) < \frac{d}{2} \exp(-\beta c_0). \qquad (3.5)$$

Proof A straightforward check ensures that $x = 0$ solves the fixed-point equation $f_N(x) = x$, with f_N as in (3.4).

In order to show that $x = 0$ is the unique steady state it suffices to observe that f_N is positive if and only if x is negative.

The stability condition follows by imposing that $f_N'(0) \in (-1, 1)$. By direct computations, we have

$$f_N'(0) = \frac{-2\beta s^2 \sum_{n=1}^{N} b_n^2 \exp\left(\frac{-\beta s b_n^2}{2} - \beta c_n \right)}{d \left(\exp(-\beta c_0) + 2 \sum_{n=1}^{N} \exp\left(\frac{-\beta s b_n^2}{2} - \beta c_n \right) \right)}.$$

Since $f_N'(0)$ is always negative, the stability of $x = 0$ is guaranteed by $f_N'(0) > -1$, which is equivalent to (3.5). □

Due to the several possibilities which may arise from an increase in the various parameter values, rather than presenting the results in propositions, we prefer to gather all the comments hereinafter, as this choice allows for a larger freedom.

Since c_0 reduces the term on the right-hand side of (3.5), increasing such parameter may have a destabilizing effect on the equilibrium when the left-hand side of (3.5) is positive, and this happens for instance when $b_1 > \sqrt{d/(\beta s^2)}$. Indeed, when the biases of optimists and pessimists are large enough, raising the information cost of unbiased fundamentalists makes the share of agents opting for that strategy decrease, due to the lower fitness in terms of profits, not only for prices far from the equilibrium, but also in a neighborhood of the steady state. Since fundamentalists have a stabilizing effect on the equilibrium, a reduction in their share makes more difficult for prices to converge towards the fundamental value.

As concerns the effect of an increase in the information costs of biased fundamentalists on the equilibrium stability, by (3.5) it follows that it is not univocal and depends on the intensity of the corresponding bias. Namely, if $b_m > \sqrt{d/(\beta s^2)}$ (if $b_m < \sqrt{d/(\beta s^2)}$) for some $m \in \{1, \ldots, N\}$, then raising c_m may have a stabilizing (destabilizing) effect on the equilibrium, since the left-hand side of (3.5) decreases (increases). Indeed, if the bias of a group of optimists or pessimists is excessively large, increasing the corresponding information cost makes the profits of the agents opting for that strategy decrease also when prices are far from the equilibrium and thus their share falls. The consequent raise in the shares of agents with a lower bias and of fundamentalists makes prices more likely converge towards the fundamental value.

Similarly to what happens with information costs of optimists and pessimists, we observe that increasing the biases has not a univocal effect. Namely, rewriting (3.5) as

$$b_1^2 < \frac{d\left(2 + \exp\left(\frac{\beta s b_1^2}{2} - \beta(c_0 - c_1)\right)\right) + 2\left(\sum_{n=2}^{N}\left(d - \beta s^2 b_n^2\right)\exp\left(\frac{-\beta s}{2}(b_n^2 - b_1^2) - \beta(c_n - c_1)\right)\right)}{2\beta s^2}. \quad (3.6)$$

and comparing such condition with (2.9), when identifying b_1 and c_1 with b and c, respectively, we conclude that the introduction of further coupled groups of biased fundamentalists may have either a destabilizing or a stabilizing effect on the equilibrium according to the sign of

$$\sum_{n=2}^{N}\left(d - \beta s^2 b_n^2\right)\exp\left(-\beta s b_n^2/2 - \beta c_n\right). \quad (3.7)$$

Indeed, when the latter term is negative (positive), the values of b_1 which satisfy (3.6) are a subset (superset) of the values of b which satisfy (2.9). We stress that term in (3.7) is negative e.g. when the biases of the newly introduced coupled groups are excessively large.

We finally notice that drawing conclusions on the role of β on the system stability is very difficult, due to the several occurrences of β in (3.5).

As explained after (3.4), since the map f_N is decreasing for every $N \geq 1$, no complex dynamics can occur. However, in addition to the phenomena portrayed in Figs. 1 and 2, when considering further coupled groups of optimists and pessimists, multiple coexisting locally stable period-two cycles can emerge, one for each added couple of groups. In particular, in Fig. 1 in Hommes and Wagener (2010) such phenomenon is illustrated when $N = 2$ for the setting without information costs in the case $x = 0$ is locally stable, since in that framework the steady state can not lose stability. On the contrary, thanks to the introduction of information costs, dealing with at least two couples of groups of symmetrically biased fundamentalists, we can observe the coexistence of multiple locally stable period-two cycles also when $x = 0$ is unstable. Indeed, in the presence of information costs, for $N = 2$ we show up to two coexisting locally stable period-two cycles, both when the steady state is locally stable (in Fig. 3) and when it is unstable (in Fig. 4), for the following parameter configuration: $A = 8$, $s = 0.95$, $d = 1$, $\beta = 10$, $b_1 = 1$, $c_1 = 0.2$, $c_2 = 0.1$, and fixing $c_0 = 0.3$ in Fig. 3 and $c_0 = 0.5$ in Fig. 4, while we will let b_2 free to vary. More precisely, in Fig. 3 we show how the shape of the second iterate of the map f_2 varies for increasing values of b_2, so that for $b_2 = 1.01$ we find in Fig. 3a the unstable steady state $x = 0$, denoted by a small circle, and a globally stable period-two cycle, denoted by black squares, while for $b_2 = 1.5$ we observe in Fig. 3c that $x = 0$, now denoted by a black dot, has become locally stable and that an unstable period-two cycle, denoted by empty squares, has emerged through a pitchfork bifurcation of the second iterate of f_2 occurring for $b_2 \approx 1.06$ and illustrated in Fig. 3b, as it happened in Fig. 2d when raising b. When increasing the bias of the second coupled groups of optimists and pessimists to $b_2 = 2$, we observe in Fig. 3e a further pair composed by a locally stable and an unstable period-two cycles, denoted respectively by black and empty stars, which are born for $b_2 \approx 1.82$ through a double fold bifurcation of the second iterate of f_2 (see Fig. 3d) without affecting the local stability of $x = 0$. Raising the information cost faced by unbiased fundamentalists from $c_0 = 0.3$ to $c_0 = 0.5$, in Fig. 4a, like in Fig. 3a, we find for $b_2 = 1.5$ the unstable steady state $x = 0$ and a globally stable period-two cycle. On the other hand, when increasing the value of the bias of the second coupled groups of optimists and pessimists to $b_2 = 2$, in Fig. 4c we observe also a locally stable and an unstable period-two cycles, born for $b_2 \approx 1.83$ through a double fold bifurcation of the second iterate of f_2 (see Fig. 4b) while $x = 0$ is still unstable.

Hence, the main difference between the frameworks portrayed in Figs. 3 and 4 lies in the absence in the latter, due to a larger value of the information cost of unbiased fundamentalists, of the pitchfork bifurcation of f_2^2 depicted in Fig. 3b through which $x = 0$ becomes locally stable.

We stress that the mechanism which allows for the emergence of the second couple of locally stable and unstable period-two cycles in Fig. 1 in Hommes and Wagener (2010)—obtained for the same parameter configuration we are dealing with, except for the zero information costs and for the consideration in that work of increasing values of the intensity of choice parameter, rather than of the bias—is very similar to that which leads from Fig. 3c to Fig. 3e, and indeed it is based on a double fold bifurcation of the second iterate of the map governing the dynamics, too. We

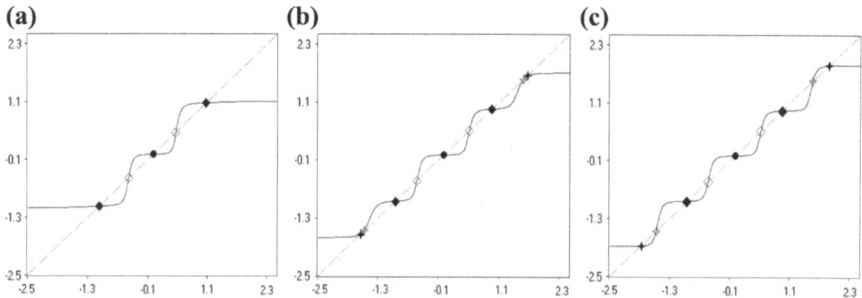

Fig. 5 The graph of the second iterate of f_2 for $c = c_0 = 0$, and $b_2 = 1.01$ in **a**, $b_2 = 1.84$ in **b** and $b_2 = 2$ in **c**

illustrate the steps related to the framework in Fig. 1 in Hommes and Wagener (2010) in Fig. 5 below for $c = c_0 = 0$, where we progressively increase b_2. For $b_2 = 1.01$ we observe in Fig. 5a the locally stable steady state $x = 0$, denoted by a black dot, and a locally stable period-two cycle, denoted by black squares, together with the unstable period-two cycle, denoted by empty squares, separating their basins of attraction. For $b_2 = 2$, in Fig. 5c we find a further pair composed by a locally stable and an unstable period-two cycles, denoted respectively by black and empty stars, which are born for $b_2 \approx 1.84$ through a double fold bifurcation of the second iterate of f_2 (see Fig. 5b), similar to the one we observed in Fig. 3d. Nonetheless, while in Fig. 3 the couple of locally stable and unstable period-two cycles emerges in the interval bounded by the already existing locally stable period-two points, for the parameter configuration considered in Fig. 1 in Hommes and Wagener (2010) and in our Fig. 5 the new couple of locally stable and unstable period-two cycles arises externally with respect to the already existing locally stable period-two cycle. In either case, the emergence of such couple of locally stable and unstable period-two cycles does not alter the stability of the preexisting attractors.

Similar phenomena to those just described occur when adding further couples of groups of biased fundamentalists into the economy. Namely, both with and without information costs, the introduction of extra pairs of groups of optimists and pessimists leads, for sufficiently high values of their bias, to the emergence of new couples of locally stable and unstable period-two cycles through a double fold bifurcation of the second iterate of the map governing the dynamics. In more detail, just one extra couple composed by a locally stable and an unstable period-two cycles emerges due to the introduction into the economy of a new pair of groups of symmetrically biased fundamentalists. As already seen with $N = 2$, also with several couples of groups of biased agents it still holds that the birth of a new pair of locally stable and unstable period-two cycles does not affect the stability of the other attractors. In particular, for $N = 3$, starting from the frameworks considered in Figs. 3 and 4 and setting $c_3 = 0.05$, we observe the emergence for $b_3 \approx 2.8$, through a double fold bifurcation of f_3^2, of a new couple of locally stable and unstable period-two cycles, externally with respect to the already existing cycles. This is still true when dealing with the parameter configurations just described, but in the absence of information costs. Namely, in such case, the double fold bifurcation of f_3^2, which allows for the emergence of a new couple of locally stable and unstable period-two cycles, occurs for $b_3 \approx 2.95$. We do not report the corresponding graphs of f_3^2, because they would be difficult to read due to the reduced size of the pictures and the relatively high number of period-two cycles.

We just conclude by remarking that, if information costs are taken into account and for $s/d < 1$, i.e., when the Muthian model is eductively stable in the sense of Guesnerie (2002), the new couples of locally stable and unstable period-two cycles may emerge—with the increase in the number of groups of biased fundamentalists, for sufficiently high values of their bias—both when $x = 0$ is locally stable, like in Fig. 3, and when it is unstable, like in Fig. 4. This is indeed the main difference between our findings and those in Hommes and Wagener (2010), where, due to the absence of information costs, the steady state is always locally stable. Hence, we can infer that eductive stability may not imply evolutionary stability when information costs are taken into account, also in the presence of several couples of groups of biased fundamentalists.

4 Conclusion

We believe that the setting analyzed here and in Naimzada and Pireddu (2020a), as well as the original framework in Hommes and Wagener (2010), can be the starting point for other research developments. Indeed, we recall that the final sentence in Hommes and Wagener (2010) reads as follows: "The study of the stability of evolutionary systems with many trader types in various market settings and with more complicated strategies remains an important topic for future work". In such perspective, the analyzed framework with heterogeneous producers could be further extended by dealing with a richer set of forecasting rules, including e.g. rational expectations agents like in Naimzada and Pireddu (2020b), in view of investigating whether dynamic phenomena more complex than period-two cycles can emerge when the fundamental steady state loses stability or even when it is stable.

A different modification of the model, which could possibly lead to rich dynamic outcomes, would consist in introducing memory in the share updating mechanism, so that agents, in choosing the heuristics to adopt, rather than taking into account just the most recently realized net profit, would consider the performance of the various forecasting rules in terms of realized profits in the recent past.

References

Anufriev, M., Assenza, T., Hommes, C., & Massaro, D. (2013). Interest rate rules and macroeconomic stability under heterogeneous expectations. *Macroeconomic Dynamics, 17*, 1574–1604.

Brock, W. A., & Hommes, C. H. (1997). A rational route to randomness. *Econometrica, 65*, 1059–1095.

Cavalli, F., Naimzada, A., & Pireddu, M. (2017). An evolutive financial market model with animal spirits: Imitation and endogenous beliefs. *Journal of Evolutionary Economics, 27*, 1007–1040.

De Grauwe, P., & Rovira Kaltwasser, P. (2012). Animal spirits in the foreign exchange market. *Journal of Economic Dynamics and Control, 36*, 1176–1192.

Ezekiel, M. (1938). The cobweb theorem. *Quarterly Journal of Economics, 52*, 255–280.

Guesnerie, R. (2002). Anchoring economic predictions in common knowledge. *Econometrica, 70*, 439–480.

Hommes, C. (2013). *Behavioral rationality and heterogeneous expectations in complex economic systems*. Cambridge: Cambridge University Press.

Hommes, C., & Wagener, F. (2010). Does eductive stability imply evolutionary stability? *Journal of Economic Behavior and Organization, 75*, 25–39.

Naimzada, A., & Pireddu, M. (2015). A financial market model with endogenous fundamental values through imitative behavior. *Chaos, 25*, 073110. https://doi.org/10.1063/1.4926326.

Naimzada, A., & Pireddu, M. (2020a). Eductive stability may not imply evolutionary stability in the presence of information costs. *Economics Letters, 186*, 108513. https://doi.org/10.1016/j.econlet.2019.06.019.

Naimzada, A., & Pireddu, M. (2020b). Rational expectations (may) lead to complex dynamics in a Muthian cobweb model with heterogeneous agents, submitted.

Naimzada, A., & Ricchiuti, G. (2008). Heterogeneous fundamentalists and imitative processes. *Applied Mathematics and Computation, 199*, 171–180.

Naimzada, A., & Ricchiuti, G. (2009). The dynamic effect of increasing heterogeneity in financial markets. *Chaos, Solitons and Fractals, 41*, 1764–1772.

Nöther's Theorem and the Lie Symmetries in the Goodwin-Model

József Móczár

Abstract The dynamic behavior of a physical system can concisely be described by the least action principle. In the centrum of its mathematical presentations is a specific function of the coordinates and velocities, i.e., the Lagrangian. If the integral of a Lagrangian is stationary, then the system is moving along an extremal path through the phase space. All Lie symmetries of a Lagrangian correspond to a conserved quantity, and the conservation principle is explained by variation symmetry. Briefly, that is the meaning of Nöther's theorem. After showing that Goodwin's cyclical growth model has a Lagrangian, we introduce the *generalized Nöther's theorem* and apply it to Goodwin's 2D model in order to get its Hamiltonian. We prove that the cyclical motion in his model derives from its dynamic Lie symmetries. These cyclical trajectories are extremal trajectories in the phase space, and along these trajectories the first integral of the model's Lagrangian is stationary, which by the principle of least action also means that they satisfy the first-order necessary conditions. The optimality still needs to satisfy the sufficient condition. Since the Legendre's second-order sufficient condition is not applicable here, we show it satisfies another systems of sufficient conditions, the local convex surface of Lagrangian with the minimum non-trivial fixed-point and contour lines. Our conclusion is that all systems' solutions described by the other first-order nonlinear ordinary differential equations are optimal if they have a Lagrangian that satisfies the sufficient and necessary conditions.

Keywords Generalized Nöther's theorem · Optimal control theory · Goodwin model · Lie symmetries · Legendre sufficient condition

J. Móczár (✉)
Department of Mathematical Economics and Economic Analysis, Corvinus University of Budapest, Fővám tér 8, Budapest 1093, Hungary
e-mail: jozsef.moczar@uni-corvinus.hu

© Springer Nature Singapore Pte Ltd. 2020
F. Szidarovszky and G. I. Bischi (eds.), *Games and Dynamics in Economics*,
https://doi.org/10.1007/978-981-15-3623-6_8

143

1 Introduction

Here we apply the *generalized Nöther's theorem* and the *Lie symmetries* to Goodwin's economic growth cycle model in conjunction with Lagrangian theory.[1] Exploring and characterizing the behavior of the Lie symmetries are the subject and result of most recent papers in this field. For example, Sen and Tabor (1990) examined the three dimensional model of Lorenz (1963) that is essential in complex chaotic dynamics; Baumann and Freyberger (1991) scrutinized the two-dimensional scaled Lotka-Volterra dynamic system (Lotka 1925; Volterra 1931), while Almeida and Moreira (1992) looked at the three-wave interaction problem using the Lie symmetries.

As is also known, the solution curves of the Goodwin-model (Goodwin 1967) are closed trajectories with a dynamic equilibrium that can also be given directly based on the Lyapunov function (see Hirsch et al. 2004). Although Goodwin used the first integral concept to define the solutions of his economic model, due to the lack of sufficient explanation its connection to physics and the roots in the Lagrangian remained in obscurity all the time to economists. Neither in his paper of 1967 nor in any of his later works, neither Goodwin himself nor anyone else demonstrated that his non-linear dynamical system has a Lagrangian and that the first integral can be produced with the application of a proper Lie symmetry, or, as is more widely known, of the Hamiltonian, which is a conserved quantity of the dynamical system. Nevertheless, this approach through the *generalized Nöther's theorem* shows not *only* relevance but also elegance since we can freely use the language of both classical (non-relativistic) mechanics and mathematical control theory and their concept systems too. In Sect. 2 we concisely introduce Nöther's theorem and in general the Lagrangian. We show the *Euler-Lagrange differential equation* then we introduce the *generalized Nöther's theorem*, which will be used in Sect. 4 for the investigation of Goodwin's non-linear dynamic system. In Sect. 3 we assume that the readers are familiar with *Goodwin's original growth cycle model and economics*; so we emphasize only his model's non-linear dynamic system and give some economic interpretations of the variables and parameters. Section 4 defines the *Lagrangian and Hamiltonian* by using the generalized Nöther's theorem, and Sect. 5 determines the *Lie symmetries* of Goodwin's non-linear dynamic system. Finally in Sect. 6 we show that the Goodwinian cyclical motion is generated by the model's Lie symmetries, more specifically its dynamic symmetries. The first integral, i.e., the constant conserved quantity derived from the model's Lagrangian, results this inner symmetry. We will also show that Nöther's generalized theorem plays an important role in the mechanics of Goodwin's model.

---

[1] This approach is an improved version of that introduced earlier in Móczár (2010), and Móczár and Márkus (2011). The fundamentals of the mathematical principles used in this paper can be found in Hydon (2000), Nutku (1990), Olver (1993), and Velan and Lakshmanan (1995).

2 The Generalized Nöther's Theorem and the Conservation Laws

Nöther (1918) was the first who formulated a universal theorem for the connection between the dynamic symmetries and the conservation laws. In the spirit of her renowned theorem, generally, it is also true that the symmetry of a Lagrangian corresponds to a conserved quantity, and vice versa. We can illustrate this statement with a simple example. Let us consider the classical Lagrangian of a free particle with mass m which simply is $L = (1/2)m\dot{x}^2$, where $\dot{x} = dx/dt$. It can be seen that L is dependent only on the velocity of x, i.e., on $\dot{x}$ and independent of the position of x which means that x is invariant under spatial translation. Thus, $dL/dx = 0$, i.e., L is symmetric by x. From this we can show, by using the appropriate Euler-Lagrange equation that $\partial L/\partial \dot{x} = m\dot{x} = p$ is constant, which means that impulse p is a conserved quantity.

The Lagrangian defines the trajectories of any dynamical system. The corresponding Lie symmetries are such that, under infinitesimal transformations of the dependent and independent variables of the integral-functional, they leave the total structure invariant. In the literature this invariance is sometimes referred to as variation symmetry. In our discussion we pay special attention to the so-called dynamical symmetries which are not connected to the geometric symmetries of coordinate transformation, and it will play a crucial role in understanding the generalized Nöther's theorem. We must note, that a Lagrangian cannot be given to all dynamic systems and thus in those cases (generalized) Nöther's theorem cannot be applied (Wigner 1954). At the same time several different Lagrangians can be written up to a dynamic system, but their integral functionals behave differently under the same infinitesimal transformation (Morandi et al. 1990).

From our studies in physics, we know that Nöther's theorem can be useful in other disciplines too, if the problem can be formulated by using the calculus of variations. The theorem connects the invariant characteristics of integral-functional

$$S = \int_{t_1}^{t_2} L(t, q(t), \dot{q}(t))dt \qquad (1)$$

with the conservation laws, i.e., the integral of the appropriate Euler-Lagrange or Hamiltonian differential equations (see Nagy 1981). Thus the conservation of impulse (motion quantity) and angular momentum in mechanics correspond to a spatial translation and rotational invariance of the integral-functional while the invariance under the translation with respect to time corresponds to the conservation of energy. The translational and rotational invariances with respect to both spatial and time variables are also called geometric invariances. A further and deeper understanding of these invariances can be obtained if we explore (find out) the appropriate Lagrangian and the symmetries of the equations of motion derived from it. Each, independent symmetry provides a further conversation law of the process.

The extremal principles in mechanics are of prime importance in both physics and optimal control theory. They describe the motion between the initial q_1 and the final q_2 states on the time interval $[t_1, t_2]$. The object is the calculation of the extremal path. By mathematics, if the action function is extremal to the optimal (real) path, then the integral-functional in (1) cannot take its extremal under the perturbed path $q(t) + \delta q(t)$ with an infinitesimally small δ, even if it is very close to the extremal. It is said, that we vary the action function, i.e., we consider the first variance of the integral:

$$\delta S = \int_{t_1}^{t_2} [L(t, q(t) + \delta q(t), \dot{q}(t) + \delta\dot{q}(t) - L(t, q(t), \dot{q}(t)))]dt. \qquad (2)$$

All this is used to find the necessary conditions of extremality under which the equation $\delta S = 0$ is satisfied. This is the *least action principle*. Applying the steps of the calculus of variations (see Budó 1965), we get the *Euler-Lagrange differential equation*:

$$\frac{d}{dt}\frac{\partial L}{\partial \dot{q}} = \frac{\partial L}{\partial q}, \qquad (3)$$

whose explicit form is a second-order ordinary differential equation if $\partial L/\partial q \neq 0$. The solution of this equation provides the extremal motion path.

Proceeding on our way towards the presentation of the *generalized Nöther's theorem*, we consider a system with the coordinates, $q_1, q_2, \ldots, q_n$. The Lagrangian is now a function of all variables and their derivatives with respect to time. In this generalized case the derivation of Hamiltonian is as follows.

At first let us consider the total derivative of $L(q_1, q_2, \ldots, q_n, \dot{q}_1, \dot{q}_2, \ldots, \dot{q}_n)$ with respect to time:

$$\frac{dL}{dt} = \sum_{i=1}^{n} \left(\frac{\partial L}{\partial q_i}\frac{dq_i}{dt} + \frac{\partial L}{\partial \dot{q}_i}\frac{d\dot{q}_i}{dt} \right).$$

The Euler-Lagrange equations for the generalized system enable us to substitute for the partials of L with respect to q_i, so we have:

$$\frac{dL}{dt} - \sum_{i=1}^{n} \left(\dot{q}_i \frac{d}{dt}\left(\frac{\partial L}{\partial \dot{q}_i} \right) + \frac{\partial L}{\partial \dot{q}_i}\frac{d\dot{q}_i}{dt} \right) = 0,$$

which can be rewritten as follows:

$$\frac{d}{dt}\left[\sum_{i=1}^{n} \left(\frac{\partial L}{\partial \dot{q}_i}\dot{q}_i \right) - L \right] = 0.$$

Thus the quantity inside the square brackets (i.e., the Hamiltonian or first integral) is constant:

$$H = \sum_{i=1}^{n} \left(\frac{\partial L}{\partial \dot{q}_i} \dot{q}_i \right) - L = const. \tag{4}$$

This is called the *generalized Nöther's theorem* which will be used for $n = 2$ in Sect. 4.

3 Goodwin's Non-linear Dynamic System

Goodwin's growth cycle model of 1967 is a simple dynamic model of the distribution ratios of gross output and employment. The trajectories of employment and the workers' share of the total income are given by the following non-linear dynamic system:

$$\dot{v} = [(1/\sigma) - (\alpha + \beta) - (1/\sigma)u]v \tag{5}$$

$$\dot{u} = \left[-(\alpha + \gamma) + \rho v \right] u \tag{6}$$

where v is the employment rate and u is the workers' share of total income, $\dot{v}$ and $\dot{u}$ are their derivatives with respect to time, σ is the fixed capital-output ratio, α is the growth rate of labor productivity which is constant assuming an unchanged rate of non-embodied technical progress. β is the fixed growth rate of the labor force as well as $-\gamma + \rho v = \dot{w}/w$ is a linearized Phillips curve where w is the wage rate, γ and ρ are constants with appropriately high positive values. We note that (5) and (6) form a 2D first-order nonlinear autonomous homogeneous differential equations system.

If u is absent from Eq. (5), then the employment rises at a constant rate, i.e., $\dot{v}/v = (1/\sigma) - (\alpha + \beta)$ where $(1/\sigma) > (\alpha + \beta)$, which means that the efficiency of capital is greater than the sum of the growth rate of productivity and the growth rate of the labor force. To put it differently, we can say that the employment rate rises if the productivity of capital is higher than the sum of the extensive and intensive factors of economic growth. Likewise, if v is absent from Eq. (6), then the workers' share of the total income declines at a steady rate, $\dot{u}/u = -(\alpha + \gamma)$. This means that the workers' share of total income declines at such a rate that is precisely equal to the sum of the growth rate of labor productivity and the autonomous growth parameter of wages. In both cases the trajectory is an exponential function, in the former case an increasing one, while in the latter a decreasing one. But, if we also consider the other variable u in (5), then the capital efficiency through variable u decreases in the growth rate of employment while in (6) the increasing pace of wage rate to a unit of employment rate will, through the variable v, cause the decrease of the rate by which the workers' share of the total income declines.

Thereafter, Goodwin eliminates the time from Eqs. (5) and (6). To produce the closed trajectories with the dynamic equilibrium of his model, Goodwin applied the same simple method as Andronov et al. (1966, pp. 143–145) did to the Lotka-Volterra system. The time-variance, just like in classical mechanics, gives the first integral:

$$[(1/\sigma) - (\alpha + \beta)] \ln u + (\alpha + \gamma) \ln v - (1/\sigma)u - \rho v = const. \qquad (7)$$

While Goodwin said nothing about Eq. (7), we will now demonstrate that it is the Hamiltonian or else conserved quantity of his dynamic system given by (5) and (6).

4 The Lagrangian and Hamiltonian of Goodwin's Non-linear Dynamic System

At first, to produce the Lagrangian of the Goodwin model, we use the results in Fernandez-Nuñez (1998). We will denote it with L' which can be given as follows:

$$L'(v, \dot{v}, u, \dot{u}) = \frac{1}{2} \frac{\ln u}{v} \dot{v} + \frac{1}{2} \frac{\ln v}{u} \dot{u} -$$
$$(((1/\sigma) - (\alpha + \beta)) \ln u + (\alpha + \gamma) \ln v - (1/\sigma)u - \rho v) \qquad (8)$$

If we take the Euler-Lagrange equations of the Lagrangian L' with respect to u and v, we get back Eqs. (5) and (6) of the Goodwin model. Thus we have shown that the Goodwin model also has the Lagrange structure.

According to *the generalized Nöther's theorem* (4), we can show by a simple calculation that the Hamiltonian of the Goodwin model is the following:

$$H' = \dot{v} \frac{\partial L'}{\partial \dot{v}} + \dot{u} \frac{\partial L'}{\partial \dot{u}} - L'$$
$$= ((1/\sigma) - (\alpha + \beta)) \ln u + (\alpha + \gamma) \ln v - (1/\sigma)u - \rho v, \qquad (9)$$

which is the conserved quantity, i.e., the first integral, which corresponds to Eq. (7). If we consider the corresponding energy E' from the mechanics the first integral can be calculated as follows:

$$I' \left(= e^{E'} \right) = v^{\alpha + \gamma} u^{(1/\sigma) - (\alpha + \beta)} e^{-(\rho v + (1/\sigma)u)} \qquad (10)$$

5 The Lie Symmetries of Goodwin's Non-linear Dynamic System

Let us take the infinitesimal transformations of the state variables of the model:

$$v' = v + \varsigma_1(v, u) \tag{11}$$

$$u' = u + \varsigma_2(v, u) \tag{12}$$

and

$$t' = t \tag{13}$$

where functions $\varsigma_1(v, u)$ and $\varsigma_2(v, u)$ are directly independent of time, i.e., $\partial \varsigma_1/\partial t = 0$ and $\partial \varsigma_2/\partial t = 0$. Substituting Eqs. (11), (12) and (13) into Eqs. (5) and (6), after simplifying the received equations, we get the following *partial differential equations*:

$$\begin{aligned} &(((1/\sigma) - (\alpha + \beta))v - (1/\sigma)uv)\frac{\partial \varsigma_1}{\partial v} + (\rho vu - (\alpha + \gamma)u)\frac{\partial \varsigma_1}{\partial u} \\ &+ ((1/\sigma)u - (1/\sigma) + (\alpha + \beta))\varsigma_1 + (1/\sigma)v\varsigma_2 = 0 \end{aligned} \tag{14}$$

$$\begin{aligned} &(((1/\sigma) - (\alpha + \beta))v - (1/\sigma)uv)\frac{\partial \varsigma_2}{\partial v} \\ &+ (\rho vu - (\alpha + \gamma)u)\frac{\partial \varsigma_2}{\partial u} - \rho u\varsigma_1 + (\alpha + \gamma - \rho v)\varsigma_2 = 0 \end{aligned} \tag{15}$$

We have obtained a connected partial differential equations system which has a possible solution as follows:

$$\varsigma_1 = ((1/\sigma) - (\alpha + \beta))v - (1/\sigma)uv \tag{16}$$

$$\varsigma_2 = \rho vu - (\alpha + \gamma)u \tag{17}$$

By using this solution, we can now determine the relevant generator, i.e., the *vector of Lie symmetries*:

$$X_1' = (((1/\sigma) - (\alpha + \beta))v - (1/\sigma)uv)\frac{\partial}{\partial v} + (\rho vu - (\alpha + \gamma)u)\frac{\partial}{\partial u} \tag{18}$$

Another possible solution to (13) and (14) can be given as:

$$\varsigma_1 = \frac{v^{\alpha + \gamma} u^{(1/\sigma) - (\alpha + \beta)}}{e^{\rho v + (1/\sigma)u}}((1/\sigma) - (\alpha + \beta) - (1/\sigma)u)v \tag{19}$$

$$\varsigma_2 = \frac{v^{\alpha+\gamma}u^{(1/\sigma)-(\alpha+\beta)}}{e^{\rho v+(1/\sigma)u}}(\rho v - (\alpha + \gamma))u \tag{20}$$

The infinitesimal generator can now be given as follows:

$$X_2' = \frac{v^{\alpha+\gamma}u^{(1/\sigma)-(\alpha+\beta)}}{e^{\rho v+(1/\sigma)u}}((1/\sigma)v - (\alpha + \beta) - (1/\sigma)uv)\frac{\partial}{\partial v}$$
$$+ \frac{v^{\alpha+\gamma}u^{(1/\sigma-(\alpha+\beta))}}{e^{\rho v+(1/\sigma)u}}(\rho vu - (\alpha + \gamma)u)\frac{\partial}{\partial u} \tag{21}$$

If we compare the generators defined in Eqs. (19) and (20), then we can easily observe the following relations between them:

$$X_2' = \frac{v^{\alpha+\gamma}u^{(1/\sigma)-(\alpha+\beta)}}{e^{\rho v+(1/\sigma)u}}X_1' \tag{22}$$

i.e., the generators differ in one factor only. Since these generators share the same structure too, the factor

$$I' = \frac{v^{\alpha+\gamma}u^{(1/\sigma)-(\alpha+\beta)}}{e^{\rho v+(1/\sigma)u}} \tag{23}$$

must show the permanent quantity of motion. This is precisely equal to the conserved quantity derived from the Lagrangian in Eq. (10), which refers to the dynamic symmetries of the Goodwin model.

6 Numerical Study

The Goodwin model examines the interdependence between the accumulation of capital and the distribution of income. Its solution curves reply to the question of how the accumulation of capital changes cyclically with periodicity

$$T = 2\pi/((\alpha + \beta)((1/\sigma) - (\alpha + \beta)))^{1/2}$$

around the non-trivial equilibrium fixed-point whose coordinates are

$$u^* = 1 - (\alpha + \beta)\sigma$$

$$v^* = (\alpha + \gamma)/\rho.$$

The parameters in the model can be estimated by using econometric methods from the appropriate time-series. We can also think of these calculations as

Harvie (2000) did in his paper by testing the model with real data, whereby he makes some interesting observations.

The following graphs were made by choosing the following parameter values: $\alpha = 0.058$, $\beta = 0.1$, $\gamma = 15$, $\sigma = 0.08$ and $\rho = 15$. The initial values are $v(0) = 0.095$ and $u(0) = 0.9$. The direction of motion on the closed curve is a clockwise rotation. Figures 1 and 2 show the periodic behavior of $v(t)$ and $u(t)$, while Fig. 3 illustrates these periodicities with the phase diagram in the $(u(t), v(t))$ space.

The cyclical trajectories of the Goodwin model—as we have proved—are the extremal trajectories in the phase space, and along these trajectories the first integral of the model's Lagrangian is stationary, which by the principle of least action also means that they satisfy the first-order necessary condition. For the optimality, satisfying a second-order sufficient condition still needs, namely, the *Legendre condition*. For this purpose, first define

$$|\Delta| = \begin{vmatrix} L'_{\dot{v}\dot{v}} & L'_{\dot{v}\dot{u}} \\ L'_{\dot{u}\dot{v}} & L'_{\dot{u}\dot{u}} \end{vmatrix} = \begin{vmatrix} 0 & 0 \\ 0 & 0 \end{vmatrix}$$

where the second-order derivatives are to be evaluated along the extremals, and L' was defined in (8). Since both principal minors are zero, the Legendre condition does not work in this case.

Fig. 1 Periodic behavior of $v(t)$

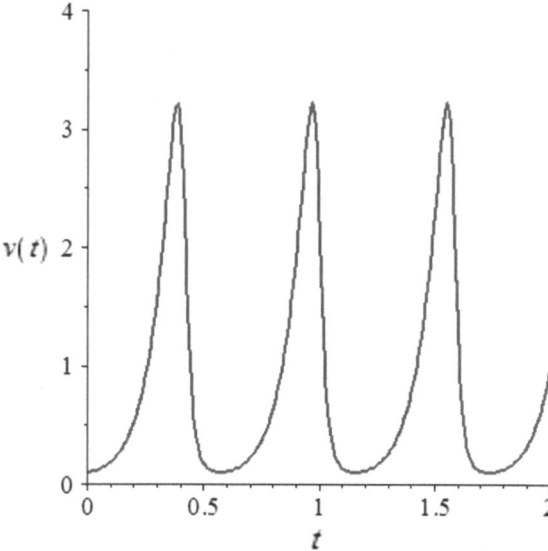

Fig. 2 Periodic behavior of
$u(t)$

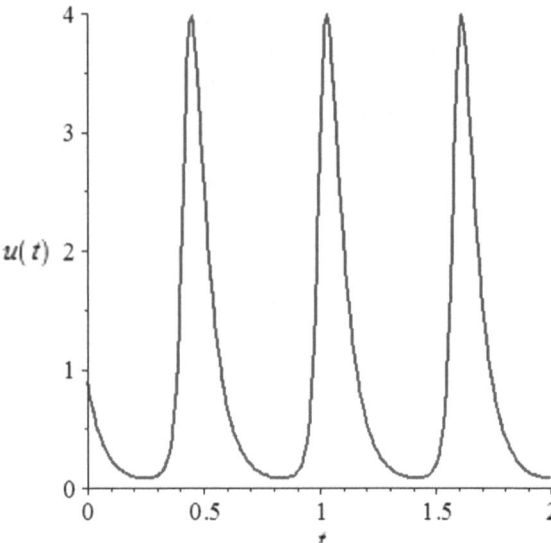

Fig. 3 Phase portrait $u(t) -$
$v(t)$

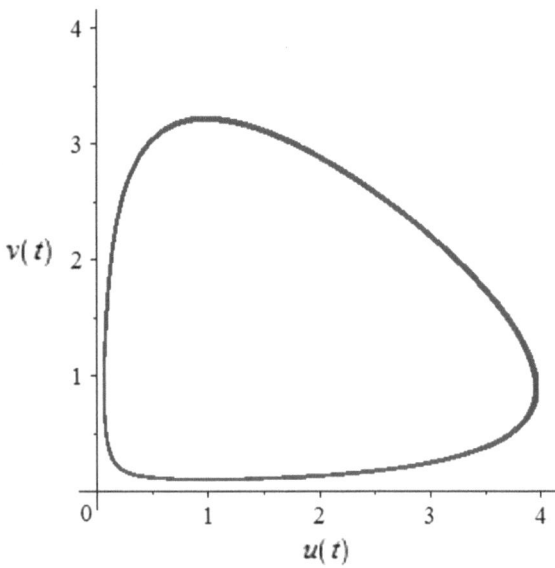

Thus we use another condition, namely, the second-order sufficient condition, the convexity of the Lagrangian. Figure 4 shows the *local convexity* of the Lagrangian with the non-trivial equilibrium fixed-point and contour lines of extremal trajectories.

This means that the change of the employment rate $v(t)$ and the workers' share of total income $u(t)$ are optimal along the extremal trajectories in the local sense, which results in optimal growth cycles.

Fig. 4 Lagrangian of the Godwin-model

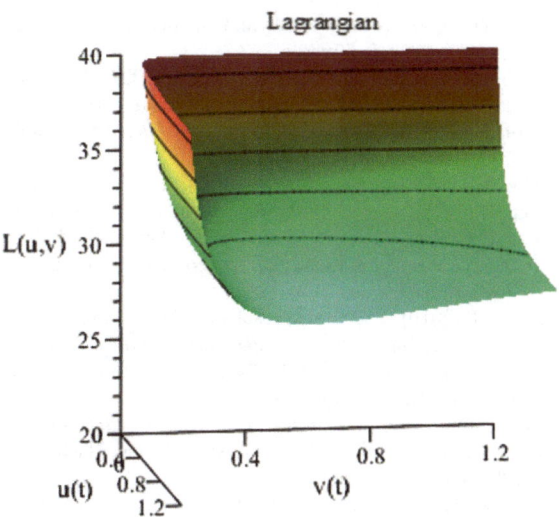

7 Conclusions

The results of this paper can be used to refine the descriptions of the solution curves of the nonlinear first-order ordinary differential equations systems by allowing us to formulate a general rule: the trajectory of the dynamic system is optimal if it has a Lagrangian that satisfies the first-order necessary and second-order sufficient conditions.

By our investigations we have shown that the cyclical motion is generated by the Goodwin model's Lie symmetries, more specifically by its dynamic symmetries. The first integral, i.e., the constant conserved quantity derived from the model's Lagrangian, results this inner symmetry. This also means that Nöther's generalized theorem plays an important role in the mechanics of Goodwin's model.

References

Almeida, M. A., & Moreira, I. C. (1992). Lie symmetries for the reduced three-wave interaction problem. *Journal of Physics A: Mathematical and General, 25*, 669–672.

Andronov, A. A., Vitt, A. A., & Khaikin, S. E. (1966). *Theory of oscillators*. Oxford: Pergamon Press.

Baumann, G., & Freyberger, M. (1991). Generalized symmetries and conserved quantities of the Lotka-Volterra model. *Physics Letters A, 156*(9), 488–490.

Budó, Á. (1965). *Mechanics (in Hungarian)*. Budapest: Tankönyvkiadó.

Fernandez-Nuñez, J. (1998). Lagrangian structure of the two-dimensional Lotka-Volterra system. *International Journal of Theoretical Physics, 37*(9), 2457–2462.

Goodwin, H. R. (1967). A growth cycle. In C. H. Feinstein (Ed.), *Socialism, capitalism and economic growth*. Cambridge: Cambridge University Press.

Harvie, D. (2000). Testing Goodwin: Growth cycles in ten OECD countries. *Cambridge Journal of Economics, 24,* 349–376.

Hirsch, M. W., Smale, S., & Devaney, L. R. (2004). *Differential equations, dynamical systems, & an introduction to chaos.* New York: Academic Press.

Hydon, P. E. (2000). *Symmetry methods for differential equations: A beginner's guide.* Cambridge, UK: Cambridge University Press.

Lotka, A. J. (1925). *Elements of mathematical (physical) biology.* New York: Dover Publication.

Lorenz, E. (1963). Deterministic non-period flows. *Journal of Atmospheric Sciences, 20,* 130–141.

Morandi, G., Ferrario, C., Lo Vecchio, G., Marmo, G., & Rubano, C. (1990). The inverse problem in the calculus of variations and the geometry of the tangent bundle. *Physics Reports, 188*(3–4), 147–284.

Móczár, J. (2010). The newest results of physical mathematics as the possible investigation tools of economics (in Hungarian). *Alkalmazott Matematikai Lapok, 27*(1), 41–77.

Móczár, J., & Márkus, F. (2011). An economic application of the Lie symmetries (in Hungarian). *Alkalmazott Matematikai Lapok, 28*(1), 1–7.

Nagy, K. (1981). *Quantum mechanics (in Hungarian).* Budapest: Tankönyvkiadó.

Nöther, E. (1918). Invariante variationsprobleme, Nachr. König. Gesellsch. Wiss. zu Göttingen, *Math-phys. Klasse*, 235–257.

Nutku, Y. (1990). Hamiltonian structure of the Lotka-Volterra equations. *Physics Letters A, 145*(1), 27–28.

Olver, P. J. (1993). *Applications of lie groups to differential equations* (2nd ed.). New York: Springer-Verlag.

Sen, T., & Tabor, M. (1990). Lie symmetries of the Lorenz model. *Physics D, 44,* 313–339.

Velan, M. S., & Lakshmanan, H. (1995). Lie symmetries and infinite dimensional Lie algebras of certain nonlinear dissipative systems. *Journal of Physics A: Mathematical and General, 28,* 1929–1942.

Volterra, V. (1931). *Lecons sur la theorie mathematique de la lutte pour la vie.* Paris: Gaithier-Villars.

Wigner, E. P. (1954). Conversation laws in classical and quantum physics. *Progress of Theoretical Physics, 11,* 437–440.

Games

Mixed Duopolies with Advance Production

Tamás László Balogh and Attila Tasnádi

Abstract Production to order and production in advance have been compared in many frameworks. In this paper we investigate a production in advance version of the capacity-constrained Bertrand-Edgeworth mixed duopoly game and determine the solution of the respective timing game. We show that a pure-strategy (subgame-perfect) Nash-equilibrium exists for all possible orderings of moves. It is pointed out that unlike the production-to-order case, the equilibrium of the timing game lies at simultaneous moves. An analysis of the public firm's impact on social surplus is also carried out. All the results are compared with those of the production-to order version of the respective game and with those of the mixed duopoly timing games.

Keywords Bertrand-Edgeworth · Mixed duopoly · Timing games

1 Introduction

We can distinguish between production-in-advance (PIA) and production-to-order (PTO) concerning how the firms organize their production in order to satisfy the consumers' demand.[1] In the former case production takes place before sales are realized, while in the latter one sales are determined before production takes place. Markets of perishable goods are usually mentioned as examples of advance production in a market. Phillips et al. (2001) emphasized that there are also goods which can be traded both in a PIA and in a PTO environment since PIA markets can be regarded as a kind of spot market whereas PTO markets as a kind of forward market. For example, coal and electricity are sold in both types of environments.

---

[1]The PIA game is also frequently called the price-quantity game or briefly PQ-game.

T. L. Balogh
Mathematics Connects Association, Laktanya u. 40., Debrecen 4028, Hungary
e-mail: billtm@gmail.com

A. Tasnádi (✉)
Department of Mathematics, Corvinus University of Budapest, Fővám tér 8,
Budapest 1093, Hungary
e-mail: attila.tasnadi@uni-corvinus.hu

© Springer Nature Singapore Pte Ltd. 2020
F. Szidarovszky and G. I. Bischi (eds.), *Games and Dynamics in Economics*,
https://doi.org/10.1007/978-981-15-3623-6_9

The comparison of the PIA and PTO environments has been carried out in experimental and theoretical frameworks for standard oligopolies.[2] For instance, assuming strictly increasing marginal cost functions, Mestelman et al. (1987) found that in an experimental posted offer market the firms' profits are lower in case of PIA. For more recent experimental analyses of the PIA environment we refer to Davis (2013) and Orland and Selten (2016). In a theoretical paper Shubik (1955) investigated the pure-strategy equilibrium of the PIA game and conjectured that the profits will be lower in case of PIA than in case of PTO. Levitan and Shubik (1978) and Gertner (1986) determined the mixed-strategy equilibrium for the constant unit cost case without capacity constraints.[3] Assuming constant unit costs and identical capacity constraints, Tasnádi (2004) found that profits are identical in the two environments and that prices are higher under PIA than under PTO. Zhu et al. (2014) showed for the case of strictly convex cost functions that PIA equilibrium profits are higher than PTO equilibrium profits. In addition, considering different orders of moves and asymmetric cost functions Zhu et al. (2014) demonstrated that the leader-follower PIA game leads to higher profit than the simultaneous-move PIA game.[4]

Concerning our theoretical setting, the closest paper is Tasnádi (2004) since we will investigate the constant unit cost case with capacity constraints. The main difference is that we will replace one profit-maximizing firm with a social surplus maximizing firm, that is we will consider a so-called mixed duopoly. We have already considered the PTO mixed duopoly in Balogh and Tasnádi (2012) for which we found (i) the payoff equivalence of the games with exogenously given order of moves, (ii) an increase in social surplus compared with the standard version of the game, and (iii) that an equilibrium in pure strategies always exists in contrast to the standard version of the game.[5] In this paper we demonstrate for the PIA mixed duopoly the existence of an equilibrium in pure strategies, (weakly) lower social surplus than in case of the PTO mixed duopoly and the emergence of simultaneous moves as a solution of a timing game.

It is also worthwhile to relate our paper briefly to the literature on mixed oligopolies. In a seminal paper Pal (1998) investigates for mixed oligopolies the endogenous emergence of certain orders of moves. Assuming linear demand and constant marginal costs, he shows for a quantity-setting oligopoly with one public firm that, in contrast to our result, the simultaneous-move case does not emerge.

---

[2]We call an oligopoly standard if all firms are profit maximizers, which basically means that they are privately owned.

[3]Gertner (1986) also derived some important properties of the mixed-strategy equilibrium of the PIA game for strictly convex cost functions. For more on the PIA case see also Bos and Vermeulen (2015), van den Berg and Bos (2017), and Montez and Schutz (2018).

[4]From the mentioned papers only Zhu et al. (2014) considered sequential orders of moves. For more on standard duopoly leader-follower games we refer to Boyer and Moreaux (1987), Deneckere and Kovenock (1992) and Tasnádi (2003) in the Bertrand-Edgeworth framework. Furthermore, Din and Sun (2016) extended Zhu et al. (2014) to mixed duopolies.

[5]We refer the reader also to Bakó and Tasnádi (2017) which proves the validity of the Kreps and Scheinkman (1983) result for mixed duopolies by employing the Kreps and Scheinkman tie-breaking rule at the price-setting stage.

Matsumura (2003) relaxes the assumptions of linear demand and identical marginal costs employed by Pal (1998). The case of increasing marginal costs in Pal's (1998) framework has been investigated by Tomaru and Kiyono (2010). In line with our result on the timing of moves Bárcena-Ruiz (2007) obtained the endogenous emergence of simultaneous moves for a heterogeneous goods price-setting mixed duopoly timing game. In case of emission taxes Lee and Xu (2018) find that the sequential-move (simultaneous-move) game emerges in the equilibrium of the mixed duopoly timing game under significant (insignificant) environmental externality. There is also an evolving literature on managerial mixed duopolies, for instance, Nakamura (2019) shows that in this case a sequential order of moves emerges in which the private firm with a price contract moves first, while the public firm with a quantity contract moves second.

The remainder of the paper is organized as follows. In Sect. 2 we present our framework, Sects. 3–5 contain the analysis of the three games with exogenously given order of moves, Sect. 6 solves the timing game, and we conclude in Sect. 7.

2   The Framework

The demand is given by function D on which we impose the following restrictions.

Assumption 1 The demand function D intersects the horizontal axis at quantity a and the vertical axis at price b. D is strictly decreasing, concave and twice continuously differentiable on $(0, b)$; moreover, D is right-continuous at 0, left-continuous at b and $D(p) = 0$ for all $p \geq b$.

Clearly, any price-setting firm will not set its price above b. Let us denote by P the inverse demand function. Thus, $P(q) = D^{-1}(q)$ for $0 < q \leq a$, $P(0) = b$, and $P(q) = 0$ for $q > a$.

On the producers side we have a public firm and a private firm, that is, we consider a so-called mixed duopoly. We label the public firm as 1 and the private firm as 2. Henceforth, we will also label the two firms as i and j, where $i, j \in \{1, 2\}$ and $i \neq j$. Our assumptions imposed on the firms' cost functions are as follows.

Assumption 2 The two firms have identical $c \in (0, b)$ unit costs up to the positive capacity constraints k_1, k_2 respectively.

We shall denote by p^c the market clearing price and by p^M the price set by a monopolist without capacity constraints, i.e. $p^c = P(k_1 + k_2)$ and $p^M = \arg\max_{p \in [0,b]}(p - c)D(p)$. In what follows $p_1, p_2 \in [0, b]$ and $q_1, q_2 \in [0, a]$ stand for the prices and quantities set by the firms.

For any firm i and for any quantity q_j set by its opponent j we shall denote by $p_i^m(q_j)$ the profit maximizing price on firm i's residual demand curve $D_i^r(p, q_j) = \left(D(p) - q_j\right)^+$ with respect to its capacity constraint, i.e. $p_i^m(q_j) = \arg\max_{p \in [0,b]} (p - c)\min\{D_i^r(p, q_j), k_i\}$. Clearly, p_i^m is well defined whenever $c < P(q_j)$ and

Assumptions 1–2 are satisfied. If $c \geq P(q_j)$, then $p_i^m(q_j)$ is not unique, as any price $p_i \in [0, b]$ together with quantity $q_i = 0$ results in $\pi_i = 0$ and π_i cannot be positive. For notational convenience we define $p_i^m(q_j)$ by b in case of $c \geq P(q_j)$.

For a given quantity q_j we shall denote the inverse residual demand curve of firm i by $R_i(\cdot, q_j)$. In addition, we shall denote by $q_i^m(q_j)$ the profit maximizing quantity on firm i's inverse residual demand curve subject to its capacity constraint, i.e. $q_i^m(q_j) = \arg\max_{q \in [0, k_i]}(R_i(q, q_j) - c)q$. It can be checked that $R_i(q_i, q_j) = P(q_i + q_j)$ and $q_i^m(q_j) = D_i^r\left(p_i^m(q_j), q_j\right)$.[6]

Let us denote by $p_i^d(q_j)$ the smallest price for which

$$(p_i^d(q_j) - c) \min\left\{k_i, D\left(p_i^d(q_j)\right)\right\} = (p_i^m(q_j) - c)q_i^m(q_j),$$

whenever this equation has a solution.[7] Provided that the private firm has 'sufficient' capacity, that is $\max\{p^c, c\} < p_2^m(k_1)$, then if it is a profit-maximizer, it is indifferent to whether serving residual demand at price level $p_2^m(q_1)$ or selling $\min\{k_2, D\left(p_2^d(q_1)\right)\}$ at the weakly lower price level $p_2^d(q_1)$. Observe that if $R_i(k_i, q_j) = p_i^m(q_j)$, then $p_i^d(q_j) = p_i^m(q_j)$, which can be the case for some values of q_j in case of $p^M < P(k_i)$. We shall denote by $\tilde{q}_j$ the largest quantity for which $q_i^m(\tilde{q}_j) = k_i$ in case of $p^M \leq P(k_i)$ (i.e. $q_i^m(0) = k_i$), and zero otherwise. From Deneckere and Kovenock (1992, Lemma 1) it follows that $p_i^d(\cdot)$ and $p_i^m(\cdot)$ are strictly decreasing on $[\tilde{q}_j, k_j]$. Moreover, $q_i^m(\cdot)$ is strictly decreasing on $[\tilde{q}_j, k_j]$ and constant on $[0, \tilde{q}_j]$, and therefore $\tilde{q}_j = \inf\{q_j \in [0, a] \mid q_i^m(q_j) < k_i\}$ is always uniquely defined.

We assume efficient rationing on the market, and thus, the firms' demands equal

$$\Delta_i(D, p_1, q_1, p_2, q_2) = \begin{cases} D(p_i) & \text{if } p_i < p_j, \\ T_i(p, q_1, q_2), & \text{if } p = p_i = p_j \\ \left(D(p_i) - q_j\right)^+ & \text{if } p_i > p_j, \end{cases}$$

for all $i \in \{1, 2\}$, where T_i stands for a tie-breaking rule. We will consider two sequential-move games (one with the public firm as the first mover and one with the private firm as the first-mover) and a simultaneous-move game. We employ the same tie-breaking rule as Deneckere and Kovenock (1992).

Assumption 3 If the two firms set the same price, then we assume for the sequential-move games that the demand is allocated first to the second mover[8] and for the simultaneous-move game that the demand is allocated in proportion of the firms' capacities.

---

[6]Note that $D_i^r\left(p_i^m(q_j), q_j\right) \leq k_i$ since $p_i^m(q_j) \geq P(k_i + q_j)$.

[7]The equation defining $p_i^d(q_j)$ has a solution for any $q_j \in [0, k_j]$ if, for instance, $p_i^m(q_j) \geq \max\{p^c, c\}$, which will be the case in our analysis when we refer to $p_i^d(q_j)$.

[8]This ensures for the case when the public firm moves first the existence of a subgame perfect Nash equilibrium in order to avoid the consideration of ε-equilibria implying a more difficult analysis without substantial gain.

Now we specify the firms' objective functions. The public firm aims at maximizing total surplus, that is,

$$\pi_1(p_1, q_1, p_2, q_2) = \int_0^{\min\{(D(p_j)-q_i)^+, q_j\}} R_j(q, q_i)dq + \int_0^{\min\{a, q_i\}} P(q)dq - c(q_1 + q_2)$$

$$= \begin{cases} \int_0^{\min\{D(p_j), q_1+q_2\}} P(q)dq - c(q_1 + q_2) & \text{if } D(p_j) > q_i, \\ \int_0^{\min\{D(p_i), q_i\}} P(q)dq - c(q_1 + q_2) & \text{if } D(p_j) \le q_i, \end{cases} \tag{1}$$

where $0 \le p_i \le p_j \le b$. We illustrate social surplus in Fig. 1.

The private firm is a profit maximizer, and therefore,

$$\pi_2(p_1, q_1, p_2, q_2) = p_2 \min\{q_2, \Delta_2(D, p_1, q_1, p_2, q_2)\} - cq_2. \tag{2}$$

We divide our analysis into three cases.

1. The *strong private firm case*, where we assume that $q_2^m(k_1) < k_2$ and $P(k_1) > c$. This means that the private firm's capacity is large enough to have strategic influence on the outcome and the public firm cannot capture the entire market.
2. The *weak private firm case*, where we assume that $q_2^m(k_1) = k_2$ and $P(k_1) > c$. In this case the private firm's capacity is not large enough to have strategic influence on the outcome, but it has a unique profit-maximizing price on the residual demand curve.
3. The *high unit cost case*, where we assume that $c \ge P(k_1)$. In this case if the public firm produces at its capacity level, then there is no incentive for the private firm to enter the market, because the cost level is too high.

Clearly, the three cases are well defined and disjunct from each other.

Fig. 1 Social surplus

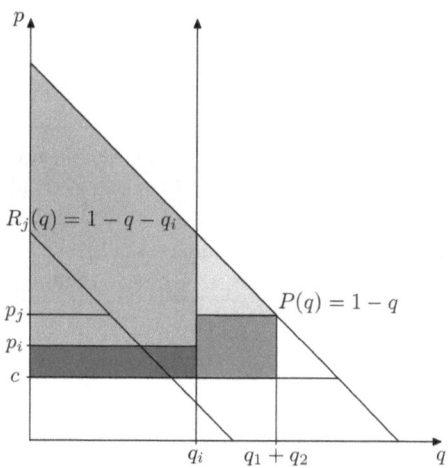

We now determine all the equilibrium strategies of both firms for the three possible orderings of moves in each of the three main cases. Within every case we begin with the simultaneous moves subcase, thereafter we focus on the public-firm-moves-first subcase, finally we analyze the private-firm-moves-first subcase. The results are always illustrated with numerical examples. For better visibility, the most interesting equilibria are depicted.

3   The Strong Private Firm Case

The following two inequalities remain true for the simultaneous moves and public leadership cases.

Lemma 1 *Under Assumptions 1–3, $q_2^m(k_1) < k_2$ and $P(k_1) > c$ we must have in case of simultaneous moves and public leadership that*

$$p_2^* \geq p_2^d(q_1^*) \tag{3}$$

in any equilibrium $(p_1^*, q_1^*, p_2^*, q_2^*)$ in which $q_1^* > 0$.

Proof We obtain the result directly from the definition of $p_2^d(q_1)$. Clearly, $p_1^* \leq p_2^m(q_1^*)$. For any $q_1 \in [0, k_1]$, the private firm is better off by setting $p_2 = p_2^m(q_1)$ and $q_2 = q_2^m(q_1)$ than by setting any price $p_2 < p_2^d(q_1)$ and any quantity $q_2 \in [0, k_2]$. □

Lemma 2 *Under Assumptions 1–3, $q_2^m(k_1) < k_2$ and $P(k_1) > c$ we have in case of simultaneous moves and public leadership that*

$$p_2^* \leq p_2^m(0) = \max\{P(k_2), p^M\} \tag{4}$$

in any equilibrium $(p_1^*, q_1^*, p_2^*, q_2^*)$.

Proof Suppose that $p_2^* > \max\{P(k_2), p^M\}$. If $p_2^* \leq p_1^*$, then the private firm would be better off by setting price $\max\{P(k_2), p^M\}$ and quantity $D\left(\max\{P(k_2), p^M\}\right)$. If $p_2^* > p_1^*$, then the private firm serves residual demand, and therefore it could benefit from switching to action $(p_2^m(q_1^*), q_2^m(q_1^*))$, $\left(\max\{P(k_2), p^M\}, D\left(\max\{P(k_2), p^M\}\right)\right)$, or $\left(p_1^* - \varepsilon, \min\left\{k_2, D\left(p_1^* - \varepsilon\right)\right\}\right)$, where ε is a sufficiently small positive value. For all three cases we have obtained a contradiction. □

3.1 Simultaneous Moves

For the case of simultaneous moves we have a pure-strategy Nash equilibrium family, which contains profiles where the private firm maximizes its profit on the residual demand choosing $p_2^* = p_2^m(q_1^*)$ and $q_2^* = q_2^m(q_1^*)$, while the public firm can choose any price level not greater than $p_2^d(q_1^*)$ and produce any non-negative amount up to its capacity. It is worth emphasizing that in case of $p_2^m(q_1^*) = p_2^d(q_1^*)$ the private firm can sell its entire capacity.

Proposition 1 *Let Assumptions 1–3, $q_2^m(k_1) < k_2$ and $P(k_1) > c$ be satisfied. A strategy profile*

$$\left(p_1^*, q_1^*, p_2^*, q_2^*\right) = \left(p_1^*, q_1^*, p_2^m\left(q_1^*\right), q_2^m\left(q_1^*\right)\right) \tag{5}$$

is for a quantity $q_1^* \in (0, k_1]$ and for any price $p_1^* \in \left[0, p_2^d\left(q_1^*\right)\right]$ or for any $q_1^* = 0$ and any $p_1^* \in [0, b]$ a Nash-equilibrium in pure strategies if and only if

$$\pi_1\left(p_2^d\left(q_1^*\right), q_1^*, p_2^m\left(q_1^*\right), q_2^m\left(q_1^*\right)\right) \geq \pi_1\left(P\left(k_1\right), k_1, p_2^m\left(q_1^*\right), q_2^m\left(q_1^*\right)\right), \tag{6}$$

where there exists a nonempty closed subset H of $[0, k_1]$ satisfying condition (6)[9] Finally, no other equilibrium in pure strategies exists.

Proof Assume that $(p_1^*, q_1^*, p_2^*, q_2^*)$ is an arbitrary equilibrium profile. We divide our analysis into three subcases. In the first case (Case A) we have $p_1^* = p_2^*$, in the second one (Case B) $p_1^* > p_2^*$ holds true, while in the remaining case we have $p_1^* < p_2^*$ (Case C).

Case A: We claim that $p_1^* = p_2^*$ implies $q_1^* + q_2^* = D(p_2^*)$. Suppose that $q_1^* + q_2^* < D(p_2^*)$. Then[10] because of $p_2^* > \max\{p^c, c\}$ by a unilateral increase in output the public firm could increase social surplus or the private firm could increase its profit; a contradiction. Suppose that $q_1^* + q_2^* > D(p_2^*)$. Then the public firm could increase social surplus by decreasing its output or if $q_1^* = 0$, the private firm could increase its profit by producing only $D(p_2^*)$; a contradiction.

We know that we must have $p_1^* = p_2^* \geq p_2^d(q_1^*)$ by Lemma 1. Assume that $q_1^* > 0$. Then we must have $q_2^* = \min\{k_2, D(p_2^*)\}$, since otherwise the private firm could benefit from reducing its price slightly and increasing its output sufficiently (in particular, by setting $p_2 = p_2^* - \varepsilon$ and $q_2 = \min\{k_2, D(p_2)\}$). Observe that $p_2^m(0) = p_2^d(0)$, $p_2^m(q_1) = p_2^d(q_1)$ for all $q_1 \in [0, \tilde{q}_1]$ and $p_2^m(q_1) > p_2^d(q_1)$ for all $q_1 \in (\tilde{q}_1, k_1]$.[11] Moreover, it can be verified by the definitions of $p_2^m(q_1^*)$ and $p_2^d(q_1^*)$ that $q_1^* + k_2 \geq D(p_2^d(q_1^*)) \geq D(p_2^*)$, where the first inequality is strict if $q_1^* > \tilde{q}_1$. Thus, $q_1^* > \tilde{q}_1$ is in contradiction with $q_2^* = \min\{k_2, D(p_2^*)\}$ since we

---

[9]In particular, there exists a subset $[\bar{q}, k_1]$ of H.

[10]Observe that by Lemma 1, the monotonicity of $p_2^d(\cdot)$, $q_2^m(k_1) < k_2$ and $P(k_1) > c$, we have $p_2^* \geq p_2^d(q_1) \geq p_2^d(k_1) > \max\{p^c, c\}$.

[11]We recall that $\tilde{q}_i$ has been defined after $p_i^d(q_j)$.

already know that $q_1^* + q_2^* = D(p_2^*)$ in Case A. Hence, an equilibrium in which both firms set the same price and the public firm's output is positive exists if and only if $p_2^m(q_1^*) = p_2^d(q_1^*)$ (i.e., $q_1^* \in (0, \tilde{q}_1)$) and (6) is satisfied. This type of equilibrium appears in (5) with $q_2^* = q_2^m(q_1^*) = k_2$.

Moreover, it can be verified that $(p_1^*, q_1^*, p_2^*, q_2^*) = (p_2^m(0), 0, p_2^m(0), q_2^m(0))$ is an equilibrium profile in pure strategies if and only if

$$\pi_1(p_2^m(0), 0, p_2^m(0), q_2^m(0)) \geq \pi_1(P(k_1), k_1, p_2^m(0), q_2^m(0)), \qquad (7)$$

where we emphasize that $p_2^m(0) = \max\{P(k_2), p^M\}$ and $q_2^m(0) = D(\max\{P(k_2), p^M\})$.

Case B: Suppose that $p_1^* > p_2^* \geq p_2^d(q_1^*)$ and $D(p_2^*) > q_2^*$. Then the public firm could increase social surplus by setting price $p_1 = p_2^*$ and $q_1 = \min\{k_1, D(p_2^*) - q_2^*\}$; a contradiction.

Assume that $p_1^* > p_2^* \geq p_2^d(q_1^*)$ and $D(p_2^*) = q_2^*$. Then in an equilibrium we must have $q_1^* = 0$, $p_2^* = p_2^m(0)$ and $q_2^* = q_2^m(0)$. Furthermore, it can be checked that these profiles specify equilibrium profiles if and only if Eq. (6) is satisfied.

Clearly, $p_1^* > p_2^* \geq p_2^d(q_1^*)$ and $D(p_2^*) < q_2^*$ cannot be the case in an equilibrium since the private firm could increase its profit by producing $q_2 = D(p_2^*)$ at price p_2^*. Finally, by Lemma 1 $p_2^* < p_2^d(q_1^*)$ cannot be the case either.

Case C: In this case $p_2^* = p_2^m(q_1^*)$ and $q_2^* = q_2^m(q_1^*)$ must hold, since otherwise the private firm's payoff would be strictly lower. In particular, if the private firm sets a price not greater than p_1^*, we are not anymore in Case C; if $q_2^* > \min\{D_2^r(p_2^*, q_1^*), k_2\}$, then the private firm either produces a superfluous amount or is capacity constrained; if $q_2^* < \min\{D_2^r(p_2^*, q_1^*), k_2\}$, then the private firm could still sell more than q_2^*; and if $q_2^* = \min\{D_2^r(p_2^*, q_1^*), k_2\}$, then the private firm will choose a price-quantity pair maximizing profits with respect to its residual demand curve $D_2^r(\cdot, q_1^*)$ subject to its capacity constraint. In addition, in order to prevent the private firm from undercutting the public firm's price we must have $p_1^* \leq p_2^d(q_1^*)$.

Clearly, for the given values p_1^*, p_2^* and q_2^* from our equilibrium profile the public firm has to choose a quantity $q_1' \in [0, k_1]$, which maximizes function $f(q_1) = \pi_1(p_1^*, q_1, p_2^*, q_2^*)$ on $[0, k_1]$. We show that $q_1' = q_1^*$ must be the case. Obviously, it does not make sense for the public firm to produce less than q_1^* since this would result in unsatisfied consumers. Observe that for all $q_1 \in [q_1^*, \min\{D(p_2^*), k_1\}]$

$$f(q_1) = \int_0^{D(p_2^*)-q_1} (R_2(q, q_1) - c)\, dq + \int_0^{q_1} (P(q) - c)\, dq - c(q_1 - q_1^*) =$$

$$= \int_0^{D(p_2^*)} P(q)\, dq - D(p_2^*)c - c(q_1 - q_1^*). \qquad (8)$$

Since only $-c(q_1 - q_1^*)$ is a function of q_1 we see that f is strictly decreasing on $[q_1^*, \min\{D(p_2^*), k_1\}]$.

Subase (i): In case of $k_1 \leq D(p_2^*)$ we have already established that q_1^* maximizes f on $[0, k_1]$. Moreover, (p_1^*, q_1^*) maximizes $\pi_1(p_1, q_1, p_2^*, q_2^*)$ on $[0, p_2^*] \times [0, k_1]$

since Eq. (8) is not a function of p_1^*. Hence, for any $p_1 \leq p_2^d\left(q_1^*\right)$ such that $p_1 < p_2^*$ we have that $\left(p_1, q_1^*, p_2^m\left(q_1^*\right), q_2^m\left(q_1^*\right)\right)$ specifies a Nash equilibrium for any $q_1 \in [0, k_1]$ satisfying $k_1 \leq D\left(p_2^m\left(q_1^*\right)\right)$. However, note that in case of $q_1^* \in [0, \tilde{q}_1]$ and $p_1 = p_2^d\left(q_1^*\right)$ we are leaving Case C and obtain a Case A Nash equilibrium.

Observe that $p_2^m\left(k_1\right) > \max\{p^c, c\}$ implies that $k_1 < D\left(p_2^m\left(k_1\right)\right)$, and therefore we always have Subcase (i) equilibrium profiles. Since $D\left(p_2^m\left(\cdot\right)\right)$ is a continuous and strictly increasing function, interval $\left[\tilde{q}_1, k_1\right] \cap (0, k_1]$ determines the set of quantities yielding an equilibrium for Subcase (i).

Subase (ii): Turning to the more complicated case of $k_1 > D\left(p_2^*\right)$, we also have to investigate function f above the interval $\left[D\left(p_2^*\right), k_1\right]$ in which region the private firm does not sell anything at all at price p_2^* and

$$f(q_1) = \int_0^{\min\{q_1, D(p_1^*)\}} (P(q) - c)\,dq - cq_2^* - c\left(q_1 - D\left(p_1^*\right)\right)^+. \quad (9)$$

Observe that we must have $P(k_1) < p_2^*$. If the public firm is already producing quantity $q_1 = D\left(p_2^*\right)$, the private firm does not sell anything at all and contributes to a social loss of cq_2^*. Therefore, $f(q)$ is increasing on $\left[D\left(p_2^*\right), \min\left\{D\left(p_1^*\right), k_1\right\}\right]$.

Assume that $k_1 \leq D\left(p_1^*\right)$. Then for any $p_1 \leq p_2^d\left(q_1^*\right)$ we get that $\left(p_1, q_1^*,\right.$ $\left. p_2^m\left(q_1^*\right), q_2^m\left(q_1^*\right)\right)$ is a Nash equilibrium if and only if

$$\pi_1\left(p_2^d\left(q_1^*\right), q_1^*, p_2^m\left(q_1^*\right), q_2^m\left(q_1^*\right)\right) \geq \pi_1\left(p_2^d\left(q_1^*\right), k_1, p_2^m\left(q_1^*\right), q_2^m\left(q_1^*\right)\right) =$$
$$= \pi_1\left(P\left(k_1\right), k_1, p_2^m\left(q_1^*\right), q_2^m\left(q_1^*\right)\right), \quad (10)$$

where the last equality follows from the inequalities $p_1^* \leq P(k_1) \leq p_2^*$ valid for this case and the fact that social surplus is maximized in function of (p_1, q_1) subject to the constraint that the private firm does not sell anything at all if the public firm sets an arbitrary price not greater than $P(k_1)$ and produces k_1.

Assume that $k_1 > D\left(p_1^*\right)$. Therefore, $f(q)$ would be decreasing on $\left[D\left(p_1^*\right), k_1\right]$. However, it can be checked that the public firm could increase social surplus by switching to strategy $(P(k_1), k_1)$ from strategy $\left(p_1^*, D\left(p_1^*\right)\right)$. In addition, any strategy (p_1, k_1) with $p_1 \leq P(k_1)$ maximizes social surplus subject to the constraint that the private firm does not sell anything at all. Therefore, $\left(p_2^d\left(q_1^*\right), q_1^*, p_2^m\left(q_1^*\right), q_2^m\left(q_1^*\right)\right)$ is a Nash equilibrium if and only if condition (6) is satisfied. Comparing Eq. (10) with Eq. (6), we can observe that we have derived the same necessary and sufficient condition for a strategy profile being a Nash equilibrium, which is valid for Subcase (ii).

So far we have established that there exists a function g, which uniquely determines the highest equilibrium price as a function of quantity q produced by the public firm. Clearly, $g(q) = p_2^d(q)$, where the domain of g is not entirely specified. At least we know from Subcase (i) that the domain of g contains $\left[\tilde{q}_1, k_1\right]$. Observe also that the equilibrium profiles of Subcase (i) satisfy condition (6). Let $u\left(q_1\right) = \pi_1\left(p_2^d\left(q_1\right), q_1, p_2^m\left(q_1\right), q_2^m\left(q_1\right)\right)$ and $v\left(q_1\right) = \pi_1\left(P\left(k_1\right), k_1, p_2^m\left(q_1\right), q_2^m\left(q_1\right)\right)$. Hence, q_1 determines a Nash equilibrium profile if and only if $u(q_1) \geq v(q_1)$. It can be verified that u and v are continuous, and therefore, set $H = \{q \in [0, k_1] \mid u(q) \geq v(q)\}$ is a closed set containing $[\tilde{q}_1, k_1]$.   □

For the illustration of the Nash equilibrium profile mentioned in the statement we consider the following example.

Example 1 The firms moves simultaneously. Let $D(p) = 1 - p$, $k_1 = 0.5$, $k_2 = 0.4$, and $c = 0.1$.

The following values can be calculated directly from the exogenously given data for Example 1: $p^c = 0.1$, $p_2^m(k_1) = 0.3$, $q_2^m(k_1) = 0.2$, $p_2^d(k_1) = 0.2$, $\tilde{q}_1 = 0.1$. It can be verified that (6) is satisfied with equality at $q' = 1 - (0.1 + \sqrt{0.24}) - k_2 \approx 0.010102$ resulting in an equilibrium price $p' = 0.1 + \sqrt{0.24} \approx 0.589898$ and equilibrium profit $\sqrt{0.24}k_2 \approx 0.195959$ for the private firm. In particular, (5) takes the form

$$(p_1^*, q_1^*, p_2^*, q_2^*) = \left(p_1^*, q_1^*, \max\left\{ \frac{1 - q_1^* + c}{2}, P(q_1^* + k_2) \right\}, \min\left\{ \frac{1 - q_1 - c}{2}, k_2 \right\} \right)$$

in equilibrium, where $q_1^* \in [q', 0.5]$ and $p_1^* \in [0, p_2^d(q_1^*)]$.

In particular, if $q_1^* = k_1 = 0.5$ and $p_1^* = p_2^d(k_1) = 0.2$, then $p_2^* = 0.3$ and $q_2^* = 0.2$ (see Fig. 2). Calculating the social surplus (the sum of dark gray and light gray areas in Fig. 2) and the private firm's profit (the light gray area indicated by π_2), we obtain $\pi_1 = 0.435$ and $\pi_2 = 0.04$. It is easy to check that for this profile the necessary condition (6) is satisfied.

Clearly, p_1^* and q_1^* can vary within the given ranges. Decreasing p_1^* results in lower producer surplus for the public firm, but in an equally large increase in consumer surplus. Thus, payoffs remain the same. Altering q_1^* shifts the residual demand curve, and results in varying payoffs. The possible payoff intervals can also be calculated for Example 1: $\pi_1 \in [0.285, 0.435]$ and $\pi_2 \in [0.04, 0.196]$.

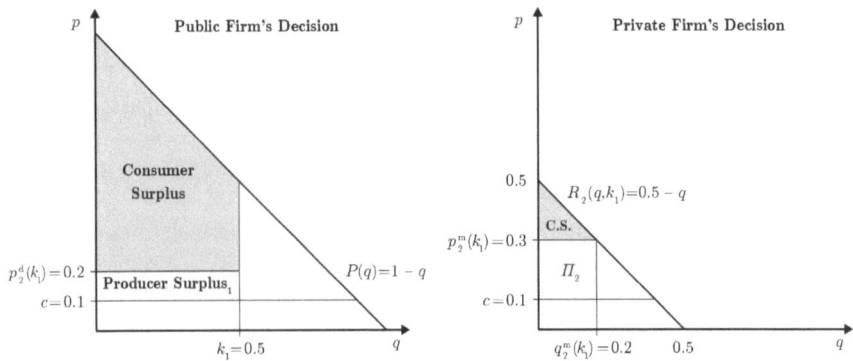

Fig. 2 The strong private firm case

3.2 Public Firm Moves First

We continue with the case of public leadership. Here, we have a unique family of pure-strategy subgame-perfect Nash equilibria, where the public firm produces its capacity limit at a price not greater than $p_2^d(k_1)$. The private firm serves residual demand and acts as a monopolist on the residual demand curve, as presented in the following proposition.

Proposition 2 *Let Assumptions 1–3, $q_2^m(k_1) < k_2$ and $P(k_1) > c$ be satisfied. Then the set of SPNE prices and quantities are given by*

$$\left(p_1^*, q_1^*, p_2^*, q_2^*\right) = \left(p_1, k_1, p_2^m(k_1), q_2^m(k_1)\right) \tag{11}$$

for any $p_1 \leq p_2^d(k_1)$.

Proof First, we determine the best reply $BR_2 = (p_2^*(\cdot, \cdot), q_2^*(\cdot, \cdot))$ of the private firm. Observe that the private firm's best response correspondence can be obtained from the proof of Proposition 1. $BR_2(p_1, q_1) =$

$$\begin{cases} \left\{\left(p_2^m(q_1), q_2^m(q_1)\right)\right\} & \text{if } p_1 < p_2^d(q_1); \\ \left\{\left(p_2^m(q_1), q_2^m(q_1)\right), (p_1, \min\{k_2, D(p_1)\})\right\} & \text{if } p_1 = p_2^d(q_1); \\ \left\{(p_1, \min\{k_2, D(p_1)\})\right\} & \text{if } p_2^d(q_1) < p_1 \leq p_2^m(0); \\ \left\{\left(p_2^m(0), q_2^m(0)\right)\right\} & \text{if } p_2^m(0) < p_1. \end{cases}$$

Though there are two possible best replies for the private firm to the public firm's first-period action $\left(p_2^d(q_1), q_1\right)$, in an SPNE the private firm must respond with $\left(p_2^m(q_1), q_2^m(q_1)\right)$ because otherwise, there will not be an optimal first-period action for the public firm. Hence, the public firm maximizes social surplus in the first period by choosing price $p_1^* = p_2^d(k_1)$ and quantity k_1. Then the private firm follows with price $p_2^* = p_2^m(k_1)$ and quantity $q_2^* = q_2^m(k_1)$. $\qquad\square$

Continuing with the example of linear demand $D(p) = 1 - p$, we focus on the simultaneous-move outcome, which matches the SPNE emerging in case of public leadership.

Example 2 The public firm moves first. Let $D(p) = 1 - p$, the capacities and the unit cost be $k_1 = 0.5$, $k_2 = 0.4$ and $c = 0.1$, respectively.

Then the actions associated with the only subgame-perfect Nash equilibrium profile are

$$\left(p_1^*, q_1^*, p_2^*, q_2^*\right) = \left(p_1^*, 0.5, 0.3, 0.2\right).$$

where $p_1^* \in [0, 0.3]$. The social surplus and the private firm's profit are equal to $\pi_1 = 0.435$ and $\pi_2 = 0.04$.

3.3 Private Firm Moves First

Now we consider the case of private leadership. In this case, there exists one type of subgame-perfect Nash equilibria in which the private firm produces on the original demand curve at the highest price level not above its monopoly price for which it is still of the public firm's interest to remain on the residual demand curve and produce less than it would produce on the original demand curve. Formally, the private firm sets price $\tilde{p}_2 =$

$$\max \left\{ p_2 \in [c, p_2^m(0)] \mid \pi_1(p_2, \min \left\{ D_1^r(p_2, \min\{D(p_2), k_2\}), k_1 \right\}, p_2, \min\{D(p_2), k_2\}) \right.$$
$$\left. \geq \pi_1(P(k_1), k_1, p_2, \min\{D(p_2), k_2\}) \right\}$$

in the first stage. The equilibrium profiles with their necessary conditions are given formally in the following proposition and the existence of the price $\tilde{p}_2$ is shown in its proof.

Proposition 3 *Let Assumptions 1–3, $q_2^m(k_1) < k_2$ and $P(k_1) > c$ be satisfied. The equilibrium actions of the firms associated with an SPNE are the following ones*

$$\left(p_1^*, q_1^*, p_2^*, q_2^* \right) = (p_1, \min \left\{ D_1^r(\tilde{p}_2, \min\{D(\tilde{p}_2), k_2\}), k_1 \right\}, \tilde{p}_2, \min\{D(\tilde{p}_2), k_2\}) \tag{12}$$

where $p_1 \in [0, \tilde{p}_2]$ can be an arbitrary price; furthermore, $p_1 \in (\tilde{p}_2, b]$ are also equilibrium prices in case of $q_1^* = 0$.

Proof We determine the SPNE of the private leadership game by backwards induction without explicitly referring to the proof of Proposition 1. For any given first-stage action (p_2, q_2) of the private firm the public firm never produces less than $\min\{D_1^r(p_2, q_2), k_1\}$ in the second stage. Moreover, if the public firm does not capture the entire market (i.e. the private firm's sales are positive), it never produces more than $\min\{D_1^r(p_2, q_2), k_1\}$. If

$$\pi_1(p_2, \min\{D_1^r(p_2, q_2), k_1\}, p_2, q_2) \geq \pi_1(P(k_1), k_1, p_2, q_2) \tag{13}$$

is satisfied at a price $p_2 \in [c, b]$ and a quantity $q_2 \in (0, k_2]$, then the private firm, by choosing its first-stage action (p_2, q_2), becomes a monopolist on the market (in case of $q_2 \geq D(p_2)$) or sells its entire production (in case of $q_2 < D(p_2)$) since the public firm cannot increase social surplus by setting a lower price than p_2 and it definitely does not set a price above p_2. To be more precise if (13) is satisfied with equality the public firm could also respond with price $P(k_1)$ and quantity k_1; however, as it can be verified later in an SPNE the public firm does not choose the latter response. Clearly, if (13) is violated, the public firm responds with price $P(k_1)$ and quantity k_1. Therefore, we get $BR_1(p_2, q_2) =$

$$
\begin{cases}
\left\{\left(p_1, \min\left\{D_1^r(p_2, q_2), k_1\right\}\right) \mid p_1 \le p_2\right)\right\} & \text{if } \pi_1\left(p_1, \min\left\{D_1^r(p_2, q_2), p_2, q_2\right), k_1\right\} \\
& > \pi_1(P(k_1), k_1, p_2, q_2); \\[2ex]
\left\{(p_1, k_1) \mid p_1 \le P(k_1)\right\} & \text{if } \pi_1\left(p_1, \min\left\{D_1^r(p_2, q_2), p_2, q_2\right), k_1\right\} \\
& < \pi_1(P(k_1), k_1, p_2, q_2); \\[2ex]
\left\{\left(p_1, \min\left\{D_1^r(p_2, q_2), k_1\right\}\right) \mid p_1 \le p_2\right)\right\} \cup & \\
\left\{(p_1, k_1) \mid p_1 \le P(k_1)\right\} & \text{if } \pi_1\left(p_1, \min\left\{D_1^r(p_2, q_2), p_2, q_2\right), k_1\right\} \\
& = \pi_1(P(k_1), k_1, p_2, q_2);
\end{cases}
$$

Clearly, the private firm does not set a price below c jointly with a positive quantity. Furthermore, the private firm can make positive profits because of $q_2^m(k_1) < k_2$ and $P(k_1) > c$, and therefore it sets a price above c. For any given $p_2 > c$ the private firm will never produce less than $\min\{D_2^r(p_2, k_1), k_2\}$ and the left hand side of (13) is constant in q_2 on $\left[\min\{D_2^r(p_2, k_1), k_2\}, \min\{D(p_2), k_2\}\right]$, while the profits of the private firm are strictly increasing in q_2 on the latter interval. Therefore, the private firm produces $q_2 = \min\{D(p_2), k_2\}$ if it produces at all. Henceforth, we substitute $q_2 = \min\{D(p_2), k_2\}$ in Eq. (13). Then the private firm would like to set price $p_2^m(0)$ if (13) is satisfied at this price level, otherwise it sets the highest price still satisfying (13). Note that (13) is definitely satisfied at price $p_2^m(0)$ if $P(k_1) \ge p_2^m(0)$, and otherwise the LHS of (13) is larger than its RHS at price $P(k_1)$, the LHS is strictly decreasing and continuous, while the RHS is strictly increasing and continuous on $[P(k_1), p_2^m(0)]$, and therefore if (13) is not satisfied at $p_2^m(0)$, there exists a unique price $\tilde{p} \in [P(k_1), p_2^m(0))$ such that (13) is satisfied with equality at price $\tilde{p}$. In the former case the private firm sets price $p_2^m(0)$, while in the latter case price $\tilde{p}$ in the SPNE. □

To illustrate Proposition 3 take again the linear demand curve $D(p) = 1 - p$.

Example 3 The private firm moves first and let $D(p) = 1 - p$, $k_1 = 0.5$, $k_2 = 0.4$ and $c = 0.1$.

The following values can be calculated directly from the exogenously given data and the result of the best possible outcome for the private firm from the set of equilibria determined in Example 1. Thus, the actions associated with the SPNE in this case are for all $p_1 \in [0, 0.589898]$:

$$
\left(p_1^*, q_1^*, p_2^*, q_2^*\right) = (p_1, 0.010102, 0.589898, 0.4)
$$

The respective payoffs are as follows: $\pi_1 = 0.285$ and $\pi_2 = 0.196$.

4 The Weak Private Firm Case

The main assumption throughout this section is that the private firm does not have sufficient capacity to influence the market strategically, that is why we call the private firm weak. Formally, $q_2^m(k_1) = k_2$, and in addition $P(k_1) > c$, which in turn implies

$p^c > c$. We begin the analysis with the following lemma which dictates that the private firm is not intended to set any price below the market clearing price.

Lemma 3 *Assume that Assumptions 1–3, $q_2^m(k_1) = k_2$ and $P(k_1) > c$ hold true. Given any strategy (p_1, q_1) of the public firm, the private firm's strategies (p_2, q_2) with price level $p_2 < \max\{p^c, c\}$ and any quantity $q_2 > 0$ are strictly dominated, for instance by a strategy with $p_2 = \max\{p^c, c\}$ and $q_2 > 0$, in all three possible orderings.*

Proof If $p_2 < \max\{p^c, c\}$, then the private firm can sell its entire capacity or makes losses, independently from the public firm's strategy. Clearly, given any (p_1, q_1) and $q_2 > 0$, replacing the private firm's price level by $p_2 = \max\{p^c, c\}$, π_2 increases, thus, the private firm's strategies with lower price levels become strictly dominated. ☐

4.1 Simultaneous Moves

Here, we have two main types of subgame-perfect Nash equilibria. In the first type the private firm sets the highest price level at which it can still produce on the original demand curve. As a particular case of this equilibrium, clearing the market may emerge. The second type contains profiles for which the private firm is a monopolist on the original demand curve.

Proposition 4 *Assume that Assumptions 1–3, $q_2^m(k_1) = k_2$ and $P(k_1) > c$ hold. A strategy profile*

$$\left(p_1^*, q_1^*, p_2^*, q_2^*\right) = \left(p_1^*, \min\left\{D_1^r(\hat{p}, \min\left\{k_2, D(\hat{p})\right\}), k_1\right\}, \hat{p}, \min\left\{k_2, D(\hat{p})\right\}\right) \tag{14}$$

where $p_1^* \in [0, \hat{p}]$ in case of $q_1^* > 0$ and $p_1^* \in [0, b]$ in case of $q_1^* = 0$, defines a Nash equilibrium family in pure strategies if and only if all of the following conditions hold:

$$p_2^m(0) \geq \hat{p} \geq p_2^m(q_1^*) \tag{15}$$

and

$$\pi_1(p^c, k_1, \hat{p}, q_2^*) \leq \pi_1(p_1^*, q_1^*, \hat{p}, q_2^*). \tag{16}$$

In particular, if $\hat{p} = p^c$, then $\left(p_1^*, q_1^*, p_2^*, q_2^*\right) = \left(p_1^*, k_1, p^c, k_2\right)$ is a Nash equilibrium.

Proof Assume that $(p_1^*, q_1^*, p_2^*, q_2^*)$ is an arbitrary equilibrium profile. It can be verified that $q_1^* + q_2^* = D(p')$, where p' stands for the highest price from p_1^*, p_2^* at which at least one firm sells a positive amount. Like in the analysis of the strong private firm case, we divide our analysis into three subcases. In the first case (Case A) we have $p_1^* = p_2^*$, in the second one (Case B) $p_1^* > p_2^*$ holds, while in the remaining case we have $p_1^* < p_2^*$ (Case C).

Case A: By Lemma 3 we have $p_1^* = p_2^* \geq p^c$. First, we verify that the strategy profile given by (14) is a Nash-equilibrium profile for any $\hat{p} \geq p^c$ if (15) and (16) are satisfied. Hence, firms set quantities $q_2^* = \min\{k_2, D(\hat{p})\}$ and $q_1^* = D_1^r(\hat{p}, \min\{k_2, D(\hat{p})\})$. By the second inequality in (15), the private firm has no incentive to increase its price. If $D(\hat{p}) \geq k_2$, then decreasing p_2 is trivially irrational for the private firm that already sells its entire capacity. In case $k_2 > D(\hat{p})$, we obtain a particular equilibrium $\left(p_1^*, q_1^*, p_2^*, q_2^*\right) = (p^*, 0, \hat{p}, D(\hat{p}))$, which means that the public firm is not present on the market, and therefore, by the first inequality in (15) the private firm has no incentive to decrease its price.

Now we consider the public firm's actions. Clearly, increasing the public firm's price would not increase, but in fact reduce total surplus if $q_1^* > 0$. Moreover, prices $p_1^* = p_2^* = p^c$ with quantities $q_1^* = D_1^r(\hat{p}, \min\{k_2, D(p^c)\}) = k_1$ and $q_2^* = \min\{k_2, D(p^c)\} = k_2$ would result in the best possible outcome for the public firm. Hence, we still have to investigate the effect of a potential price decrease by the public firm in case of $p_1^* = p_2^* > p^c$. If the public firm reduces its price without increasing its quantity, obviously total surplus cannot increase. To analyze the case in which the public firm decreases its price and increases its quantity at the same time, observe that the sum of consumer surplus and the two firms' revenues (which equals $\pi_1(p_1, q_1, p_2, q_2) + c(q_1 + q_2)$) is only a function of the highest price at which sales are still positive. Therefore, total surplus is strictly decreasing in q_1 on $\left(q_1^*, D(\hat{p})\right)$ and strictly increasing in q_1 on $\left[D(\hat{p}), k_1\right]$ for a given $p_1 < p_1^*$. To see the latter statement notice that within $\left[D(\hat{p}), k_1\right]$ the superfluous production of the private firm remains the same, that is its entire production. Hence, we have shown that the benchmark action of the public firm in order to determine whether the public firm has an incentive to reduce its price is (p^c, k_1), which is in line with (16).

Turning to the case where (15) is violated, we show that (14) cannot be a Nash-equilibrium profile. If $\hat{p} < p_2^m(q_1^*)$ the private firm will increase its price until $p_2^m(q_1^*)$ to become a monopolist on the residual demand curve, where we are not anymore in Case A of our analysis. Note that any $p_1^* \in [0, \hat{p}]$ results in the same outcome, but if $p_1^* \neq p_2^*$, we are again either in Case B or in Case C. If $p_2^m(0) < \hat{p}$, the private firm will switch to price $p_2^m(0)$.

As a special case of $\hat{p} = p^c$, clearing the market is always a Nash equilibrium for the following reason: by $p^c \geq p_2^m(k_1)$ the private firm cannot be better off by unilaterally increasing its price even by reducing its quantity, accordingly. Note that the market-clearing equilibrium ensures that an equilibrium in pure strategies always exists in the weak private firm case.

Now we show that no other equilibrium exists given that $p_1^* = p_2^* \geq p^c$. Assume that $q_2^* < \min\{k_2, D(p_1^*)\}$. In such cases the private firm gets better off by slightly undercutting p_1^* and selling $q_2^* = \min\{k_2, D(p_1^* - \varepsilon)\}$. Now assume that $q_1^* \neq D_1^r(p_1^*, \min\{k_2, D(p_1^*)\})$. If the left hand side is larger, then there is superfluous production that results in surplus loss; if the left hand side is smaller, then there is a loss in consumer surplus. Thus, there are no more equilibria, if $p_1^* = p_2^*$.

Case B: By Lemma 3 $p_1^* > p_2^* \geq p^c$. By decreasing p_1 to p_2^*, the public firm can always increase social surplus, unless $q_1^* = 0$. In the extreme case of $q_1^* = 0$, p_1^* can obviously be any nonnegative amount. Besides, if $k_2 \geq D(p_2^*)$ and (16) holds, we arrive to the Nash equilibria in which the private firm sets price $p_2^m(0)$. If $k_2 < D(p_2^*)$, then the public firm can increase social surplus by setting price $p_1 = p_2^*$ and quantity $q_1^* = D(p_2^*) - k_2$.

Case C: Now we have $p_2^* > p_1^*$. As already shown in Case A, this case emerges in equilibrium if $\left(p_1^*, q_1^*, p_2^*, q_2^*\right) = \left(p^*, D_1^r(\hat{p}, \min\{k_2, D(\hat{p})\}), \hat{p}, \min\{k_2, D(\hat{p})\}\right)$, and $p_1^* < \hat{p}$, that is, we have the Nash equilibrium mentioned in the statement. It remains to show that there is no other possible equilibrium in this case. If $p_2^* > p_1^*$, then $p_2^* = p_2^m(q_1^*)$ and $q_2^* = \min\{D_2^r(p_2^*, q_1^*), k_1\} = q_2^m(q_1^*)$ must hold, since otherwise the private firm's payoff would be strictly lower. The arguments for this are analogous to those mentioned in the strong private firm case.[12] As $q_2^m(k_1) = k_2$, due to the fact that $q_2^m(\cdot)$ is decreasing[13] in q_1, for any $q_1 < k_1$, $q_2^m(q_1) > q_2^m(k_1) = k_2$. Thus, q_2^* must equal k_2. It is easy to see that for this case the only possible type of equilibrium is characterized in the statement. $\qquad\square$

The weak private firm case emerges in the following example.

Example 4 The firms moves simultaneously. Let $D(p) = 1 - p$, $k_1 = 0.9$, $k_2 = 0.02$, $c = 0.01$.

For the exogenously given values in Example 4 we get $p^c = 0.08$ and $\hat{p}_{max} = 0.102$ as the highest possible price satisfying (15) and (16). In this case we have several Nash equilibrium profiles, which are not payoff equivalent. For all $\hat{p} \in [0.08, 0.102]$ and any $p_1 \in [0, \hat{p}]$,

$$\left(p_1^*, q_1^*, p_2^*, q_2^*\right) = (p_1, 0.98 - \hat{p}, \hat{p}, 0.02)$$

defines the family of Nash equilibrium profiles. In particular, if $\hat{p} = p^c$, then

$$\left(p_1^*, q_1^*, p_2^*, q_2^*\right) = (p_1, 0.9, 0.08, 0.02)$$

and the social surplus associated to the market clearing equilibrium is $\pi_1 = 0.4876$, while the private firm's profit is $\pi_2 = 0.0014$.

In the case in which the firms do not choose the market clearing price, let $\hat{p} = 0.102$ (see Fig. 3). Then the equilibrium profile is

$$\left(p_1^*, q_1^*, p_2^*, q_2^*\right) = (p_1, 0.878, 0.102, 0.02),$$

the corresponding payoffs are $\pi_1 = 0.4858$ (the sum of dark and light gray areas) and $\pi_2 = 0.0018$ (the light gray area indicated by π_2).

---

[12]In particular, if the private firm sets a price not greater than p_1^*, we are not anymore in Case C; if $q_2^* > \min\{D_2^r(p_2^*, q_1^*), k_2\}$, then the private firm produces a superfluous amount; if $q_2^* < \min\{D_2^r(p_2^*, q_1^*), k_2\}$, then the private firm could still sell more than q_2^*; and if $q_2^* = D_2^r(p_2^*, q_1^*)$, then the private firm will choose a price-quantity pair maximizing profits with respect to its residual demand curve $D_2^r(\cdot, q_1^*)$.

[13]Because $p_2^m(\cdot)$ is a decreasing function in q_1.

Fig. 3 The weak private
firm case—both firms have
positive output

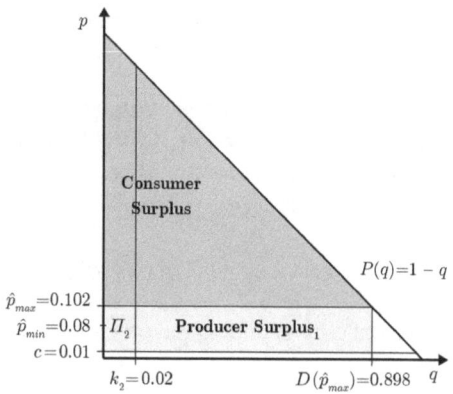

Clearly, for the equilibrium family $\pi_2(\cdot)$ is increasing in $\hat{p}$, while $\pi_1(\cdot)$ is decreasing in $\hat{p}$. The payoff intervals can also be calculated, in particular, $\pi_1 \in [0.4858, 0.4876]$, $\pi_2 \in [0.0014, 0.0018]$.

4.2 Public Firm Moves First

The case of public leadership is somewhat simpler. Namely, the firms clear the market in the only equilibrium family.[14] The results of public leadership are collected in the following proposition.

Proposition 5 *Assume that Assumptions 1–3, $q_2^m(k_1) \geq k_2$ and $P(k_1) > c$ hold. Then the prices and quantities associated with the pure strategy SPNE are*

$$\left(p_1^*, q_1^*, p_2^*, q_2^*\right) = \left(p^*, k_1, p^c, k_2\right)$$

where $p^* \in [0, p^c]$.[15]

Proof We determine the reaction function $BR_2 = (p_2^*(\cdot, \cdot), q_2^*(\cdot, \cdot))$ of the private firm. Like in the strong private firm case, the private firm's best response correspondence can be obtained from the proof of Proposition 4, the corresponding simultaneous case.

$$BR_2(p_1, q_1) = \begin{cases} (p_1, \min\{k_2, D_2^r(p_1, q_1)\}) & \text{if } p_2^m(q_1) \leq p_1; \\ (p_2^m(q_1), q_2^m(q_1)) & \text{if } p_2^m(q_1) > p_1. \end{cases} \quad (17)$$

The reaction function dictates that the public firm maximizes social surplus in the first period by choosing any price level $p_1^* \leq p^c$ and quantity k_1. □

---

[14] We speak about family, because p_1^* can vary within a given range.

[15] As mentioned earlier in the weak private firm case we have $p^c > c$.

Example 5 The public firm moves first. Let $D(p) = 1 - p$, the capacities and the unit cost be $k_1 = 0.9$, $k_2 = 0.02$ and $c = 0.01$.

Then $p^c = 0.08$. The public firm will sell its entire capacity at a $p_1^* \in [0, p^c]$ market clearing price. The private firm will react with the market clearing price, and will also sell its entire capacity. This ensures the highest possible social surplus in this setting. Thus, for all $p_1 \in [0, 0.08]$ the actions associated with the SPNE are

$$\left(p_1^*, q_1^*, p_2^*, q_2^*\right) = (p_1, 0.9, 0.08, 0.02),$$

where the corresponding payoffs are $\pi_1 = 0.4876$ and $\pi_2 = 0.0014$.

4.3 Private Firm Moves First

Finally, we consider the case of private leadership. The only pure-strategy equilibrium family of this case also appears in the simultaneous-moves subcase of the weak private firm case. Namely, the private firm produces on the original demand curve at the highest possible price level for which it is still in the public firm's interest to allow the private firm to do so. The equilibrium family is given formally in the following proposition.

Proposition 6 *Assume that Assumptions 1–3, $q_2^m(k_1) \geq k_2$ and $P(k_1) > c$ hold. Then the prices and quantities associated with the pure strategy SPNE are*

$$\left(p_1^*, q_1^*, p_2^*, q_2^*\right) = (p^*, D_1^r(\hat{p}, \min\{k_2, D(\hat{p})\}), \hat{p}, \min\{k_2, D(\hat{p})\})$$

where $p^* \in [0, \hat{p}]$, if and only if $\hat{p} \geq p^c$ and $p_2^* = \hat{p}$ is the highest price level for which

$$\pi_1(p^c, k_1, \hat{p}, \min\{k_2, D(\hat{p})\}) \leq \pi_1(p^*, D_1^r(\hat{p}, \min\{k_2, D(\hat{p})\}), \hat{p}, \min\{k_2, D(\hat{p})\}) \tag{18}$$

Proof We determine the reaction function $BR_1 = (p_1^*(\cdot, \cdot), q_1^*(\cdot, \cdot))$ of the public firm. The public firm's best response correspondence can also be obtained from the proof of Proposition 4, the corresponding simultaneous-move case. $BR_1(p_2, q_2) =$

$$\begin{cases} (p_1, k_1) & \text{if } \pi_1(p^c, k_1, p_2, q_2) \\ & > \pi_1(p_1, D_1^r(p_2, \min\{k_2, D(p_2)\}), p_2, \min\{k_2, D(p_2)\}); \\ (p_2, D_1^r(p_2, q_2)) & \text{if } \pi_1(p^c, k_1, p_2, q_2) \\ & \leq \pi_1(p_1, D_1^r(p_2, \min\{k_2, D(p_2)\}), p_2, \min\{k_2, D(p_2)\}), \end{cases}$$

where $p_1 \in [0, \hat{p}]$.

The reaction function dictates that the private firm maximizes its profit in the first period by choosing the highest possible price level, where the public firm is still better off (i.e. the social surplus is higher) by reacting with the same price and serving residual demand, than by undercutting p_2.[16] A highest price level $\hat{p}$ exists for every demand function, because if both firms choose price level p^c and sell their entire capacities (i.e. they clear the market), then (18) always holds.   □

The following example illustrates Proposition 6.

Example 6 The private firm moves first. Let $D(p) = 1 - p$, the capacities and the unit cost be $k_1 = 0.9$, $k_2 = 0.02$ and $c = 0.01$.

For the value of Example 6 we get $\hat{p} = 0.102$. The private firm will choose $p_2^* = \hat{p}$ and sells its entire capacity. The public firm will serve residual demand as it is not worth to undercutting the private firm's price which would cause superfluous production. Thus, for all $p_1 \in [0, 0.102]$ the actions associated with the only SPNE are

$$\left(p_1^*, q_1^*, p_2^*, q_2^*\right) = (p_1, 0.878, 0.102, 0.02),$$

where the corresponding payoffs are $\pi_1 = 0.4858$ and $\pi_2 = 0.0018$.

5   The High Unit Cost Case

The main assumption of this case is $c \geq P(k_1)$. In this case if the public firm produces at its capacity level, then the private firm will not enter the market because of the high cost level.

5.1   Simultaneous Moves

In this subcase we have two types of pure-strategy Nash equilibria. The first type consists of profiles in which the private firm sets a price and produces a quantity on the residual demand curve, where in the particular case when the public firm does not produce anything in equilibrium, the residual demand curve coincides with the demand curve. In the second type, the public firm produces its capacity limit, while the private firm does not enter the market.

Proposition 7 *Assume that $c \geq P(k_1)$ and Assumptions 1–3 hold. A strategy profile NE_1*

$$\left(p_1^*, q_1^*, p_2^*, q_2^*\right) = \left(p_1^*, q_1^*, p_2^m\left(q_1^*\right), q_2^m\left(q_1^*\right)\right)$$

---

[16]Depending on the parameters, it can also occur that the public firm has zero output on the residual demand curve.

is for any price-quantity pair

$$(p_1^*, q_1^*) \in \left\{ (p_1, q_1) \mid 0 < q_1 < D(c), 0 \le p_1 \le p_2^d(q_1) \right\} \bigcup \qquad (19)$$

$$\{ (p_1, q_1) \mid q_1 = 0, 0 \le p_1 \le b \} \qquad (20)$$

a Nash-equilibrium in pure strategies[17] if and only if

$$\pi_1(0, D(c), p_2^m(q_1^*), q_2^m(q_1^*)) \le \pi_1(p_1^*, q_1^*, p_2^m(q_1^*), q_2^m(q_1^*)). \qquad (21)$$

A strategy profile NE_2

$$\left(p_1^*, q_1^*, p_2^*, q_2^* \right) = \left(p_1^*, D(c), p_2^*, 0 \right)$$

where $p_1^* \in [0, c]$, *and* $p_2^* \in [0, b]$, *also defines a Nash equilibrium family. Finally, no other equilibrium exists in pure strategies.*

Proof Assume that $(p_1^*, q_1^*, p_2^*, q_2^*)$ is an arbitrary equilibrium profile. We divide our analysis into two subcases. In the first case the private firm is inactive (i.e. $q_2^* = 0$), while in the second case it is active on the market (i.e. $q_2^* > 0$).

Case A: Assume that $q_2^* = 0$, which means that only the public firm's production is positive, and since $c > P(k_1)$ it sets a price $p_1^* \le c$ and quantity $q_1^* = D(c)$ in order to maximize social surplus. Therefore, only NE_2 type equilibria can emerge. We verify that indeed NE_2 specifies equilibrium profiles. Clearly, the public firm would reduce social surplus by switching unilaterally from its NE_2 strategy to a non NE_2 one. The private firm makes losses when producing a positive amount at a price $p_2^* < c$. In addition, $D_2^r(p_2^*, D(c)) = 0$ for all prices $p_2^* \ge c$ by $c > P(k_1)$ if the public firm plays an NE_2 strategy, and thus once again the private firm will just make losses if it produces a positive amount at a price $p_2^* \ge c$.

Case B: Assume that $q_2^* > 0$, which implies $p_2^* \ge c$ since otherwise the private firm would make losses. We divide our analysis into four subcases.

Subcase (i): Assume that $p_1^* = p_2^* > c$. Clearly, we cannot have $q_1^* + q_2^* < D(p_1^*)$ since otherwise the public firm could increase social surplus by increasing its production because of $c > P(k_1)$. Obviously, we cannot have $q_1^* + q_2^* > D(p_1^*)$ since then the public firm would have an incentive to reduce its production if $q_1^* > 0$ or the private firm could gain from decreasing its production if $q_1^* = 0$. In case of $q_1^* + q_2^* = D(p_1^*)$ we must have

$$q_2^* = \min \left\{ k_2, D(p_2^*) \right\} \text{ and } q_1^* = D_1^r(p_2^*, \min \{ k_2, D(p_2^*) \}) \qquad (22)$$

since otherwise the private firm could radically increase its sales by a unilateral and sufficiently small price decrease.

Now we investigate when a strategy profile with prices $p_1^* = p_2^* > c$ and quantities given by (22) constitutes a Nash equilibrium profile. The private firm can benefit from

---

[17]Recall that $q_1 < D(c) \Leftrightarrow P(q_1) > c$. In addition, $q_1 > 0$ implies $c < p_2^d(q_1) < p_2^m(q_1)$.

setting higher prices if and only if $p_2^* < p_2^m(q_1^*)$. Moreover, the private firm can benefit from setting lower prices if and only if $p_2^* > p_2^m(q_1^*)$, which in fact can only be the case[18] if $q_1^* = 0$, because the private firm is not constrained by the production of the public firm by (22). Therefore, in a Subcase (i) equilibrium profile we must have $p_1^* = p_2^* = p_2^m(q_1^*) = p_2^d(q_1^*) > c$.[19] Clearly, if $q_1^* > 0$, then the public firm would decrease social surplus by a price increase (independently of a simultaneous quantity adjustment). If $q_1^* = 0$, then the public firm still will not benefit from setting higher prices. In addition, the public firm would not gain from setting a lower price if and only if (21) is satisfied.

To summarize, Subcase (i) admits those price-quantity pairs (p_1^*, q_1^*) from the set specified by (19) for which $p_1^* = p_2^d(q_1^*)$ results in equal prices.

Subcase (ii): Assume that $p_1^* = p_2^* = c$. As shown in Subcase (i) we must have $q_1^* + q_2^* = D(p_1^*)$. In addition, it can be easily checked that the private firm can benefit from a unilateral deviation if and only if $p_2^m(q_1^*) \in (c, a)$. Since $q_1^* < D(c)$ implies $p_2^m(q_1^*) \in (c, a)$ it follows that $q_1^* = D(c)$ should be the case, which would imply $q_2^* = 0$, leading to a departure from Case B. Hence, a Subcase (ii) equilibrium does not exist.

Subcase (iii): Assume that $p_1^* > p_2^* \geq c$. Then there cannot be an equilibrium in which $q_1^* > 0$ because the public firm could increase social surplus by switching to price p_2^* and quantity $\left(D(p_2^*) - q_2^*\right)^+$. Furthermore, in case of $q_1^* = 0$ we must have $q_2^* = D(p_2^*) \leq k_2$ since otherwise the public firm could again increase social surplus by switching to price p_2^* and quantity $\left(D(p_2^*) - q_2^*\right)^+$. Therefore, in a Subcase (iii) type equilibrium the private firm behaves as a monopolist, and thus $p_2^* = p_2^m(0)$ must be the case, which in turn is an equilibrium if and only if the public firm has no incentive to enter the market, that is (21) is satisfied.

Observe that the derived equilibrium is an NE_1 type equilibrium and the respective price-quantity pairs (p_1^*, q_1^*) are a subset of the set specified by (20).

Subcase (iv): Assume that $p_1^* < p_2^*$ and $p_2^* \geq c$. In case of $D_2^r(c, q_1^*) = 0$ we must have $q_2^* = 0$, which has been already investigated in Case A. Therefore, in what follows we can assume that $D_2^r(c, q_1^*) > 0$, which in turn implies that $p_2^m(q_1^*) \in (c, a)$ and that $p_2^d(q_1^*) \in (c, p_2^m(q_1^*))$ is well defined. Observe that we must have $q_1^* + q_2^* = D(p_2^*)$ since otherwise, for instance, the public firm could increase social surplus by either increasing or decreasing its output. It can be checked that the private firm does not undercut the public firm's price if and only if $p_1^* \leq p_2^d(q_1^*)$. Moreover, if the private firm does not undercut the public firm's price, then it will set price $p_2^m(q_1^*)$ and quantity $q_2^m(q_1^*)$. The derived strategy profile constitutes a Nash equilibrium profile if and only if the public firm has no incentive to deviate, that is (21) is satisfied.

---

[18]If $k_2 \leq D(p_2^*)$, a price decrease cannot increase the private firm's profit, and if $k_2 > D(p_2^*)$, $q_1^* = 0$.

[19]Observe that this also implies $P(q_1^*) > c$.

It can be checked that we have determined an NE_1 type equilibrium and the respective price-quantity pairs (p_1^*, q_1^*) lie in the set specified by (19), where $q_1^* > 0$ and $p_1^* \in [0, p_2^d(0)] \subset [0, p_2^m(0)]$ resulting in a higher price for the private firm.[20]   □

Example 7 The firms move simultaneously. Pick the capacities and unit cost levels $k_1 = 0.5$, $k_2 = 0.1$, $c = 0.6$ and let $D(p) = 1 - p$, which lead to the high unit cost case.

Focusing on NE_1 type equilibria, from the values of Example 7 we get that any $q_1 \in [0, 0.4]$ leads to a Nash equilibrium. Let us fix $q_1 = 0.3$. Now $p_2^m(0.3) = 0.65$ an $p_2^d(0.3) = 0.325$. Thus, $p_1^* \in [0, 0.325]$; $q_1^* = 0.3$; $p_2^* = 0.65$; $q_2^* = 0.05$. In this case, $\pi_1 = 0.0787$ and $\pi_2 = 0.0013$. Depending on q_1, profit levels can vary in the following intervals: $\pi_1 \in [0.06, 0.08]$ and $\pi_2 \in [0, 0.04]$.[21]

5.2 Public Firm Moves First

In the high unit cost case with public leadership we obtain that the private firm does not enter the market, while the public firm's output equals its capacity. This result is formalized in the following proposition.

Proposition 8 Assume that $c > P(k_1)$ and Assumptions 1–3 hold. Then the prices and quantities associated with the pure strategy SPNE are

$$\left(p_1^*, q_1^*, p_2^*, q_2^*\right) = \left(p^*, D(c), p_2^*, 0\right)$$

where $p^* \in [0, c]$ and $p_2^* \in [0, b]$.

Proof We determine the reaction function $BR_2 = (p_2^*(\cdot, \cdot), q_2^*(\cdot, \cdot))$ of the private firm. The private firm's best response correspondence can be obtained from the proof of Proposition 7, the corresponding simultaneous-move case.

$$BR_2(p_1, q_1) = \begin{cases} \{(p, 0) \mid p \in [0, b]\} & \text{if } D(c) \leq q_1 \leq k_1 \text{ and } p_1 \leq c; \\ \{(p_1, \min\{k_2, D(p_1)\})\} & \text{if } p_2^d(q_1) < p_1 \text{ and } p_1 > c; \\ \{(p_1, \min\{k_2, D(p_1)\})\} \cup & \\ \quad \{(p_2^m(q_1), q_2^m(q_1))\} & \text{if } p_2^d(q_1) = p_1 \text{ and } p_1 > c; \\ \{(p_2^m(q_1), q_2^m(q_1))\} & \text{if } p_2^d(q_1) \geq p_1 \text{ and } p_1 > c. \end{cases} \quad (23)$$

Note that the above four areas partition $[0, b] \times [0, k_1]$ since $q_1 < D(c)$ implies $p_2^d(q_1) > c$. From the derived reaction function it follows that the public firm max-

---

[20]It can be verified that we have obtained all NE_1 type equilibria.

[21]We note that here $p_1^* < c$, still, it is of the public firms interest to produce a positive amount, as this action leads to a positive change in consumer surplus. This is the reason why there is no producer surplus indicated on Fig. 4.

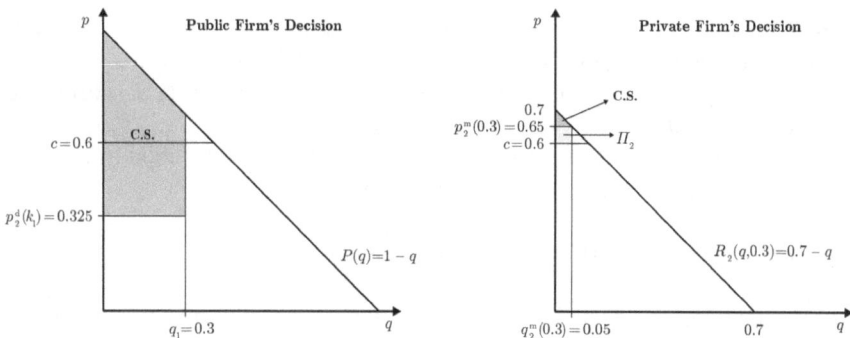

Fig. 4 The high unit cost case

imizes social surplus in the first period by choosing any price level $p^* \in [0, c]$ and quantity k_1. ☐

Example 8 The public firm moves first. Let $k_1 = 0.5$, $k_2 = 0.1$, and $c = 0.6$, and pick demand curve $D(p) = 1 - p$.

For the values given in Example 8 the private firm is not present on the market, and we obtain $p_1^* \in [0, 0.5]$, $q_1^* = 0.5$, $p_2^* \in [0, 1]$, and $q_2^* = 0$. Payoffs equal $\pi_1 = 0.08$ and $\pi_2 = 0$.

5.3 Private Firm Moves First

Finally, we consider the case of private leadership. We will establish for this case that in equilibrium the private firm chooses the highest price level at which the public firm does not capture the entire market at price c or smaller. The respective price is determined either as the price at which the public firm is indifferent between matching the private firm's price and capturing the entire market at a price less than or equal to c and producing $D(c)$, despite the fact that the production of the private firm may be wasted, or by the private firm's monopoly price.

Proposition 9 *Assume that $c \geq P(k_1)$ and Assumptions 1–3 hold. Then there exists a unique price $\hat{p} \in (c, p_2^m(0)]$ such that the prices and quantities associated with the pure strategy SPNE are*

$$(p_1^*, q_1^*, p_2^*, q_2^*) = (p_1^*, D_1^r(\hat{p}, \min\{k_2, D(\hat{p})\}), \hat{p}, \min\{k_2, D(\hat{p})\})$$

where $p_1^* \in [0, \hat{p}]$, and $\hat{p}$ is defined as the smallest price satisfying

$$\pi_1(0, D(c), \hat{p}, \min\{k_2, D(\hat{p})\}) \leq \pi_1(p_1^*, D_1^r(\hat{p}, \min\{k_2, D(\hat{p})\}), \hat{p}, \min\{k_2, D(\hat{p})\}).$$

Proof Clearly, if the private firm does not produce anything, i.e. $q_2 = 0$, then the public firm follows with $(p_1, D(c))$ such that $p_1 \leq c$. If the private firm's production is positive, i.e. $q_2 > 0$, then we must have $p_2 \geq c$. Furthermore, the private firm never produces more than $D(p_2)$.

Focusing on the SPNE, we determine the best replies of the public firm only to the first-stage actions of the private firm lying in

$$A = \{(p_2, 0) \mid p_2 \in [0, b]\} \cup \{(p_2, q_2) \mid p_2 \in [c, b] \text{ and } q_2 \in (0, D(p_2)]\}.$$

For a given $(p_2, q_2) \in A$ such that $q_2 \in (0, D(p_2)]$ the public firm never sets a price above p_2 if it decides to produce at all, i.e. $q_1 > 0$. Moreover, in the latter case the public firm's production has to equal $q_1 = D_1^r(p_2, q_2)$, since if it does not capture the entire market, social surplus will be determined at price p_2 and superfluous production decreases social surplus. Therefore, the response of the public firm is determined by inequality

$$\pi_1(0, D(c), p_2, q_2) \leq \pi_1(c, D_1^r(p_2, q_2), p_2, q_2), \qquad (24)$$

where its response equals $BR_1(p_2, q_2) = \{(p_1, D_1^r(p_2, q_2)) \mid p_1 \leq c\}$ if $q_2 > 0$ and (24) is satisfied, and $BR_1(p_2, q_2) = \{(p_1, D(c)) \mid p_1 \leq c\}$ if $q_2 = 0$ or (24) is violated.[22]

Taking the best responses of the public firm into consideration, the private firm will produce $q_2 = \min\{k_2, D(p_2)\}$ at price p_2 if (24) is satisfied.[23] By substituting $q_2 = \min\{k_2, D(p_2)\}$ into (24) it follows that the right-hand side of (24) is continuous, strictly decreasing in p_2 on $[c, p_2^m(0)]$, and it is larger than its left-hand side at price $p_2 = c$. Since the private firm does not set a price above $p_2^m(0)$ it will either set the price in $(c, p_2^m(0))$ for which (24) is satisfied with equality or price $p_2^m(0)$. □

Example 9 The private firm moves first. Pick linear demand and let the capacities and the unit costs be $k_1 = 0.5$, $k_2 = 0.1$, $c = 0.6$.

It can be determined for the values given in Example 6 that $\hat{p} = 0.8$. This leads us to $p_1^* \in [0, 0.8]$; $q_1^* = 0.1$; $p_2^* = 0.8$; $q_2^* = 0.1$, which implies $\pi_1 = 0.06$; $\pi_2 = 0.04$.

6 Solution of the Timing Game

We consider a timing in which the firms in stage 1 can choose between two periods for the announcement of their price and quantity decision. Thereafter, knowing each others timing decision, the firms in stage 2 set their prices and quantities in the

---

[22]To be precise if (24) is satisfied with equality, then both mentioned types are best responses; however, as it can be verified in a SPNE only the former type can be selected.

[23]Note that the distribution of production between the two firms does not effect (24).

Table 1 Example payoff levels for the demand function $D(p) = 1 - p$

Cases	Strong private firm	Weak private firm	High unit cost
k_1	0.5	0.9	0.5
k_2	0.4	0.02	0.1
c	0.1	0.01	0.6
π_1: *Public firm's equilibrium payoff (social surplus)*			
sim. moves	$\in [0.285, 0.435]$	$\in [0.4858, 0.4876]$	$\in [0.06, 0.08]$
As leader	0.435	0.4876	0.08
As follower	0.285	0.4858	0.06
π_2: *Private firm's equilibrium payoff (profit)*			
sim. moves	$\in [0.04, 0.196]$	$\in [0.0014, 0.0018]$	$\in [0, 0.04]$
As leader	0.196	0.0018	0.04
As follower	0.04	0.0014	0

selected periods. Hence, we investigate a timing game with observable delay à la Hamilton and Slutsky (1990).

Before we turn to the solution of the timing game, we provide a summary of the payoffs that were calculated in the numerical Examples 1–9, respectively. Table 1 provides numerical evidence of the solution of the timing game for the particular demand function $D(p) = 1 - p$, with exogenously given capacities and cost levels.

It is easy to see from Table 1 that in all the three main cases any firm has the highest payoff with certainty in case it is the first mover. Thus, intuitively as every firm wants to become the leader and there cannot be two leaders at the same time, the outcome of the timing game is simultaneous moves. To be more precise the firms strictly prefer the role of the leader to moving simultaneously and they prefer moving simultaneously to the role of the follower if in the simultaneous-move game neither the best one nor the worst one is realized from the continuum of pure-strategy equilibria. The equilibrium of the timing game for any concave, twice continuously differentiable demand function is precisely stated in the following proposition.

Proposition 10 *Under Assumptions 1–3 and that in the simultaneous-move game from the set of all possible pure-strategy equilibria for both firms neither the best one nor the worst one is realized,*[24] *in the subgame-perfect Nash equilibrium of the timing game both firms choose to set their price-quantity pairs in the first period, and therefore the firms play the simultaneous-move game in the second stage.*

Proof The equilibrium of the timing game can be derived from Propositions 1–9, by comparing the payoffs of both firms for different orderings of moves.

Let us focus on the the strong private firm case, hence we have to consider Propositions 1–3. If both firms select to chose their price-quantity decisions in the first period, then neither of them would benefit from moving in the second period. Since

---

[24]The latter assumption can be explained by risk dominance.

we assumed that in the simultaneous-move game not the worst possible equilibrium outcome will be chosen, moving second would be definitely worse. From this it also follows that none of the sequential games can be an equilibrium outcome of the timing game. If both firms select to chose their price-quantity decisions in the second period, then unilaterally becoming the leader is strictly better than moving simultaneously since we have assumed that in the simultaneous-move game the best possible equilibrium outcome will not be chosen.

The same argument applies to the weak private firm and the high unit cost cases.   □

7 Corollaries and Concluding Remarks

Our main results are collected in the following corollaries. We focus on the differences between the production-to-order case—which was investigated in earlier work—and the production-in-advance case from the point of view of equilibrium strategies, social surplus effects and equilibrium analysis of the timing game. The first corollary focuses on the public firm's influence on social surplus. One can carry out a comparison with the results for the production-to-order case presented in Balogh and Tasnádi (2012). In the PIA case the social surplus becomes lower—let them play any pure-strategy Nash equilibria—than that of the PTO case. This result is put down in the next corollary.

Corollary 1 *When playing the production-in-advance type of the Bertrand-Edgeworth game, the equilibrium strategies lead to a decrease in social surplus compared to the PTO case.*

The second main result of the paper is implicitly given in Sect. 5: independently from the parameters and the orderings of firms' decisions, the production-in-advance type Bertrand-Edgeworth mixed duopoly always has at least one pure-strategy Nash equilibrium. This result remained the same as that of the mixed PTO case. However, we emphasize that in case of standard Bertrand-Edgeworth duopolies, there is a lack of pure-strategy equilibria (see e.g. Deneckere and Kovenock 1992). We state the existence of a pure-strategy equilibrium in the third corollary.

Corollary 2 *We have at least one pure-strategy (subgame-perfect) Nash equilibrium in all three analyzed cases and for all three orderings of moves.*

These results are summarized in Table 2.

The results suggest that it is by far not all the same whether a public firm has some influence on an oligopoly market. Further research directions may include the application of our model to markets with asymmetric information, partial public ownership, and oligopolies with more than two firms. One can notice that our assumptions were quite general in the present paper. However, to present plausible results in the mentioned topics, more strict assumptions may be needed.

Table 2 Comparison of the PTO and PIA cases

	Production-to-order	Production-in-advance
Equilibrium in pure strategies	Yes	Yes
Timing game equilibrium	All possible orderings	Simultaneous moves
Public firms's social surplus effect	Positive	Negative[a]

[a]Compared to the mixed PTO case

Acknowledgements This research is granted by the Pallas Athéné Domus Spientiae Foundation Leading Researcher Program.

References

Bakó, B., & Tasnádi, A. (2017). The kreps-scheinkman game in mixed duopolies. *Journal of Institutional and Theoretical Economics, 173*, 753–768.

Balogh, T., & Tasnádi, A. (2012). Price leadership in a duopoly with capacity constraints and product differentiation. *Journal of Economics (Zeitschrift für Nationalökonomie), 106*, 233–249.

Bárcena-Ruiz, J. C. (2007). Endogenous timing in a mixed duopoly: Price competition. *Journal of Economics (Zeitschrift für Nationalökonomie), 91*, 263–272.

van den Berg, A., & Bos, I. (2017). Collusion in a price-quantity oligopoly. *International Journal of Industrial Organization, 50*, 159–185.

Bos, I., & Vermeulen, D. (2015). *On pure-strategy nash equilibria in price-quantity games* (p. 018). GSBE Research Memoranda, No: Maastricht University.

Boyer, M., & Moreaux, M. (1987). Being a leader or a follower: Reflections on the distribution of roles in duopoly. *International Journal of Industrial Organization, 5*, 175–192.

Davis, D. (2013). Advance production, inventories and market power: An experimental investigation. *Economic Inquiry, 51*, 941–958.

Deneckere, R., & Kovenock, D. (1992). Price leadership. *Review of Economic Studies, 59*, 143–162.

Din, H. R., & Sun, C. H. (2016). Combining the endogenous choice of timing and competition version in a mixed duopoly. *Journal of Economics (Zeitschrift für Nationalökonomie), 118*, 141–166.

Gertner, R. H. (1986). *Essays in theoretical industrial organization*. Massachusetts Institute of Technology, Ph.D. thesis.

Hamilton, J., & Slutsky, S. (1990). Endogenous timing in duopoly games: Stackelberg or cournot equilibria. *Games and Economic Behavior, 2*, 29–46.

Kreps, D. M., & Scheinkman, J. A. (1983). Quantity precommitment and bertrand competition yield cournot outcomes. *Bell Journal of Economics, 14*, 326–337.

Lee, S. H., & Xu, L. (2018). Endogenous timing in private and mixed duopolies with emission taxes. *Journal of Economics (Zeitschrift für Nationalökonomie), 124*, 175–201.

Levitan, R., & Shubik, M. (1978). Duopoly with price and quantity as strategic variables. *International Journal of Game Theory, 7*, 1–11.

Matsumura, T. (2003). Endogenous role in mixed markets: A two production period model. *Southern Economic Journal, 70*, 403–413.

Mestelman, S., Welland, D., & Welland, D. (1987). Advance production in posted offer markets. *Journal of Economic Behavior and Organization, 8*, 249–264.

Montez, J., & Schutz, N. (2018). All-pay oligopolies: Price competition with unobservable inventory choices. Collaborative Research Center Transregio 224, Discussion Paper Series—CRC TR 224, No 20.

Nakamura, Y. (2019). Combining the endogenous choice of the timing of setting the levels of strategic contracts and their contents in a managerial mixed duopoly. *Journal of Industry, Competition & Trade, 19*, 235–261.

Orland, A., & Selten, R. (2016). Buyer power in bilateral oligopolies with advance production: Experimental evidence. *Journal of Economic Behavior and Organization, 122*, 31–42.

Pal, D. (1998). Endogenous timing in a mixed oligopoly. *Economics Letters, 61*, 181–185.

Phillips, O., Menkhaus, D., & Krogmeier, J. (2001). Production-to-order or production-to-stock: The endogenous choice of institution in experimental auction markets. *Journal of Economic Behavior and Organization, 44*, 333–345.

Shubik, M. (1955). A comparison of treatments of a duopoly problem, Part II. *Econometrica, 23*, 417–431.

Tasnádi, A. (2003). Endogenous timing of moves in an asymmetric price-setting duopoly. *Portuguese Economic Journal, 2*, 23–35.

Tasnádi, A. (2004). Production in advance versus production to order. *Journal of Economic Behavior and Organization, 54*, 191–204.

Tomaru, Y., & Kiyono, K. (2010). Endogenous timing in mixed duopoly with increasing marginal costs. *Journal of Institutional and Theoretical Economics, 166*, 591–613.

Zhu, Q. T., Wu, X. W., & Sun, L. (2014). A generalized framework for endogenous timing in duopoly games and an application to price-quantity competition. *Journal of Economics (Zeitschrift für Nationalökonomie), 112*, 137–164.

Necessary Conditions for Concave and Cournot Oligopoly Games

Ferenc Forgó and Zoltán Kánnai

Abstract Necessary conditions for the existence of pure Nash equilibria introduced by Joó (A note on minimax theorems, Annales Univ. Sci. Budapest, **39**(1996), 175–179) for concave non-cooperative games are generalized and then applied to Cournot oligopoly games. If for a specified class of games there always exists a pure Nash equilibrium, then cost functions of the firms must be convex. Analogously, if for another specified class of games there always exists a pure Nash equilibrium, then revenue functions of the firms must be concave in their respective variables.

Keywords Concave games · Cournot oligopoly · Necessary conditions

JEL-code: L13

1 Introduction

Oligopoly is a market structure where a few competing firms are present and their individual decisions about production and/or selling price influence not only their own profit but everybody else's as well. Thus oligopoly lends itself to being modelled as a non-cooperative game where the players are the firms and payoffs are determined by profit functions usually defined as revenues less costs. Oligopolies have long been in the focus of economic research and practical market design. The ground breaking work of Cournot (1838) had laid down the foundations but intensive research only began when game theory became available to provide the necessary tools for deep analysis. Our focus will be on classical Cournot games where firms make

F. Forgó (✉)
Department of Operations Research and Actuarial Sciences, Corvinus University of Budapest, Fővám tér 8, 1093 Budapest, Hungary
e-mail: ferenc.forgo@uni-corvinus.hu

Z. Kánnai
Department of Mathematics, Corvinus University of Budapest, Fővám tér 8, 1093 Budapest, Hungary
e-mail: kannai@uni-corvinus.hu

© Springer Nature Singapore Pte Ltd. 2020
F. Szidarovszky and G. I. Bischi (eds.), *Games and Dynamics in Economics*,
https://doi.org/10.1007/978-981-15-3623-6_10

decisions on the production volume of a single homogeneous product subject to capacity constraints.

Among many other aspects the existence and uniqueness of equilibria of non-cooperative games as defined by Nash (1950) has drawn much attention. Beyond direct application of game theoretic existence theorems many papers utilized the special features of an oligopoly game. Excellent reference books on the subject are available e.g. Friedman (1977), Okuguchi and Szidarovszky (1990). In game theory much effort has been devoted to weakening conditions imposed on strategy sets/payoff functions to ensure the existence of a (pure) Nash equilibrium point (NEP for short). Staying in finite dimensional spaces, this endeavor is demonstrated by the series of papers marked by the milestone results of von Neumann (1928), Nash (1950), Nikaido and Isoda (1955), Friedman (1977).

These results of course translate to oligopoly games but sufficient conditions directly imposed on the primitives (demand, inverse demand and cost functions) are preferable since their interpretation is more direct and closely related to economic phenomena thus readily embraced by economists. It was realized early that there are limits to generalizations of revenue and/or cost functions if we do not want to lose the desirable property of the existence of a pure NEP. There are examples, a few of them analyzed in Novshek (1985), for Cournot games without pure NEP's. These are, however, only examples but not necessary conditions. Necessary conditions in relation to oligopoly games are quite rare.

In this paper we will study and apply to the Cournot game a special class of necessary conditions first formulated and proved by Joó (1986, 1996) for general concave games. The main message of our analysis, in loose terms, is that if for a special class of revenue functions there always exists a pure NEP, then the cost functions need to be convex in their respective variables. This can also be reversed: if for a special class of cost functions there always exists a pure NEP, then the revenue functions must be concave in their respective variables.

The paper is organized as follows. In Sect. 2 we set forth a special class of necessary conditions applicable in mathematical programming and game theory. In Sect. 3 we study and generalize necessary conditions for concave games due to Joó (1986, 1996). In Sect. 4 these conditions are discussed and applied to generalized Cournot games. Section 5 concludes.

2   A General Framework for Necessary Conditions in Mathematical programming and Game Theory

In mathematical programming a major line of research has aimed at extending the power of efficient solution methods to problems where assumptions about objective functions and/or constraints (feasible sets) are weaker. This has led e.g. to replacing concave (convex) objective functions with quasi-concave (quasiconvex) functions (see e.g. Diewert et al. 1981) while maintaining the power of solution methods based on local search.

The natural question emerges: where are the meaningful bounds for these generalizations? While sufficient conditions giving way to generalizations abound in the literature, necessary conditions are hard to find. A rare exception is the work of Kolstad and Mathiesen (1987) addressing the uniqueness of the NEP. The general framework set forth in this paper for necessary conditions is inspired by Martos (1975) in mathematical programming and Joó (1986, 1996) in game theory. While Martos' results are well known those of Joó's have remained unnoticed. This is mainly due to the titles not giving any orientation about what it is really all about. Especially the title of Joó (1996) is misleading.

Define a general mathematical programming problem $P(f, L)$ as

$$\max f(x) \tag{1}$$

$$\text{subject to } x \in L$$

where $L \subset \mathbb{R}^n$ is the feasible set, $f : \mathbb{R}^n \to \mathbb{R}$ is the objective function.

Let T be a particular property of $P(f, L)$. Let furthermore $\mathcal{L}$ be a family of feasible sets and $\mathcal{F}$ a family of objective functions. The following two statements are said to be Martos-type necessary conditions:

(i) If property T holds for any $P(f, L)$, $L \in \mathcal{L}$, then $f \in \mathcal{F}$.
(ii) If property T holds for any $P(f, L)$, $f \in \mathcal{F}$, then $L \in \mathcal{L}$.

An example of a Martos-type (i) condition is the following.

Theorem 1 *Martos (1975) Let L' be a compact, convex subset of $\mathbb{R}^n$. If for any compact, convex set $L \subset L'$ problem (1) has the property that every local maximum point is also a global maximum point, then f is quasiconcave on L'.*

Here $\mathcal{L}$ is the family of all compact, convex subsets of L', $\mathcal{F}$ is the family of quasiconcave functions defined on L', and property T is all local maximum points being also global on a compact, convex set.

We will consider games in normal (strategic) form: $G = \{S_1, ..., S_n; f_1, ..., f_n\}$ or briefly $G = \{S; f\}$ where $S = S_1 \times ... \times S_n$ is the set of strategy profiles and $f : S \to \mathbb{R}^n$ is the profile of payoff functions. Let T be a property of $G = \{S; f\}$. Let Σ be a family of strategy profiles and F a family of payoff profiles. The following two statements are said to be Joó-type necessary conditions:

(i) If property T holds for any $G = \{S; f\}$, $S \in \Sigma$, then $f \in F$.
(ii) If property T holds for any $G = \{S; f\}$, $f \in F$, then $S \in \Sigma$.

We will call a function f defined on a convex, compact set $C \subset \mathbb{R}^n$ partially concave if it is concave in each of its variables if the rest of the variables are held fixed. An example of a Joó-type necessary condition is the following.

Theorem 2 (Theorem 1 in Joó 1996) *Let* $f_k : [0, 1]^n \to \mathbb{R}$ $(k = 1, ..., n)$ *be continuous functions, and* $f = f_1 \times ... \times f_n$. *Let T be the following property: If* $f'_k :$ $[0, 1]^n \to \mathbb{R}$ $(k = 1, ..., n)$ *is continuous and partially concave in the k-th variable, then the game* $G = \{[0, 1]^n, f + f'\}$ *has at least one NEP, where* $f' = f'_1 \times ... \times f'_n$. *If property T holds, then each function* f_k $(k = 1, ..., n)$ *is partially concave in its k-th variable.*

Theorem 2 was extended to games with convex, compact strategy sets.

Theorem 3 (Theorem 2 in Joó 1996) *Let* $K_1, ..., K_n$ *be convex, compact subsets of finite dimensional euclidean spaces,* $f_k : K_1 \times ... \times K_n \to \mathbb{R}$ $(k = 1, ..., n)$ *be continuous functions and* $f = f_1 \times ... \times f_n$. *Let T be the following property: If* $f'_k :$ $K_1 \times ... \times K_n \to \mathbb{R}$ $(k = 1, ..., n)$ *is continuous and partially concave in the k-th variable, then the game* $G = \{K_1, ..., K_n; f + f'\}$ *has at least one NEP, where* $f' =$ $f'_1 \times ... \times f'_n$. *If property T holds, then each function* f_k $(k = 1, ..., n)$ *is partially concave in its k-th variable.*

3   Joó-type Necessary Conditions for Concave Games

One of the standard existence theorems in noncooperative game theory is due to Nikaido and Isoda (1955):

Theorem 4 *Let* $G = \{S_1, ..., S_n; f_1, ..., f_n\}$ *be a game in normal form. If*

  (i) *the strategy sets* $S_1, ..., S_n$ *are non-empty, compact, convex sets of finite dimensional euclidean spaces,*

 (ii) *the payoff functions* $f_k : \times_{j=1}^n S_j \to \mathbb{R}$ $(k = 1, ..., n)$ *are continuous and partially concave in the k-th variable, then G has at least one NEP.*

Theorem 2 of Joó 1996 gives a necessary condition for the concavity of the payoff functions when the payoff function is subjected to concave perturbations. One way to generalize Theorem 2 is through requiring of the payoff functions less then continuity. Key to the generalization is a characterization of concave functions which we will give in the form of a lemma. We need two propositions to prove the lemma.

Proposition 1 *If the function* $f : [a, b] \to \mathbb{R}$ *is bounded from above, then the function*

$$\Phi : \mathbb{R} \to \mathbb{R}$$
$$\Phi(c) := \sup_{t \in [a,b]} \left(f(t) + c \cdot t \right)$$

is Lipschitz continuous.

Proof For any $c, d \in \mathbb{R}$ and $x \in [a, b]$ we have

$$f(x) + d \cdot x \leq f(x) + c \cdot x + (|a| + |b|) \cdot |c - d| \leq \sup_{t \in [a,b]} \left(f(t) + c \cdot t \right) + (|a| + |b|) \cdot |c - d|,$$

or equivalently

$$\sup_{x \in [a,b]} \left(f(x) + d \cdot x \right) \leq \sup_{t \in [a,b]} \left(f(t) + c \cdot t \right) + (|a| + |b|) \cdot |c - d|.$$

Using the definition of Φ and rearranging we obtain

$$\Phi(d) - \Phi(c) \leq (|a| + |b|) \cdot |d - c|.$$

Changing the role of c and d we get

$$|\Phi(d) - \Phi(c)| \leq (|a| + |b|) \cdot |d - c|$$

which was to be proved. $\square$

Proposition 2 *Let* $f : [a, b] \to \mathbb{R}$ *be a function bounded from above and* $a < x < b$. *Then there exist* $c, d \in \mathbb{R}$ *such that*

$$\sup_{t \in [a,x]} \left(f(t) + c \cdot t \right) \leq \sup_{t \in [x,b]} \left(f(t) + c \cdot t \right)$$

$$\sup_{t \in [a,x]} \left(f(t) + d \cdot t \right) \geq \sup_{t \in [x,b]} \left(f(t) + d \cdot t \right).$$

Proof Define

$$c := \max \left\{ 0, \; \frac{\sup\limits_{[a,x]} f - f(b)}{b - x} \right\}.$$

Then for every $t \in [a, x]$ we have

$$c \cdot (b - t) \geq c \cdot (b - x) \geq \sup_{[a,x]} f - f(b) \geq f(t) - f(b),$$

implying

$$f(t) + c \cdot t \leq f(b) + c \cdot b,$$

or equivalently

$$\sup_{t \in [a,x]} \left(f(t) + c \cdot t \right) \leq f(b) + c \cdot b,$$

from which we get the first assertion of the proposition. Define

$$d := \min \left\{ 0, \ \frac{f(a) - \sup_{[x,b]} f}{x - a} \right\}.$$

By similar reasoning as before we will arrive at

$$\sup_{t \in [x,b]} \left(f(t) + d \cdot t \right) \le f(a) + d \cdot a,$$

leading to the second assertion of the proposition. □

Lemma 1 *Let $f : [a, b] \to \mathbb{R}$ be an upper semicontinuous function. If for any $c \in \mathbb{R}$ the set*

$$\left\{ x \in [a, b] : f(x) + c \cdot x = \max_{t \in [a,x]} (f(t) + c \cdot t) \right\}$$

is a closed interval, then f is concave.

Proof We will show that at any point $a < x_0 < b$ there is a line supporting f from above. Consider the function

$$\Psi : \mathbb{R} \to \mathbb{R},$$

$$\Psi(c) := \max_{t \in [a,x_0]} \left(f(t) + c \cdot t \right) - \max_{t \in [x_0,b]} \left(f(t) + c \cdot t \right).$$

Ψ is continuous by Proposition 1, and by Proposition 2 there are numbers $c, d \in \mathbb{R}$ such that $\Psi(c) \le 0 \le \Psi(d)$. Thus by Bolzano's theorem there is a number c_* for which $\Psi(c_*) = 0$, i.e.

$$\max_{t \in [a,x_0]} \left(f(t) + c_* \cdot t \right) = \max_{t \in [x_0,b]} (f(t) + c_* \cdot t).$$

This common maximum is also the maximum of the function $t \to f(t) + c_* \cdot t$ on the interval $[a, b]$. Therefore there are numbers $a \le x_1 \le x_0 \le x_2 \le b$ such that

$$f(x_1) + c_* \cdot x_1 = \max_{t \in [a,b]} \left(f(t) + c_* \cdot t \right) = f(x_2) + c_* \cdot x_2. \qquad (2)$$

By the assumption, the level set H belonging to the maximum of the function $t \to f(t) + c_* \cdot t$ is a closed interval and by (2) $x_1, x_2 \in H$. Thus by $x_1 \le x_0 \le x_2$ we have $x_0 \in H$. Therefore for any $t \in [a, b]$,

$$f(x_0) + c_* \cdot x_0 \ge f(t) + c_* \cdot t$$

holds. After rearrangement we get

$$f(t) \leq f(x_0) - c_* \cdot (t - x_0). \tag{3}$$

The expression on the right-hand side of (3) is a straight line which supports f from above at x_0. □

Remarks

1. It is known that concave, upper semicontinuous functions over a closed interval are continuous. Joó (1986) stated Lemma 1 for continuous functions but gave no proof. Though a posterior we know that these functions are continuous but a priori we only need to assume upper semicontinuity. This is why we think that Lemma 1 is a generalization of Joó's lemma.

2. Since in the proof of Theorem 2 Lemma 1 plays a crucial role and continuity of the functions $f_k : [0, 1]^n \rightarrow \mathbb{R}$ $(k = 1, ..., n)$ is basically needed to ensure that the set of maximum points of $f_k(y_1, ..., y_{k-1}, x_k, y_{k+1}, ..., y_n)$ in x_k for any fixed $y_1, ..., y_{k-1}, y_{k+1}..., y_n$ is a closed interval, the continuity of f_k in x_k can be weakened to upper semicontinuity resulting in a generalization of Theorem 2.

3. It is not clear but seems probable that in Theorem 3 continuity, a priori, can also be relaxed.

4. Similar characterization of quasiconcave functions based on the nature of the set of maximum points was given by Forgó (1996): Let $X \subset \mathbb{R}^n$ be a non-empty convex set and $f : \mathbb{R}^n \rightarrow \mathbb{R}$ a continuous function. Then f is quasiconcave on X if and only if for any closed interval $I \subset X$ the set of maximum points of f over I is a closed interval. Interestingly, the continuity assumption cannot be relaxed to upper semicontinuity.

4 Necessary Conditions for Cournot Oligopoly Games to Have a Pure Nash Equilibrium

In Cournot oligopolies firms make decisions about the volume of production of a homogeneous product. Production may have capacity bounds other than the natural lower bound 0. Selling price is determined by the production of the entire industry via an inverse demand function. Cost of production may vary from firm to firm. Gross profit is defined as revenue (volume times selling price) minus cost. This model gives rise to a game, called the Cournot game, defined by strategy sets $S_i = [a_i, b_i]$ for firm $i = 1, ..., n$ ($b_i = \infty$ is allowed for some or all i), payoff (profit) functions $f_i(q) = q_i P(1^T q) - C_i(q_i)$, where $P : \mathbb{R}_+ \rightarrow \mathbb{R}_+$ is the inverse demand function assigning to total industry output the highest price the market clears at, 1 denotes a vector of $1's$, $C_i \rightarrow \mathbb{R}$ is the cost function assigning to the production q_i of firm i the total cost incurred at that level of production. So the Cournot game G in normal form is given as $G = \{S_1, ..., S_n; f_1, ..., f_n\}$.

It has long been a major line of research in economics in general and industrial organization in particular, to give ever weaker sufficient conditions imposed on the ingredients of the Cournot game that ensure the existence (uniqueness) of a pure NEP. From an ocean of contributions we only mention three landmarks:

(i) The classical works of Szidarovszky and Yakovitz (1977, 1982) where inverse demand is assumed to be concave and the cost functions convex.
(ii) The paper of Novshek (1985) where concavity/convexity assumptions are considerably relaxed: concavity of inverse demand is replaced by the condition that each firm's marginal revenue be a declining function of the total output of the others, convexity of the cost function is abandoned altogether and it is only assumed that it be a nondecreasing lower semicontinuous function.
(iii) Ewerhart (2014) brings many known sufficient conditions under the umbrella of biconcavity.

In the efforts to get ever weaker sufficient conditions, after Novshek's result where the cost function is very general, attention has been focused on the inverse demand function and more generally on the revenue function. It turns out that if we allow more general revenue functions not just the conventional "quantity times price" form, then the existence of a pure NEP necessitates the convexity of the cost function.

Let us redefine the Cournot oligopoly game $G = \{S_1, ..., S_n; f_1, ..., f_n\}$ where $S_i = [0, 1]$, $f_i(x) = R_i(x) - C_i(x), i = 1, ..., n$. Here $R_i, C_i : S = \times_{j=1}^{j=n} S_j \to \mathbb{R}$ are the (generalized) revenue and cost functions. Notice that in this set-up revenues and costs of each firm may depend on the industry production profile. Revenue in the classical model does depend on the production profile of the industry, specifically on the firm's own level of production and the total industry production. In case of a generalized revenue function this is not necessarily so, other functional dependence of the revenue on the production profile of the industry is allowed. For cost functions, as opposed to the classical form, the cost of each firm may depend not only on its own production volume but on the production profile of the whole industry.

The general revenue function allows for getting different levels of revenue for two production profiles with the same total production. Indeed, an evenly distributed production profile gives less chance for the firm to get extra leverage by utilizing its position marked by a dominant market share. Also, a general revenue function can take into account other market forces than price (discounts, all sorts of promotions, etc.). By not assuming anything a priori about the monotonicity and the shape of the inverse demand function, unusual markets, such as markets of Giffen and Veblen goods (see Varian 1922) can be studied in the same model.

Costs can also depend on the whole production profile. Overuse of natural resources may incur costs that increase much faster as industry output increases compared to the situation when only an individual firm uses more of the resource. Even monotonicity can be violated in special cases. In some countries zero-level production in agriculture is rewarded by subsidies which disappear as production moves away from zero. This is also an example of the presence of discontinuities as well.

The following theorem emphasizes the importance of convexity of the cost functions if we want to ensure the existence of a pure NEP.

Theorem 5 *Let all the cost functions C_i of a generalized Cournot game be continuous. If the generalized Cournot oligopoly game $G = \{S_1, ..., S_n, R_1 - C_1, ..., R_n - C_n\}$ has a pure NEP for any partially concave continuous revenue function R_i $i = 1, ..., n$, then all C_i $i = 1, ..., n$ are partially convex.*

Proof By Theorem 2 $-C_i$ is partially concave implying that C_i is partially convex for all $i = 1, ..., n$. □

The following question comes naturally to mind: If we only consider Cournot games (not generalized!) which means that we only require the existence of a pure NEP for a special class of revenue functions, what can be said about the cost function? Surely less than convexity. Maybe quasi-convexity?

The role of the revenue and cost functions can be reversed in a natural way. We then obtain the following necessary condition.

Theorem 6 *Let all the revenue functions R_i of a generalized Cournot game be continuous. If the generalized Cournot oligopoly game $G = \{S_1, ..., S_n, R_1 - C_1, ..., R_n - C_n\}$ has a pure NEP for any partially convex continuous cost function C_i $i = 1, ..., n$, then all R_i $i = 1, ..., n$ are partially concave.*

Theorems similar to Theorems 5 and 6 can be stated for multiproduct oligopolies as defined in Forgó (1999) page 67–72. In this case Theorem 3 has to be invoked in order to arrive at the same results.

5 Conclusion

Necessary conditions for the existence of pure NEP's were derived for generalized Cournot oligopoly games. If for all revenue functions there exists at least one pure NEP for a fixed continuous cost function, then the cost function must be convex. The question of how to characterize cost functions within the framework of the classical Cournot game where revenues are calculated as the product of volume and price determined by the total production of the industry through an appropriately conditioned inverse demand function remains open.

Acknowledgements Research was done in the framework of Grant NKFI K-1 119930.

References

Cournot A A (1838) Recherches sur les Principes Mathématiques de la Th éorie des Richesses. L. Hachette. English edition (translated by N.Bacon): Researches into the Mathematical Principles of the Theory of Wealth. Macmillan, New York, 1897

Diewert, W. E., Avriel, M., & Zang, I. (1981). Nine kinds of quasiconcavity. *Journal of Economic Theory*, 25, 397–420.

Ewerhart, C. (2014). Cournot games with biconcave demand. *Games and Economic Behavior, 85,* 37–47.

Forgó, F. (1996). On Béla Martos' contribution to mathematical programming. *Szigma, 27,* 1–9. (in Hungarian).

Forgó, F., Szép, J., & Szidarovszky, F. (1999). *Introduction to the theory of games, concepts, methods, applications.* Dordrecht/Boston/London: Kluwer Academic Publishers.

Friedman, J. W. (1977). *Oligopoly and the theory of games.* Amsterdam: North-Holland.

Joó, I. (1986). Answer to a problem of M. Horváth and A. Sövegjártó. *Annales Univ. Sci. Budapest Sectio Math., 29,* 203–207.

Joó, I. (1996). A note on minimax theorems Annales Univ. *Sci. Budapest Sectio Math., 39,* 175–179.

Kolstad, C. D., & Mathiesen, L. (1987). Necessary and sufficient conditions for uniqueness of a Cournot equilibrium. *Review of Economic Studies, 54,* 681–690.

Martos, B. (1975). *Nonlinear programming theory and methods.* Budapest: Akadémiai Kiad ó.

Nash, J. F. (1950). Equilibrium points in n-person games. *Proceedings of the National Academy of Sciences, 36,* 48–49.

Nikaido, H., & Isoda, K. (1955). Note on noncooperative convex games. *Pacific Journal of Mathematics, 5,* 807–815.

Novshek, W. (1985). On the existence of Cournot equilibrium. *The Review of Economic Studies, 52,* 85–98.

Okuguchi, K., & Szidarovszky, F. (1990). *The theory of oligopoly with multi-product firms.* Berlin/Heidelberg: Springer.

Szidarovszky, F., & Yakowitz, S. (1977). A new proof of the existence and uniqueness of the Cournot equilibrium. *International Economic Review, 18,* 787–789.

Szidarovszky, F., & Yakowitz, S. (1982). Contributions to Cournot oligopoly theory. *Journal of Economic Theory, 28,* 51–70.

Varian, H. R. (1992). *Microeconomic analysis.* New York/London: W.W. Norton and Company

von Neumann, J. (1928). Zur Theorie der Gesellschaftsspiele. *Mathematische Annalen, 100,* 295–320.

Set-Valued Techniques in Dynamic Economic Models

Zoltán Kánnai, Imre Szabó and Peter Tallos

Abstract Existence of solutions in optimization problems is usually hard to prove. Mostly, deep results from set-valued analysis are employed and convexity assumptions are imposed. In this paper we show that the theory of differential inclusions can successfully be used for solving nonconvex problems, such as nonconvex optimal control problems, or implicit differential schemes. Results are applied to a generalized dynamic input-output economic model.

Mathematics Subject Classifications: 34A60 · 49K15 · 93D15

1 Introduction

Several problems in the theory of economic dynamics lead to optimal control, see Sethi and Thompson (2006) and the references within. Some other problems are formulated in terms of implicit equations, like classical input-output Leontief processes. In both cases we propose the use of set-valued analysis for verifying the existence of solutions in some types of problems.

The present paper is principally divided into two parts and is organized as follows. The first 5 Sections deal with a special type of nonconvex optimal control problems. Section 2 introduces the basic problem. In Sect. 3 we give a brief review of conjugate functions. Section 4 contains a short account on differential inclusions, and in Sect. 5 we prove the existence of optimal control. In Sects. 6, 7 and 8 we show how differential inclusions can be used for examining implicit differential equations, including a generalized dynamic input-output model.

Z. Kánnai · I. Szabó · P. Tallos (✉)
Department of Mathematics, Corvinus University of Budapest, Budapest, Hungary
e-mail: tallos@uni-corvinus.hu

© Springer Nature Singapore Pte Ltd. 2020
F. Szidarovszky and G. I. Bischi (eds.), *Games and Dynamics in Economics*,
https://doi.org/10.1007/978-981-15-3623-6_11

2   Nonconvex Control Problems

Consider the following simple optimal control problem:

$$H(x, u) = \int_0^T h(x(t), u(t)) \, dt \rightarrow \min$$
$$x'(t) = u(t), \quad x(0) = x_0 \tag{1}$$
$$u(t) \in K \quad \text{almost everywhere.}$$

The Pontryagin maximum principle (see Kánnai et al. 2014) formulates a necessary condition for the optimal control u, but provides no hint on whether or not the optimal control exists. In general, the problem of existence can be approached by exploiting rather deep results of set-valued analysis, and most works impose some sort of convexity assumptions on the functions involved. Here we refer to the classic comprehensive work of Rockafellar (1976) or the monograph by Aubin and Frankowska (1990).

Below we demonstrate that the existence of optimal control can be verified in an interesting class of optimization problems by using the technique of differential inclusions and without setting strong convexity or differentiability assumptions.

3   Conjugate Functions

Let X be a real Hilbert-space, X^* its dual and $f : X \rightarrow \mathbb{R}$ a given function.

Definition 1   The function $f^* : X^* \rightarrow \mathbb{R}$ defined by

$$f^*(p) = \sup_{x \in X} \{\langle p, x \rangle - f(x)\}$$

for every $p \in X^*$ is called the *conjuagte function* of f.

Definition 2   The *subdifferential* $\partial f(x)$ of f at $x \in X$ is defined by

$$\partial f(x) = \left\{ p \in X^* : f(x) - f(y) \leq \langle p, x - y \rangle \; \forall y \in X \right\}$$

which is a convex subset of X^*. The elements of this set are called the *subgradients* of f at the point x.

For basic properties of conjugate functions and subdifferentials we refer to the monograph by Aubin and Cellina (1984).

Proposition 1   *For every $x \in X$ and $p \in \partial f(x)$ we have $x \in \partial f^*(p)$.*

Proof In view of the definition, for an arbitrary $y \in X$ we have $\langle p, y - x \rangle \leq f(y) - f(x)$, and hence $\langle p, y \rangle - f(y) \leq \langle p, x \rangle - f(x)$. This implies $f^*(p) = \langle p, x \rangle - f(x)$, so it follows

$$f(x) = \langle p, x \rangle - f^*(p) \leq f^{**}(x).$$

Making use of the Fenchel-inequality we get $f^{**}(x) = f(x)$ and $f^{**}(x) = \langle p, x \rangle - f^*(p)$. Therefore, for any $y \in X$

$$\langle y, x \rangle - f^*(y) \leq \langle p, x \rangle - f^*(p),$$

which means $x \in \partial f^*(p)$.   □

4   Differential Inclusions

Consider a set-valued mapping F defined on $\mathbb{R}^n$, with nonempty, compact values in $\mathbb{R}^n$. Find an absolutely continuous function x with

$$x'(t) \in F(x(t)) \quad x(0) = x_0 \tag{2}$$

almost everywhere on an interval $[0, T]$. Such a function x is called a solution to the differential inclusion (2) on $[0, T]$.

A set-valued map F defined on $\mathbb{R}^n$ with nonempty values in $\mathbb{R}^n$ is said to be upper semicontinuous if for every x in the domain and every $\varepsilon > 0$ there exists a $\delta > 0$ such that

$$F(x + \delta B) \subset F(x) + \varepsilon B$$

where B stands for the unit ball in $\mathbb{R}^n$. We note that if the range of F lies in a fixed compact set, then upper semicontinuity of F is equivalent to having a closed graph. (i.e. if $y_n \in F(x_n)$ and $x_n \to x$, $y_n \to y$, then $y \in F(x)$. We refer to Aubin and Cellina (1984) for more details.)

As is well known, the differential inclusion (2) admits a solution, if the map F is upper semicontinuous and has convex values (see for instance Aubin and Cellina 1984). In fact, a fixed point argument shows that the map

$$\Phi(x) = \{y : y'(t) \in F(x(t)), \ y(0) = x_0\}$$

posesses a fixed point in the space of absolutely continuous functions. The fixed point $x \in \Phi(x)$ is then clearly a solution to the original problem.

A simple counterexample shows that the convexity assumption cannot be omitted. Indeed, if F is defined on the real line by

$$F(x) = \begin{cases} \{+1\} & \text{if } x < 0 \\ \{-1; +1\} & \text{if } x = 0 \\ \{-1\} & \text{if } x > 0 \end{cases}$$

then the differential inclusion (2) has no solution starting from the origin. The reason is the lack of convexity at $x = 0$. If F is convexified at $x = 0$, then the constant zero is the obvious solution.

However, the convexity assumption can be relaxed by assuming the existence of a convenient potential function (see Bressan et al. 1989).

Theorem 1 *Consider the (not necessarily convex-valued) set-valued F on X. Suppose that there exists a lower semicontinuous and convex function $V : X \to \mathbb{R}$ such that*

$$F(x) \subset \partial V(x) \tag{3}$$

for every $x \in X$. Then the differential inclusion (2) posesses a solution on the interval $[0, T]$.

The convexity assumption was replaced by the (weaker) lower regularity on the potential function V by Kánnai and Tallos (1998). For the extension of this result to the non-autonomous case by regularization of the set-valued map F, we refer to Tallos and Kánnai (2003). The basic constructive idea is that for the approximate solutions x_n the sequence of derivatives x_n' is not only weakly convergent, but convergent with respect to the L^2-norm as well.

5   The Existence of Optimal Control

Let K be a nonempty compact subset of the Hilbert-space X, and $c > 0$ be a given constant. Consider the real-valued functions f and g defined on X and suppose that both are continuous, and f is convex.

Introduce the following set-valued map F on X:

$$F(x) = \arg \min \{u \in K : f(u) - c\langle x, u\rangle + g(x)\} . \tag{4}$$

It is easy to verify that F has nonempty compact values in X. Moreover, in view of Berge's theorem (see Aubin and Cellina 1984) we have that F is upper semicontinuous.

Proposition 2 *For every $x \in X$*

$$F(x) \subset \partial \left(\frac{1}{c}f + \delta_K\right)^* (x)$$

where δ_K denotes the characteristic function of the set K.

Proof Take a point $u \in F(x)$ arbitrarily. Then for any $v \in K$ we obtain

$$f(u) - c\langle x, u \rangle + g(x) \leq f(v) - c\langle x, v \rangle + g(x)$$

which means

$$\langle x, v - u \rangle \leq \frac{1}{c} f(v) - \frac{1}{c} f(u).$$

Consequently, for every $v \in X$ we have

$$\langle x, v - u \rangle \leq \left(\frac{1}{c} f + \delta_K \right)(v) - \left(\frac{1}{c} f + \delta_K \right)(u).$$

In view of the definition of the subdifferential, we deduce

$$x \in \partial \left(\frac{1}{c} f + \delta_K \right)(u).$$

Now making use of Proposition 1, we conclude that

$$u \in \partial \left(\frac{1}{c} f + \delta_K \right)^{*}(x).$$

$\square$

Under the above conditions consider the following optimal control problem:

$$\int_0^T (f(u(t)) - c\langle x(t), u(t) \rangle + g(x(t))) \, dt \to \min$$
$$x'(t) = u(t), \quad x(0) = x_0$$
$$u(t) \in K \quad \text{almost everywhere},$$

which is not necessarily convex with respect to the state variable x.

Theorem 2 *The above optimal control problem has a solution.*

Proof Create the set-valued map F in (4) and consider the differential inclusion

$$x'(t) \in F(x(t))$$
$$x(0) = x_0$$

It is easy to see that every solution of this inclusion provides an optimal trajectory of the control problem. Obviously, the pointwise minimum yields the minimum value of the integral.

On the other hand, in view of Proposition 2 the potential condition (3) is fulfilled for the map F. Indeed, by introducing the function

$$V(x) = \left(\frac{1}{c}f + \delta_K\right)^*(x)$$

$(x \in X)$, the map F satisfies the conditions of Theorem 1. Therefore, the differential inclusion has a solution on the interval $[0, T]$.   □

6 Implicit Differential Schemes

Consider the differential relation

$$0 \in G(x(t), x'(t)) \tag{5}$$

where G is an upper semicontinuous map with nonempty closed images in $\mathbb{R}^n$. A solution to (5) is an absolutely continuous function x such that the inclusion holds on an interval almost everywhere.

Such problems arise naturally in economic dynamics. For instance, the classical Leontief's dynamic input-output model is governed by the implicit differential equation

$$x(t) = Ax(t) + Bx'(t) + c$$
$$x(0) = x_0$$

where A is the productivity matrix, B is the investment matrix and c stands for the final consumption. The problem can be reformulated in a more general setting:

$$x(t) - Ax(t) - Bx'(t) \in U \tag{6}$$
$$x(0) = x_0$$

where U is a given nonempty closed set in $\mathbb{R}^n$.

This implicit scheme can be regarded as a differential inclusion by introducing the set-valued map

$$H(x) = \{v \in \mathbb{R}^n : x - Ax - Bv \in U\}.$$

Then the differential relation (6) is equivalent to the explicit differential inclusion

$$x'(t) \in H(x(t)) \tag{7}$$
$$x(0) = x_0$$

This type of problem was extensively studied (including the nonautonomous case) in Kánnai and Tallos (1999). We present a more direct approach below.

7   A Set-Valued Inverse-Continuity Theorem

Let M be a nonempty convex and compact set in $\mathbb{R}^n$ and consider a set-valued map A defined on M with nonempty convex and compact images in $\mathbb{R}^n$. Set

$$S = I - A$$

where I stands for the $n \times n$ identity matrix. Introduce the set

$$V_0(S) = M \cap S(M)$$

where $S(M)$ is the range of S on M:

$$S(M) = \cup_{x \in M} S(x).$$

The effective domain of the map S is given by

$$D(S) = \{z \in M : (A(x) + z) \cap M \neq \emptyset \ \forall x \in \text{bd } M\}$$

where bd M denotes the boundary of the set M.

Consider the generalized inverse S^- of the map S that is defined by

$$S^-(z) = \{x \in M : z \in S(x)\}$$

Then S^- is a set-valued map defined on $V_0(S)$ and with closed images in $\mathbb{R}^n$. The following theorem is due to Liu and Zhang (2008) and it is an important application of the Rogalski-Cornet surjectivity theorem.

Theorem 3 *Suppose that $D(S)$ is not empty. If the set-valued map A is upper semi-continuous, then*

$$D(S) \subset V_0(S)$$

moreover S^- is upper semicontinuous with nonempty closed images.

8   Application to a Generalized Dynamic Input-Output Economic Model

Let M be a nonempty convex and compact set in $\mathbb{R}^n$ and consider a set-valued map A defined on M with nonempty convex and compact images in $\mathbb{R}^n$. The generalized dynamic input-output model is definded by the implicit differential relation

$$0 \in x(t) - Bx'(t) - A(x(t)) \tag{8}$$
$$x(0) = x_0 \in M$$

where B is an $n \times n$ matrix.

The static version of this model with $B = 0$ was studied by Liu and Zhang (2008). Following Liu and Zhang (2008), we set the following conditions.

- Feasibility condition:
$$(x - A(x)) \cap \operatorname{im} B \neq \emptyset. \tag{9}$$

- Regularity condition:

$$\exists z \in M \, \forall v \in \operatorname{bd} M \, \exists x \in M : v + z - x \in M, \; Bv \in A(v + z - x) - (v + z - x)$$

Theorem 4 *Under the above conditions the inclusion (8) has at least one solution for every x_0 in M.*

Proof First introduce the following set-valued map on M

$$F(x) = \{v \in \mathbb{R}^n : 0 \in x - Bv - A(x)\}.$$

In view of condition (9) the values of F are not empty, and they are obviously closed and convex sets. It is easy to verify that relation (8) and the explicit differential inclusion

$$x'(t) \in F(x(t)) \tag{10}$$
$$x(0) = x_0$$

have precisely the same solution set.

Now consider another set-valued map defined by

$$S(v) = \{x \in M : Bv \in x - A(x)\}$$

Then S has a closed graph, since the graph of A is closed. Further, the values are in a fixed compact set M, consequently S is upper semicontinuous. For the generalized inverse we have:
$$S^-(x) = \{v \in \mathbb{R}^n : x \in S(v)\} = F(x).$$

By using the notations of the previous section we also observe that the regularity condition and the feasibility condition (9) imply that $D(S) \neq \emptyset$. Making use of Theorem 3 we conclude that F is upper semicontinuous.

Exploiting the classical result of the theory of differential inclusions (see Aubin and Cellina 1984), we get that the initial value problem

$$x'(t) \in F(x(t))$$
$$x(0) = x_0$$

has a solution for every x_0 in M that yields a solution to (8).   □

References

Aubin, J.-P., & Cellina, A. (1984). *Differential inclusions*. Berlin, Heidelberg, New York, Tokyo: Springer-Verlag.

Aubin, J.-P., & Frankowska, H. (1990). *Set-Valued analysis*. Boston, Basel, Berlin: Birkhäuser Verlag.

Bressan, A., Cellina, A., & Colombo, G. (1989). Upper semicontinuous differential inclusions without convexity. *Proceedings of the American Mathematical Society, 106*, 771–775.

Kánnai, Z., Szabó, I., & Tallos, P. (2014). *Calculus of variations and optimal control (in Hungarian)*. Budapest: Typotex. http://www.tankonyvtar.hu.

Kánnai, Z., & Tallos, P. (1998). Potential type inclusions, Lecture Notes in Nonlinear. *Analysis, 2*, 215–222.

Kánnai, Z., & Tallos, P. (1999). Selections, differential inclusions and economic modeling. *Szigma, 29*, 213–220.

Liu, Y., & Zhang, Q. (2008). The Rogalski-Cornet theorem and a Leontief-type input-output inclusion. *Nonlinear Analysis, Theory, Methods and Applications, 69*, 425–433.

Rockafellar, T. (1976). Integral functionals, normal integrands and measurable selections. In J. Mawhin (Ed.), *Nonlinear operators and the calculus of variations*, Lecture Notes in Mathematics, vol. 543. Berlin, Heidelberg, New York: Springer.

Sethi, S., & Thompson, G. (2006). *Optimal control theory. Applications to management sciences and economics*. Berlin, Heidelberg, New York: Springer.

Tallos, P., & Kánnai, Z. (2003). Viable solutions to nonautonomous inclusions without convexity. *Central European Journal of Operations Research, 11*, 47–55.

Agent Behavior and Transitions in N-Person Social Dilemma Games

Ugo Merlone, Daren R. Sandbank and Ferenc Szidarovszky

Abstract One of the most common way of analyzing the evolution of networks of interacting agents is based on simulation. The interactions among the agents can be modeled by N-person social dilemma games, and the evolution of the network can be characterized by repeated transitions. The nature of the transitions as well as the long-term behavior of the state of the network depend on the structures of the particular games under consideration. The purpose of this paper is to present the transitions that occur between N-person social dilemma games, the agent behavior causing these transitions, and the impacts of various parameters. Formerly, N-person social dilemma games such as Prisoners' Dilemma, Chicken, Stag Hunt and Battle of the Sexes have been separately researched and analyzed. In this paper the specific behavior of Pavlovian agents are explored in a two dimensional cellular automaton environment using parameters that extend over all of these games. It is found that there are three significantly different types of agent behavior: bipartisan, partisan, and unison. Bipartisan agents stochastically decide to cooperate or defect based on a cooperating probability greater than zero and less than one, partisan agents either cooperate with certainty or defect with certainty, and unison agents cooperate 100% of the time. Each agent behavior can be associated with a set of social dilemma games and be represented as plateaus on a three dimensional graph. These plateaus themselves and the transitions between them are investigated in terms of where and why they occur. Lastly, this paper reviews the impact of initial cooperating probability, neighborhood size, learning factors, and grid size on these plateaus and transitions.

U. Merlone (✉)
Department of Psychology, Center for Logic, Language, and Cognition,
University of Torino, via Verdi 10, 10124 Torino, Italy
e-mail: ugo.merlone@unito.it

D. R. Sandbank
Systems and Industrial Engineering Department, University of Arizona,
Tucson, AZ 85721-0020, USA

F. Szidarovszky
Department of Mathematics, Corvinus University of Budapest, Fővám tér 8, Budapest
1093, Hungary

© Springer Nature Singapore Pte Ltd. 2020
F. Szidarovszky and G. I. Bischi (eds.), *Games and Dynamics in Economics*,
https://doi.org/10.1007/978-981-15-3623-6_12

205

1 Introduction

Agent-based social simulation started in the 1990s (Davidsson 2001, 2002) and this research area is growing significantly. The main advantage of agent-based simulation is that it is a bottom-up approach to analyze networks of agents where the agents' attributes may include behavioral traits, characteristics and learning capabilities. Agent-based simulation can be used to study artificial societies (Epstein and Axtell 1996) for emergence of groups with common attributes or social segregation. In this paper the specific behavior of Pavlovian agents in a two dimensional cellular automaton environment to cooperate or defect is based on reinforcement learning with linear payoff functions. This model has been used to study several real world applications in the literature. In Power (2009) the towns of Catalina, New Foundland and Labrador, Canada are modeled as a cellular automaton environment with Pavlovian agents, who's behaviors are identical to those studied in this paper in order to examine the collective communication and N-person prisoner's dilemma cooperation within a socio-geographic community. Pavlovian agents are considered also in Szilagyi (2009) in order to model and examine an N-person chicken dilemma game where agents in a large city decide to either cooperate with each other for the collective best interest and use public transportation or defect and drive their car. Other potential fields where this type of model is applicable include military expenditures, oil cartels, and climate change. In each of these areas countries or agents can be modeled in a cellular automaton environment where each must decide to cooperate for the collective best interest or defect for their own self interest. The collective best interest for military expenditures would be to limit arms production, for oil cartels to set monopolistic prices and market share, and for climate change to curb CO_2 emissions. The self interest temptation for each country or agent respectively is to raise arms production and become a dominate military force, lower price and increase profits by picking up more market share, and allow CO_2 emissions and spend less to save the environment. The worst situation is if all agents defect since in each case discussed above the country or agent would be negatively impacted without getting any competitive advantage.

Two player social dilemma games are defined with a payoff matrix shown in Table 1. The rows show the strategies of player 1, and the columns indicate the strategies of player 2. For each corresponding strategy pair the first number in each position gives the payoff of player 1, and the second value is the payoff of player 2. The parameters P, R, S and T are derived from the Prisoners' Dilemma game and are referred to as Punishment, Reward, Sucker's Bet, and Temptation respectively. The Prisoners' Dilemma game will be described in more detail in upcoming paragraphs.

A multi-person or N-person extension of the model takes into account the collective behavior in society where agents of a network may cooperate with each other for the collective best interests or defect to pursue their own self interest. This paper assumes a two dimensional cellular automaton environment where each agent is represented by a cell on a rectangular grid and may interact with its immediate neighbors or with all the agents as a collective set. An example of a cellular automaton environ-

Table 1 Payoff matrix for two player games

2 1	Cooperate	Defect
Cooperate	R R	T S
Defect	S T	P P

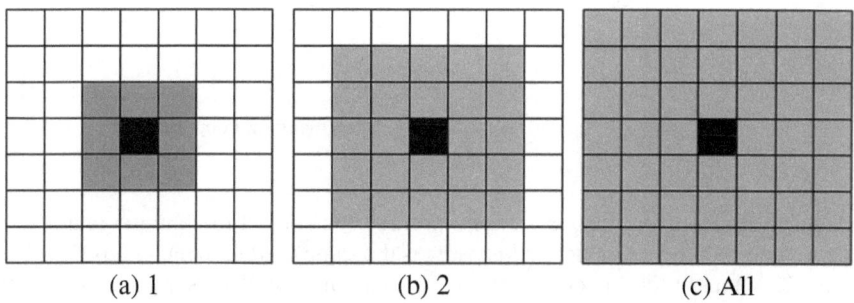

(a) 1   (b) 2   (c) All

Fig. 1 Cellular automaton environment with Moore neighborhoods

ment with Moore neighborhoods is shown in Fig. 1. In this figure the gray cells are the neighbors for the black cell with Moore neighborhoods of one, two, and all. For a discussion about spatial structures in agent-based models and how explicit spatial (lattice based) structures are related to networks, the reader may refer to Ausloos et al. (2015).

In each iteration the agents decide whether to cooperate or defect based on a certain probability distribution. After each agent chooses to cooperate or defect a reward or punishment is received that depends on the accumulated choices of the other agents in its designated neighborhood. The amount of reward or punishment an agent receives is derived from a payoff function. A typical linear payoff function is shown in Fig. 2. In this payoff function x is the percentage of cooperators, $C(x)$ is the payoff for those agents that are cooperating and $D(x)$ is the payoff for those agents that are defecting.

In an N-person social dilemma game each agent is characterized by a behavioral styles which dictates how the agent will decide to cooperate or defect. In Szilagyi (2003) several potential behavioral style are presented including Greedy, Conformist, and Pavlovian. Greedy agents duplicate the decision of the agent in their neighborhood that receives the highest payoff. Conformist agents make their decision in line with the majority in their neighborhood. Pavlovian agents base their decision on reinforced learning. This paper will deal with the Pavlovian agent type only. Reinforcement learning has been used extensively in the literature for two player repeated games (Bush and Mosteller 1955; Macy and Flache 2002; Flache and Macy 2002) and is based on Thorndike's law of conditioning based on Pavlov's experiments

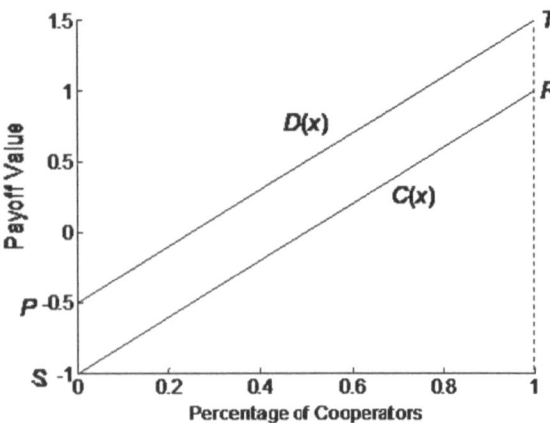

Fig. 2 Linear payoff functions for cooperators (C) and defectors (D)

where positive outcomes to an action reinforce the subject to continue that action (Thorndike 1911). The model in this paper is the same used in Szilagyi (2003), Zhao et al. (2007) and Merlone et al. (2018). They all analyze N-person social dilemma games, so the results found here can be related directly to those papers. In this model the Pavlovian agent has a certain probability p of cooperating in each time period, which changes for the next time period or iteration by a proportion of the reward or punishment received. The probability that an agent i will be cooperating at time period t can be computed as

$$
p_i(t) = \begin{cases} p_i(t-1) + \alpha C(x(t-1)) & \text{if the agent cooperated at time period } t-1 \\ p_i(t-1) - \beta D(x(t-1)) & \text{if the agent is a defector at time period } t-1 \end{cases}
$$

$$(1)$$

where $x(t-1)$ is the percentage of cooperating agents in the population at time period $t-1$, α is the proportion or learning factor for cooperators and β is the learning factor for defectors. α and β are set parameters between zero and one. Since $p_i(t)$ is a probability it must also be between zero and one. So whenever $p_i(t)$ becomes larger than one it is adjusted to be one. Likewise, whenever $p_i(t)$ becomes negative it is adjusted to be zero. A similar model extensively analyzed for two player repeated games in the literature is a variant of Bush and Mosteller (1955) linear stochastic model which was proposed by Macy and Flache (2002), Flache and Macy (2002). In these models a stimulus is calculated for any action based on a payoff and an aspiration level. This stimulus with a learning factor is used in an algorithm similar to Eq. (1) where the probability to cooperate is increased if either the cooperating agent is rewarded or defecting agent is punished; and the probability to cooperate is decreased if either the cooperating agent is punished or defecting agent is rewarded. In our model we use a similar concept with the assumption that the aspiration level is zero and the payoff values are the stimulus. Finally, it should be observed that, as in Pavlonian agents the process in which learning occurs is a function of the consequences of behavior, they are called, more correctly Skinnerian agents in Merlone et al. (2013).

The model works as follows. An instantiation for the payoff parameters (P, R, S, T) and the learning factors (α, β) are given. The initial conditions and simulation parameters required for a single simulation run are the initial cooperating probability for each agent, the cellular automaton grid size, the neighborhood size, and the number of iterations to be performed. To start the simulation each agent in the cellular automaton grid is given the initial probability to cooperate. Time is moved forward in iterations. In the first iteration agents simultaneously decide to cooperate or defect based on the assigned initial probabilities to cooperate. Then the percentage of cooperators of the entire grid is determined and used as a statistic to evaluate the state of the system. The payoffs are then computed for each agent. Each agent's payoff is attained by determining the percentage of cooperators x in that specific agent's neighborhood and implementing the proper payoff function, which is $C(x)$ if that agent cooperated in that iteration or $D(x)$ if that agent defected. Then each agent's probability to cooperate is adjusted by using Eq. (1). The second iteration then begins. In the second iteration the agents now simultaneously decide to cooperate or defect based on their own updated probabilities to cooperate. The system percentage of cooperators is determined and each agent's probability to cooperate is updated using the same methodology as described for the first iteration. This repeats for the designated number of iterations. Again, the main statistic tracked for the system is the percentage of cooperators of the entire population in each iteration.

The Prisoners' Dilemma game is an important example of a social dilemma which is frequently examined in the literature. In this game two suspects are arrested and separated by the police. If after questioning both suspects cooperate by remaining silent, each one is sentenced to minimal jail time on a lesser charge. If one suspect testifies against the other (defects) and the other stays silent (cooperates), the defector is released and the cooperator is sentenced to full jail time on the accused charge. If both suspects defect by testifying against each other, then both are sentenced to reduced jail time on the accused charge. In accordance with the above description a Prisoners' Dilemma game occurs when the parameters of the payoff functions satisfy the relation $T > R > P > S$. There are several social games other than Prisoners' Dilemma with much literature including Chicken which is characterized by the inequality $T > R > S > P$, Stag Hunt which follows when $R > T > P > S$, and Deadlock when $T > P > R > S$ is satisfied. These games have a colorful story and many real life applications including such topics as world politics, law, and other fields (Poundstone 1992; McAdams 2009; Axelrod and Keohane 1985). Other social games characterized in this paper include Battle of the Sexes (Zhao et al. 2008), Harmony, Coordination, Leader and two unnamed games. Harmony occurs when the highest payoff is received when both players or agents cooperate and the lowest payoff is received when both agents defect. This occurs when $R > S > P$ and $R > T > P$. The Coordination game is derived when agents receive higher payoffs if they coordinate their decision by either both cooperating or defecting. That is, they receive the highest payoff if both cooperate and the next highest payoff is if both defect. This occurs when $R > P > S$ and $R > P > T$. The Leader game is defined when two drivers want to merge into traffic from opposite directions when a single gap opens up. The highest payoff is for the agent who defects by entering the single gap while the other agent who cooperates by waiting gets the second highest payoff. The next highest payoff is that they both cooperate and wait for a gap large enough for both to enter. The lowest payoff is if both defect by entering the single gap and collide. Leader occurs when $S > R > P$.

In order to represent these games graphically, which will be done throughout this paper, we assume without loss of generality that $R > P$ for all games and then normalize the payoff function values such that $R = 1$ and $P = 0$. If $R < P$ for a particular game then we simply interchange the definitions of cooperation and defection to derive an equivalent game with $R > P$. For example, Battle of the Sexes is a game where a man and a woman are deciding where to go out on a date. The man prefers a sporting event while the woman prefers ballet. However, both of them prefer going out together rather than alone. Cooperation is defined as a person going to the event the other person prefers and defection is going to the event he or she prefers. If both cooperate (man goes to ballet and woman goes to sporting event) than both are most disappointed because they both are going to an activity they do not like and they are by themselves. If one person defects and the other person cooperates, then the defector is most happy since this person is going to an event he/she likes with their partner and the cooperator is somewhat happy because she/he is with the other person. If both defect and go to the event they enjoy alone, they are not as happy as if they could be if they went somewhere with their partner. Battle of the Sexes is typically characterized by the inequality $T > S > P > R$. However, if we simply interchange the definition of cooperate to be going to the event the individual prefers and defect to be to going to the activity the other individual prefers, then the inequality for the same story becomes $S > T > R > P$. This is an equivalent game with $R > P$. This method is also used for the Deadlock game discussed above where $T > P > R > S$ becomes $S > R > P > T$ with a different definition of cooperate and defect.

Using the above convention, each game can be represented by a region in the S, T plane as shown in Fig. 3. This type of graphical representation has been used to analyze various topics in the literature including the efficiency of adapting aspiration levels (Posch et al. 1999), fundamental clusters in spatial 2×2 games (Hauert 2001) and effects of space in 2×2 games (Hauert 2002). The ordering of R, S, T, and P with the normalized values $R = 1$ and $P = 0$ divides the plane into 12 regions where the various games are depicted graphically by name and color. That is, Prisoners' Dilemma is teal, Chicken is red, Leader and Battle of the Sexes is green, Stag Hunt is dark blue, Harmony is yellow, Coordination is light blue, and Deadlock is white. Unnamed games are colored brown and magenta. These color designations for specified games will be used throughout much of this paper.

As stated above, in Fig. 3 it is assumed that $R = 1$ and $P = 0$. Later in this paper this assumption will be dropped and similar graphical representations will be presented with various values of R and P. This will not change the relative location of the games in these figures as long as $R > P$, which as stated above will be assumed without loss of generality. Furthermore, for the first part of this paper it will be

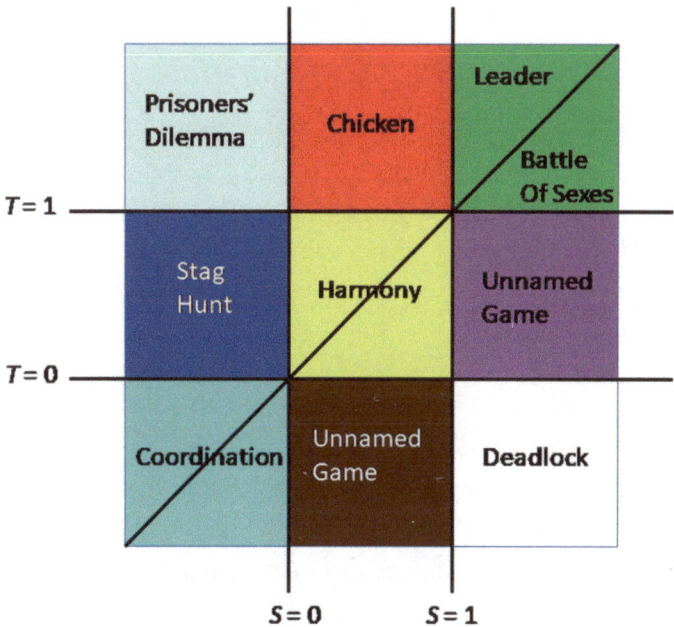

Fig. 3 Classifications of games

assumed that R is positive and P is negative. This is a reasonable assumption since R represents reward which we define as a positive payoff and P represents punishment which will be considered a negative payoff. Later in the paper, the special cases where R and P are either both positive or both negative will be investigated.

Figure 4 shows the simulation results using the linear payoff function parameters as shown in Fig. 2 ($R = 1$, $P = -0.5$, $T = 1.5$, and $S = -1$) and equal learning factors ($\alpha = \beta = 0.05$). The initial conditions and simulation parameters for this run are the initial cooperating probability 0.5, cellular automaton grid size 50×50, neighborhood size 50, and number of iterations 100. A neighborhood size 50 in this simulation essentially means the neighborhood is the entire automaton grid or all of the other agents. The figure depicts the percentage of cooperators after each iteration. The results show that the percentage of cooperators reaches a steady state just under 0.2 after about 50 iterations. In order to evaluate the dynamics of the initial conditions this simulation was repeated 1000 times. The simulation results over the 1000 runs were very similar having an average 0.1801 with a standard deviation 0.0080. This is about the same mean and standard deviation that can be seen in Fig. 4 after 50 iterations when the system becomes stable (mean 0.1798 with a standard deviation 0.0072). The variation of the final iteration of 1000 runs has a similar mean and standard deviation as a single iteration after it becomes stable.

A three dimensional simulation plot with varying payoff function parameters using the above representation will now be presented. Figure 5 shows the simulation

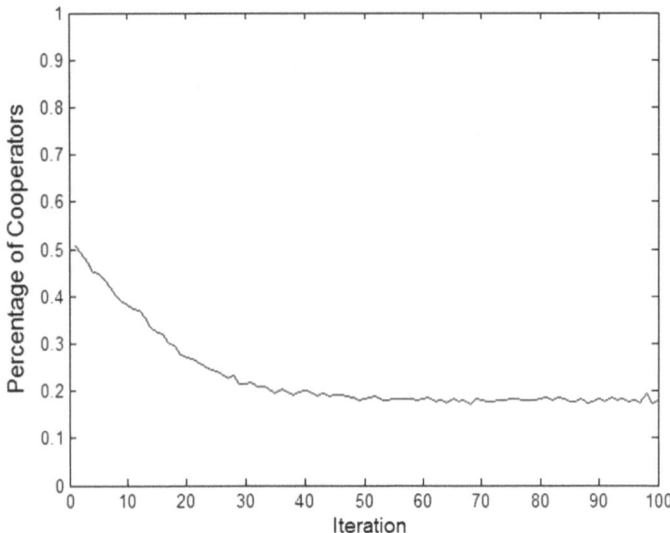

Fig. 4 Simulation results for a single set of payoff function parameters

results where $R = 1$ and $P = -1$, with T and S varying between -10 and 10. Each intersection in the mesh represents a separate simulation run on a 50×50 cellular automaton grid with each Pavlovian agent having an initial cooperating probability 0.5 and equal learning factors 0.05. The height of the plot is the percentage of cooperators after the final iteration of each simulation run. For this simulation each T and S axis is broken into 40 subdivisions. Thus Fig. 5 represents the results of 1600 single simulation runs, each with different values of T and S. Mesh intersections in close proximity have close T and S values. It can be seen in this figure that individual single runs with T and S points in close proximity to each other do yield similar results. In any case, small variances or any outliers that occur from single runs are readily seen as irregularities in the planar structure of this mesh graph. The color coding depicts the associated game as shown in Fig. 3. There are three plateaus in these simulation results. The first plateau is predominately the Prisoners' Dilemma region where T is large positive and S is large negative, but also includes portions of Chicken and Stag Hunt. In this paper this plateau is called the Prisoners' Dilemma plateau. The second plateau contains Leader and Battle of the Sexes games where both T and S are large positive. This plateau is a step higher than the Prisoners' Dilemma plateau. This plateau is called the Leader/Battle of the Sexes plateau. The third plateau includes the Harmony and Deadlock games where T is large negative and S is large positive. All of the agents are cooperating in this plateau. It is called the Harmony/Deadlock plateau. The purpose of this paper is to understand the agent

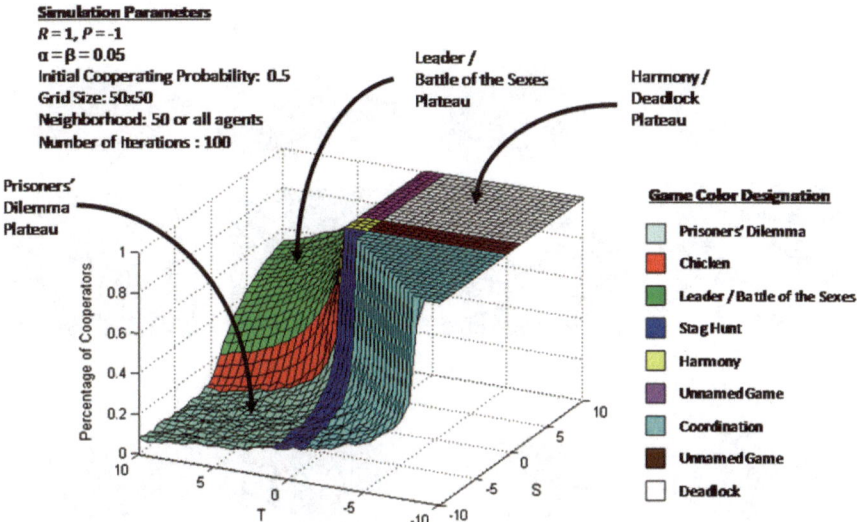

Fig. 5 Simulation result with varying *S* and *T* values

behavior in these plateaus and the transitions that occur between them. The impacts of the initial cooperating probability, neighborhood size, learning factors, and grid size on these plateaus and the transitions between them are also evaluated.

2   Agent Behavior on the Plateaus

2.1   Prisoners' Dilemma Plateau

The Prisoners' Dilemma plateau is the teal flat region in the simulation results shown in Fig. 5 where *T* is a high positive value and *S* is a high negative number. It is predominately the Prisoners' Dilemma game, but also includes portions of Chicken, Stag Hunt, and Coordination games.

Figure 6 shows the agent activity in the Prisoners' Dilemma plateau for a specific realization with $R = 1$, $P = -1$, $T = 5$, and $S = -5$. The agent simulation is run on a 10×10 cellular automaton grid with each Pavlovian agent having an initial cooperating probability 0.5 and equal learning factors 0.1. Each of the first eight iterations is shown in the figure. The black and white two dimensional charts show the decision of each agent at each iteration; white is if the agent cooperates and black is if the agent defects. In the first iteration shown in the top left hand corner it can be seen that 46 of the 100 agents are cooperating and that the number reduces to 14 in the eighth iteration shown in the bottom right hand corner. Also, it can be seen that the agents stochastically change their decision from iteration to iteration. The three

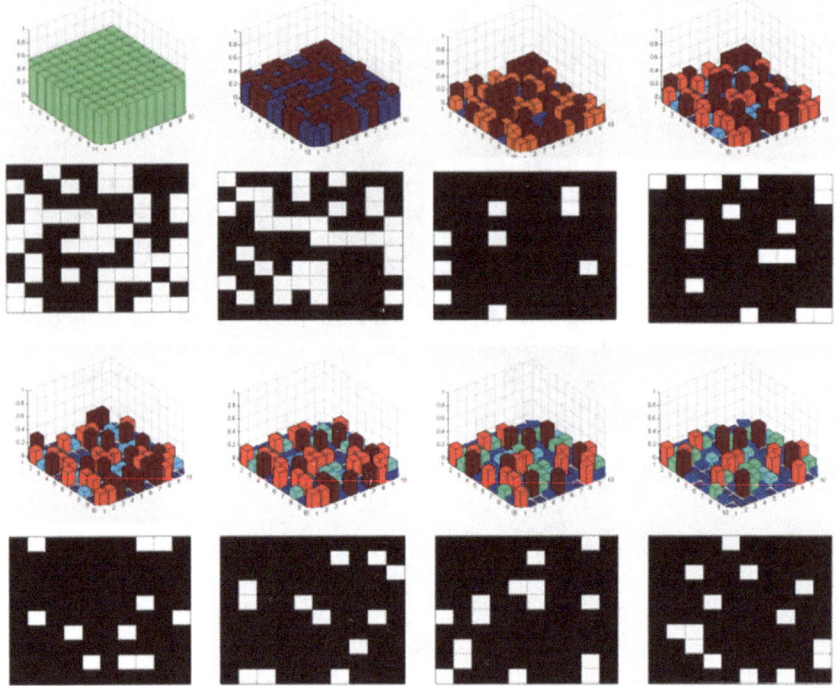

Fig. 6 Agent activity in Prisoners' Dilemma plateau

dimensional color charts show the cooperating probability for each agent at each iteration. In the first iteration, in the top left, each agent starts with an initial cooperating probability 0.50 and in the following iterations the cooperating probability fluctuates around a steady state equilibrium around 0.15. This activity of having fluctuation in the cooperating probability around a steady state equilibrium between zero and one is unique to this plateau. Agents with this unique trait are called bipartisan in this paper since they are willing to change their minds between cooperating and defecting from iteration to iteration based on their cooperating probability being greater than zero and less than one.

The steady state equilibrium for this model where agents are acting in a bipartisan manner by adjusting the cooperating probabilities around a stable value in the Prisoner's Dilemma game has been already analyzed in the literature (Szilagyi 2003; Merlone et al. 2018). In Szilagyi (2003) it was presented that the stabilization point for the percentage of cooperators in this system occurs when $x^*C(x^*) = (1 - x^*)D(x^*)$. Equal learning factors were assumed and it was described that this stabilization point occurs when the total payoff for cooperators equals the total payoff for the defectors. In Merlone et al. (2018) the stabilization point was expanded to allow for different learning factors and the equilibrium state was derived analytically by determining the percentage of cooperators from the expected value of the probability of each Pavlovian agent to cooperate and then finding the steady state equilibrium. The stabilization point in the more general system occurs when $x^* \alpha C(x^*) = (1 - x^*) \beta D(x^*)$. Note that if $\alpha = \beta \neq 0$ then this equation simplifies to $x^* C(x^*) = (1 - x^*) D(x^*)$. For linear payoff functions $x^* \alpha C(x^*) = (1 - x^*) \beta D(x^*)$ becomes the solution to a quadratic equation with zero, one or two steady states.

It is discussed in Szilagyi (2003) that the steady state solutions may be attractors or repellers. An attractor occurs when the equilibrium solution is such that the payoffs of cooperating and defecting agents are both negative. Since both types of agents are punished the equilibrium solution is an attractor because an agent's cooperating percentage decreases when cooperating and increases when defecting. In this state an agent is unlikely to cooperate or defect several iterations in a row since the cooperating percentage would change significantly in the direction that causes different decision. This leads to agents alternating between cooperating and defecting. Alternatively, a repeller equilibrium occurs when both cooperating and defecting agents are rewarded with positive payoff. A diversion from the analytical solution occurs because agents that receive reward for their behavior will have their probability to cooperate move in a direction that will make repeating that decision more likely.

Figure 7 shows the simulation results for a case in the Prisoner's Dilemma plateau. The payoff parameters are $R = 1, P = -1, S = -1.5, T = 1.5$. The parameters for this simulation are a grid size 50×50, neighborhood all, 100 iterations and equal learning factors 0.05. This plot shows the percentage of cooperators in each iteration with the initial cooperating percentages of $0.0, 0.1, 0.2, 0.3, 0.4, 0.5, 0.6, 0.7, 0.8, 0.9$ and 1.0. The simulation results for these 11 separate simulations are shown in blue and the solutions to the equilibrium equation $x^* \alpha C(x^*) = (1 - x^*) \beta D(x^*)$ are shown in red. It is clear that the steady state solution 0.2764 is attracting and the solution 0.7236 is repelling.

2.2 Leader/Battle of the Sexes Plateau

The Leader/Battle of the Sexes plateau is the green flat region in the simulation results shown in Fig. 5 where both T and S are high positive numbers. It is predominately the Leader and Battle of the Sexes games, but also includes a portion of the Chicken game.

Figure 8 shows the agent activity in the Leader/Battle of the Sexes plateau for a specific realization $R = 1, P = -1, T = 5$, and $S = 5$. The same Pavlovian agent type, initial cooperating probability, learning factors, and cellular automaton setup are used as in the case of Fig. 6. As in Fig. 6 simulation, eight iterations are shown. The simulation results for this scenario show that the cooperating probability for the agents either decreases to zero and remains there or increases to one and remains at unit level. Each agent either continually defects thereafter if its cooperating probability is zero or continually cooperates if its cooperating probability is one. This activity where agents eventually cooperate or defect continually is a significantly different case than the bipartisan behavior seen in the Prisoners' Dilemma plateau.

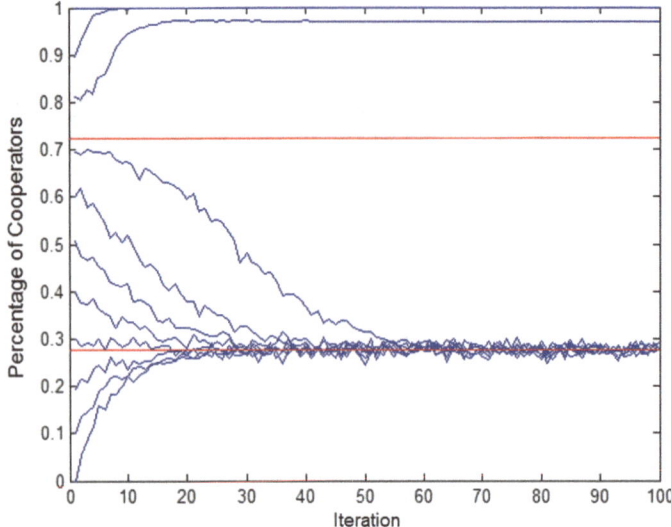

Fig. 7 Percentage of cooperators in Prisoner's Dilemma plateau with initial cooperating probabilities 0.0, 0.1, 0.2, 0.3, 0.4, 0.5, 0.6, 0.7, 0.8, 0.9 and 1.0

Agents with this trait are going to be called partisan in this paper since they will always defect or cooperate forever once their probability to cooperate reaches either zero or one, which will always be in this plateau.

In this plateau the analytical solution presented in Merlone et al. (2018) does not apply. In steady state a proportion of the agents are cooperating with certainty and the rest are defecting with certainty. The method used to derive the analytical solution in Merlone et al. (2018) does not account for this activity of limiting and adjusting the cooperating probability to zero and one. The rest of this section will present the reason why this type of activity occurs and provides an estimate of the number of partisan cooperators and defectors that will be in the system at the steady state.

In this plateau the payoff parameters R, S and T are all positive. For low percentage of cooperators this percentage will rise since cooperation is rewarded (since S and T are positive) and defection is punished (since P is negative). At some point the percentage of cooperators rises and the payoff for defecting becomes positive (since T is positive). For these higher percentages of cooperators the agents are rewarded whether they cooperate or defect. Some agents will follow a trajectory where they cooperate several iterations in a row and then cooperate with certainty and the rest of the agents will follow a trajectory where they defect several iterations in a row and then defect with certainty.

The percentage of cooperators at the end is determined by the number of agents on the continual cooperating trajectory and also by agents on the continuous defecting trajectory. In an extreme case (high S, high T) the cooperating probability for each agent will jump to zero or one in a single iteration and remains there. This means that in this extreme case the percentage of cooperators will

Fig. 8 Agent activity in Leader/Battle of the Sexes plateau

approximately be the initial cooperating probability. A simulation of this case is shown in Fig. 9 with $R = 1$, $P = -1$, $S = T = 30$. This plot shows the percentage of cooperators in each iteration given the initial cooperating percentages of $0.0, 0.1, 0.2, 0.3, 0.4, 0.5, 0.6, 0.7, 0.8, 0.9$ and 1.0. For the first iteration the initial cooperating probability and the percentage of cooperators should be approximately the same since the percentage of cooperators depends on the cooperating probability which is the same for all agents in the first iteration. It can be seen for the low initial cooperating probability that the percentage of cooperators increases and for higher initial cooperating probabilities the percentage of cooperators ends up being approximately the initial cooperating probability. This is to be expected from the above discussion.

2.3 Harmony/Deadlock Plateau

The Harmony/Deadlock plateau is the top flat region in the simulation shown in Fig. 5 where T is a high negative value and S is a high positive number. It is predom-

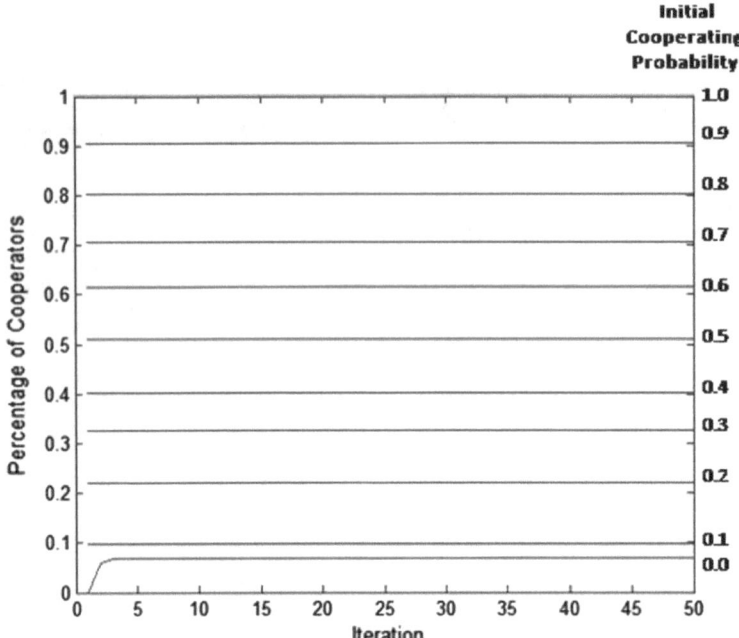

Fig. 9 Percentage of cooperators in extreme case ($S = T = 30$) in the Leader/Battle of Sexes plateau

inately the Harmony, Deadlock, and Unnamed games, but also includes a portion of Coordination game. In this plateau all of the agents cooperate 100% of the time.

For much of the region in this plateau including the Harmony and Deadlock games the payoff for cooperation is positive and the payoff for defection is negative for all percentages of cooperators. That is, cooperating is rewarded and defecting is punished. In this case each agent's probability to cooperate will increase by Eq. (1) regardless if they cooperate or defect. Eventually each agent's probability to cooperate will reach unity and each agent will cooperate with certainty thereafter.

In an unnamed and in the coordination game S can be negative meaning that the payoff for cooperation will become negative when the percentage of cooperators is small and positive when the percentage of cooperators is high. In the region where R is positive and the other payoff parameters S, T and P are negative the determination whether the state of the system results in all agents cooperating with certainty in the Harmony/Deadlock plateau or fluctuating their probability to cooperate around a stable percentage in the Prisoner's Dilemma depends on the values of the payoff parameters. This transition where this dramatic change in behavior occurs is discussed in the next section.

Figure 10 shows the simulation results for a case in the Harmony/Deadlock plateau. The payoff parameters are $R = 1$, $P = -1$, $S = 2$, $T = -2$. The parameters for this simulation are a grid size 50×50, neighborhood all, 100 iterations and equal

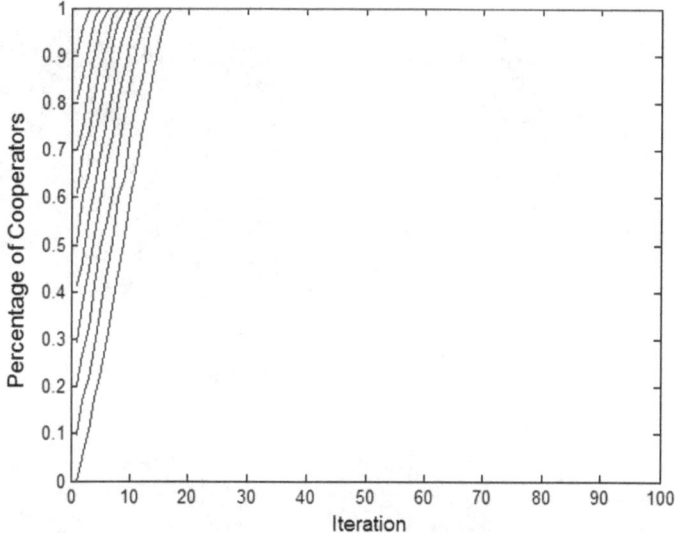

Fig. 10 Percentage of cooperators in Harmony/Deadlock plateau with initial cooperating probabilities 0.0, 0.1, 0.2, 0.3, 0.4, 0.5, 0.6, 0.7, 0.8, 0.9 and 1.0

learning factors 0.05. This plot shows the percentage of cooperators in each iteration with the initial cooperating percentages of 0.0, 0.1, 0.2, 0.3, 0.4, 0.5, 0.6, 0.7, 0.8, 0.9 and 1.0. The results for these 11 separate simulations show that the percentage of cooperators always rises to one regardless of the initial cooperating probability.

3   Analysis of Transitions Between Plateaus

This section will analyze the transitions between the three previously defined plateaus. Figure 11 shows these plateaus and transitions. These simulation results use the same parameter values as are used for Fig. 5 except with a different coloring scheme. In this figure the color is dependent on the percentage of cooperating agents at the end of the run. The left hand side plot is an isometric view and the right hand plot is a top view for the same simulation result data. In these plots the plateaus and transitions can be clearly seen.

In order to evaluate a wider solution set Fig. 12 repeats Fig. 11 for various positive *R* and negative *P* values. The results show that the existence of three plateaus with associated transitions occurs for a wide range of positive *R* and negative *P* values. The transition between the Prisoners' Dilemma and Harmony/Deadlock plateaus will first be investigated. Then the transition between the Prisoners' Dilemma and Leader/Battle of the Sexes plateaus will be examined. Finally the transition between the Leader/Battle of the Sexes and Harmony/Deadlock plateaus will be studied. After

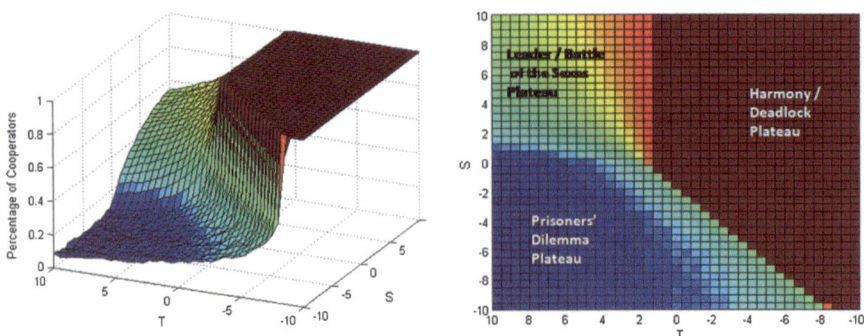

Fig. 11 Simulation result showing three plateaus and transitions

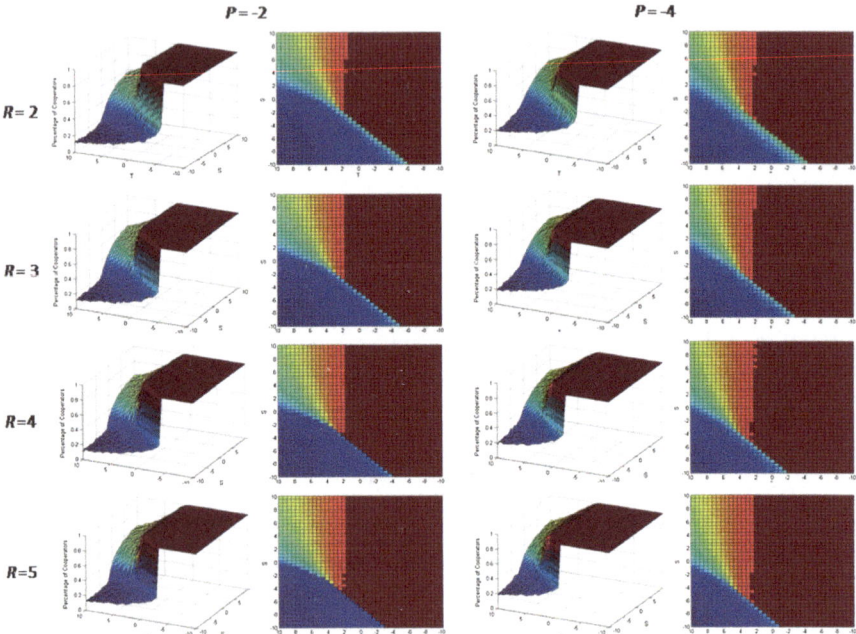

Fig. 12 Simulation results with varying P and R values

that we will look at the two special cases where R and P are either both positive or both negative.

First we will look at the transition between the Prisoners' Dilemma and Harmony/Deadlock plateaus. This is a steep transition where the agent behavior changes from bipartisan to unison. By simple observation it is apparent that this transition is a ridge or line starting in Harmony and runs though Stag Hunt and sometimes into Coordination. To evaluate where this ridge or line occurs we first have to review analytical solutions to the Prisoners' Dilemma (Szilagyi 2003; Merlone et al. 2018).

As previously discussed in these papers it is proved that for at least a portion of the Prisoners' Dilemma the analytical steady state solution x^* for Pavlovian agents in a cellular automaton environment is the solution of the equation $x^*\alpha C(x^*) = (1 - x^*)\beta D(x^*)$ which simplifies to $x^* C(x^*) = (1 - x^*)D(x^*)$ when $\alpha = \beta \neq 0$. For the linear payoff functions $C(x) = S + (R - S)x$ and $D(x) = P + (T - P)x$ the analytical steady state solution is derived by solving a quadratic equation. The solution is

$$x^* = \frac{-\alpha S - 2\beta P + \beta T \pm \sqrt{\alpha^2 S^2 - 2\alpha\beta ST + \beta^2 T^2 + 4\alpha\beta P R}}{2(\alpha R - \alpha S + \beta T - \beta P)}.$$

Since x^* is the steady-state, it has to be real meaning the value under the square root must be nonnegative. Thus

$$\alpha^2 S^2 - 2\alpha\beta ST + \beta^2 T^2 + 4\alpha\beta P R \geq 0$$

$$(\beta T - \alpha S)^2 \geq -4\alpha\beta P R$$

$$|\beta T - \alpha S| \geq 2\sqrt{-\alpha\beta P R}. \tag{2}$$

It is expected that the analytical solution will not work with complex roots, but additionally it appears that the line $\beta T - \alpha S = 2\sqrt{-\alpha\beta P R}$ is in fact the boundary in this transition region. If we set the values of α, β, P and R then this equation gives a straight line in the S, T plane. Figure 13 repeats Fig. 5 with this line shown in white on the plot. It is evident that in this example the transition is occurring when the roots to the analytical solution become complex. This is confirmed also for all examples given in Fig. 12. The other line generated from Eq. (2), $-\beta T + \alpha S = 2\sqrt{-\alpha\beta P R}$, is inside the region when cooperating probabilities are adjusted to unity, that is, where the analytical solution cannot be applied due to unison agent behavior. So this second line has no meaning and has no effect on the structure of the steady state.

Next we will look at the transition between the Prisoners' Dilemma and Leader/Battle of the Sexes plateaus. This transition is a step increase shown in Fig. 5 where the agent behavior changes from bipartisan to partisan. The reason for this transition is that the equilibrium attractor solution that occurs when cooperating and defecting are both punished as discussed for the Prisoner's Dilemma plateau disappears when transitioning into the Leader/Battle of the Sexes plateau. This specifically happens when S goes from negative to positive. The result of this is that bipartisan behavior will not work when $S > 0$ because this equilibrium attractor solution is nonexistent. Figure 13 repeats Fig. 5 with $S = 0$ as a gray line where this transition occurs. This transition is also confirmed for all examples given in Fig. 12.

Now we will look at the transition between the Leader/Battle of the Sexes and Harmony/Deadlock plateaus. This transition is a steep transition shown in Fig. 5 where the agent behavior changes from partisan to unison. The reason for this transition is that in the Harmony/Deadlock plateau cooperators are rewarded and defectors are punished for all percentages of cooperators where in the Leader/Battle of the Sexes

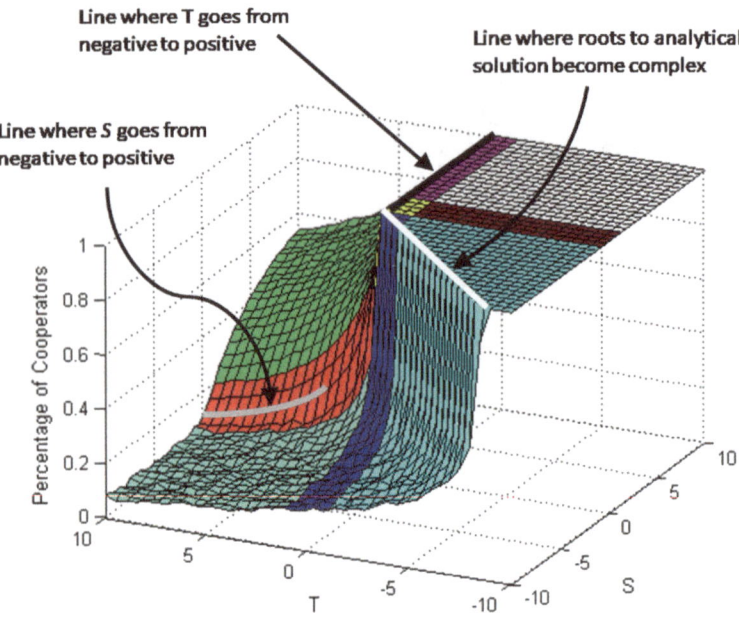

Fig. 13 Simulation with lines showing transitions

plateau there are percentage of cooperators values where defectors are rewarded and thus some agents go on a defecting trajectory with certainty. This specifically happens when T goes from negative to positive. Figure 13 shows Fig. 5 with $T = 0$ as a black line where this transition occurs. This transition is also confirmed for all examples given in Fig. 12.

Now we will look at two special cases. In Fig. 12 various positive R and negative P values were considered. Now we will examine simulation results for the case where R and P are either both positive or both negative. Figure 14 shows the simulation results using the same parameter values as are used for Fig. 5 except R and P are both positive. The results in terms of plateaus and transitions are similar to those in Fig. 12 except that all agents defect in the Prisoners' Dilemma plateau and the transition between the Leader/Battle of the Sexes and Harmony/Deadlock plateaus moves. The reason for all agents defecting in the Prisoner's Dilemma plateau is that when P changes to a positive value the equilibrium attractor solution where both cooperators and defectors are punished disappears. For low percentages of cooperators, cooperators are punished and defectors are rewarded which lowers each agent's cooperating percentage whether they cooperate or defect. This pushes all agents to defecting. The reason the transition moves between the Leader/Battle of the Sexes and Harmony/Deadlock plateaus is that when P is positive there are some scenarios with a low percentage of cooperators where defectors are rewarded. It is shown in Fig. 10 that when both P and T are negative and both R and S are positive all defectors are punished and so all agents cooperate. If either P or T turns positive

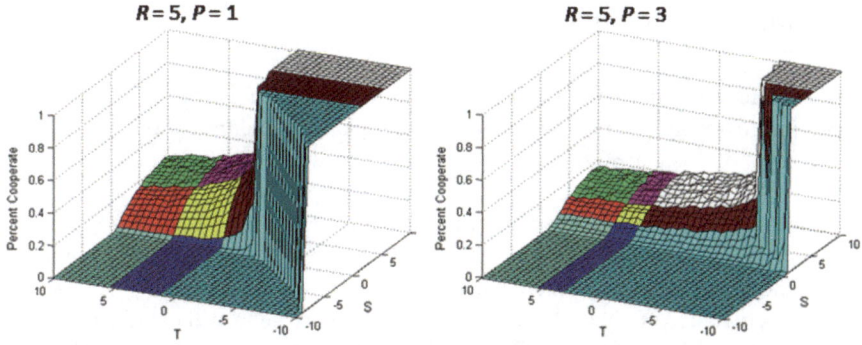

Fig. 14 Simulations when both *R* and *P* are positive

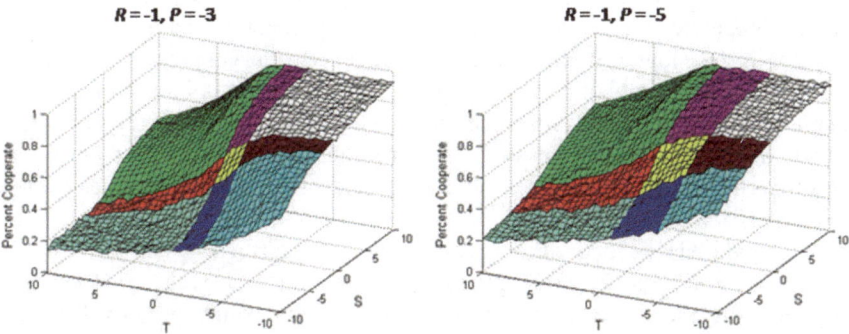

Fig. 15 Simulations when both *R* and *P* are negative

this situation changes. Figure 13 describes what happens when *T* turns positive with *P* remaining negative and Fig. 14 describes what happens when *P* turns positive.

Figure 15 shows the simulation results using the same parameter values as are used for Fig. 5 except *R* and *P* are both negative. The results are similar to those in Fig. 12 except that agents are not all cooperating with certainty in the Harmony/Deadlock plateau. This is because when *R* turns negative the equilibrium attractor solution where both cooperators and defectors are punished is existent and thus the same type of bipartisan behavior occurs as in the Prisoner's Dilemma plateau. The solution to $x^*\alpha C(x^*) = (1 - x^*)\beta D(x^*)$ now is applicable in this region since there is an equilibrium attractor solution. The partisan behavior in the Leader/Battle of the Sexes plateau still exists as in the previous discussion. Although this bipartisan behavior is not typical in the harmony and deadlock stories, making *R* and *P* both negative does appear to be a way to expand bipartisan behavior beyond the Prisoners' Dilemma plateau.

4 Impacts of Changing Parameters

In this section we will review the impact of initial cooperating probability, neighborhood size, learning factors, and grid size on the plateaus and transitions.

For most of this paper an initial cooperating probability 0.5 is assumed. Now we will review the impacts of changing this value. Figure 16 shows simulation results for the same parameter values as are used for Fig. 5 except with the initial cooperating probabilities 0.4, 0.6, and 0.8. The results show that the Leader/Battle of the Sexes plateau height or value is greater as the initial cooperating probability increases. In fact, the percentage of cooperators is approximately the same as the initial cooperating probability. This is expected since an agent with a higher cooperating probability is more likely to be on the continual cooperating trajectory as previously described for this plateau and ends up cooperating with certainty. This is because more agents will initially cooperate with a higher initial cooperating probability and fewer sequential cooperating decisions are required to move the agents to unit cooperating probability.

In all previous simulations it is assumed that each agent is interacting with all agents as a whole. In other words, each agent's neighborhood is all of the other agents. We will now look at the impacts of different neighborhood sizes on the plateaus and transitions. Figure 17 presents the simulation results using the same parameter values as are used for Fig. 5 except with Moore neighborhoods of one, two, and five. In Fig. 5 the neighborhood is defined as the entire collection of agents. The results show that the plateaus and transitions are very similar when neighborhood sizes change. There are a couple of differences when the neighborhood has close proximity or low values. One difference is that the transition between the Prisoners' Dilemma and Leader/Battle of the Sexes plateaus starts when S is slightly more negative and the transition step is slightly steeper. Also the percentage of cooperators in the Leader/Battle of the Sexes plateau is slightly higher. These differences can be explained by the fact that when the neighborhoods are in close proximity then their payoff values become more discreet making an agent's cooperating probability oscillate more. For example, in a neighborhood equal to one there are only eight other neighbors plus the agent itself in the neighborhood. This means that there are only ten possible percentages of the cooperating agents (0, 11, 22, 33, 44, 56, 67, 78, 89, and 100%) and thus only ten possible discreet payoff values. In a 50 × 50 grid with the neighborhood being all

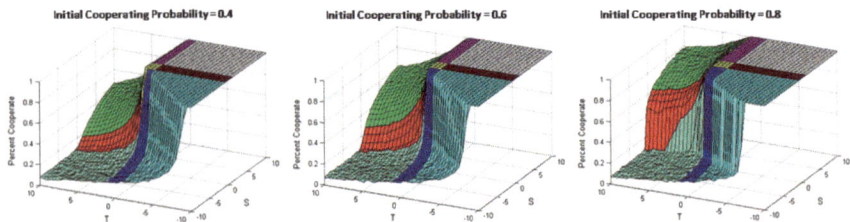

Fig. 16 Comparison of simulation results with different initial cooperating probabilities

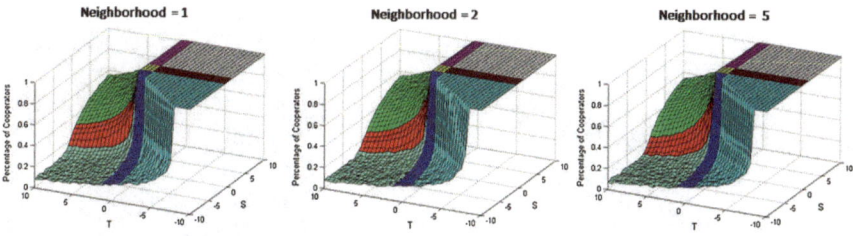

Fig. 17 Comparison of simulation results with varying neighborhoods

agents there are 2500 neighbors and possible payoff values. This is much more of a continuum or continuous case. The lower number of and wider range between payoff values in the Moore neighborhood of one will make an agent's cooperating probability oscillate more from iteration to iteration which will have an impact on this transition and plateau level. The values are slightly higher in this case because the higher oscillation results in more instances of an agent's probability being negative per Eq. (1) and having it artificially increased to zero. This periodic artificial upward adjustment in an agent's cooperating probability from a negative number to zero results in a slight raise in the percentage of cooperators in the system.

Up to this point all of the simulations have assumed that the learning factors α and β are equal. Now we will examine the impacts when $\alpha \neq \beta$. Figure 18 shows simulation results for same parameter values as are used for Fig. 5 except with different learning factors. These results show that the transition line between the Prisoners' Dilemma and Harmony/Deadlock Plateaus moves and the height or value of the Leader/Battle of the Sexes plateau changes as α and β differ. The transition line between the Prisoners' Dilemma and Harmony/Deadlock Plateaus moves as expected to where the roots to the analytical solution become complex per relation (2). It is interesting to note that the transition line does not move if both learning factors change, but remain equal. This can be seen by comparing Fig. 11 with the bottom row of Fig. 18 where the only difference is that the learning factors change from $\alpha = \beta = 0.05$ to $\alpha = \beta = 0.08$. This is because if the learning factors are equal then they cancel out in Eq. (2) regardless of their specific value. The height or value of the Leader/Battle of the Sexes plateau changes as α and β differ because these values impact how fast each agent's cooperating probability changes per Eq. (1) and thus influences the number of agents that will cooperate or defect with certainty. For example, if α is raised in a given scenario, than an agent's cooperating probability will raise more when it cooperates per Eq. (1). We know from previous discussion that in the Leader/Battle of the Sexes plateau that the percentage of cooperators at the end is determined by the number of agents on the continual cooperating trajectory and by the agents on the continual defecting trajectory. If an agent's cooperating probability raises faster, then the agent will more likely be on a continually cooperating trajectory and at the end will cooperate with certainty. This can be seen in Fig. 18 where the height or value of the Leader/Battle of the Sexes plateau is higher when $\alpha > \beta$ (middle row) and lower when $\alpha < \beta$ (top row).

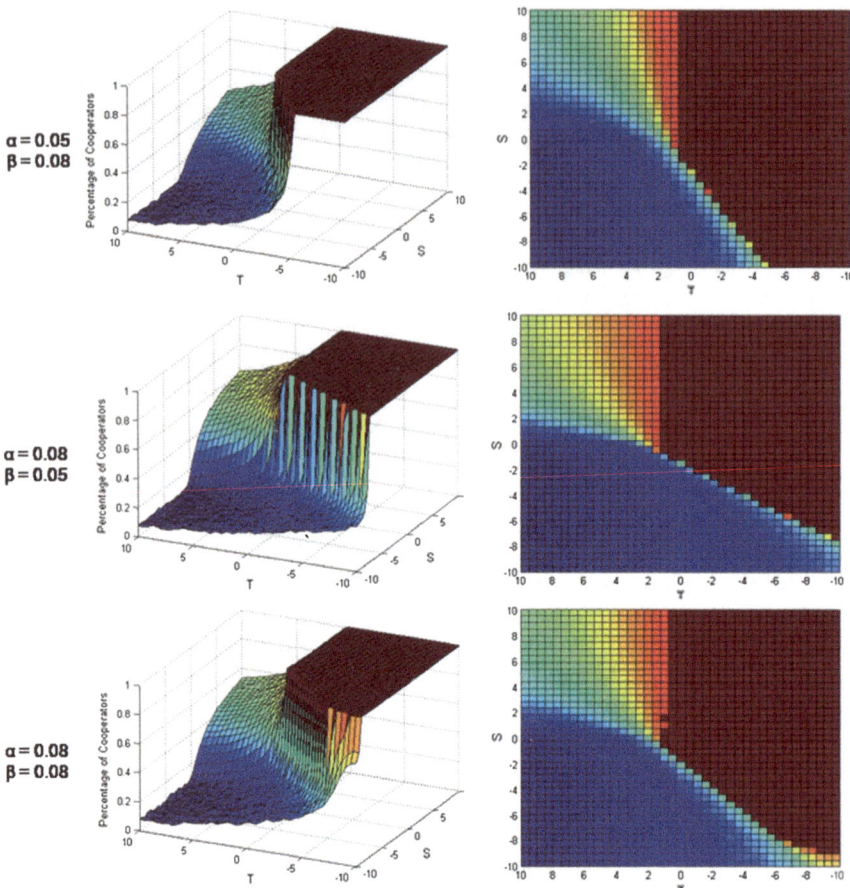

Fig. 18 Comparison of simulation results with different learning factors

For most of this paper the cellular automaton environment is assumed to be a
50×50 grid. Now we will review the impacts of smaller grid sizes. Figure 19 shows
simulations results for the same parameter values as are used for Fig. 5 except with
grid sizes 10×10 and 25×25. It is clear that there is more irregularity in the output
for smaller grid sizes in the Prisoners' Dilemma and Leader/Battle of the Sexes
plateaus. Since there is an element of randomness for the agents in these plateaus
we would expect more regular results with larger grid sizes due to the Law of Large
Numbers.

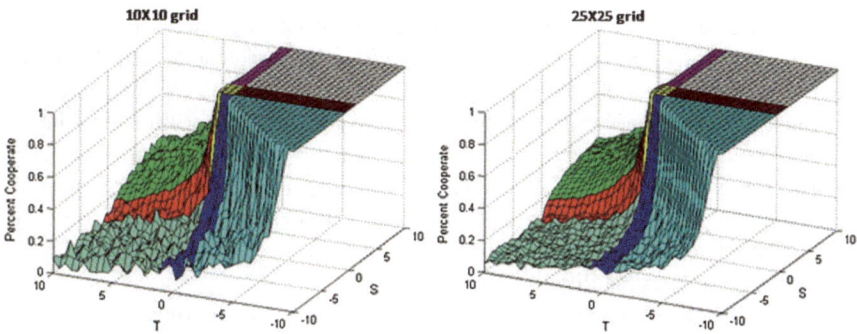

Fig. 19 Comparison of simulation results with varying different grid sizes

5 Conclusions

This paper presented the plateaus and transitions to the N-person social dilemma game assuming Pavlovian agents with linear payoff functions in a two-dimensional cellular automaton environment when the parameter R is positive and P is negative. The behavior of the agents in each of these plateaus was analyzed.

It was discovered that there are three plateaus where agents act in different manners. The first plateau consists of predominately the Prisoners' Dilemma game where each agent's cooperating probability fluctuates around a steady state equilibrium. These agents are called bipartisan agents since they are willing to change their minds from iteration to iteration. The second plateau consists of a region consisting of Leader and Battle of the Sexes games where the agents decide to either cooperate or defect and then never change their minds. These agents are called partisan because of their reluctance to change their decision. The final plateau consists of the region including Harmony and Deadlock games where all agents decide to cooperate. These agents are called unison since they all make the same decision.

The transitions between each of these three plateaus were evaluated. It was found that the transition between the Prisoners' Dilemma and Harmony/Deadlock plateaus occurs when the roots to the analytical solution become complex, that the transition between the Prisoners' Dilemma and Leader/Battle of the Sexes plateaus happens when S goes from negative to positive, and that the transition between the Leader/Battle of the Sexes and Harmony/Deadlock plateaus takes place when T goes from negative to positive. Each of these transitions was analyzed and verified using agent simulation.

The special cases where either both R and P are positive or negative were considered. When both R and P are positive all of the agents in the Prisoners' Dilemma plateau defect because when P is positive there is no equilibrium attractor solution. All agents defect in this plateau since cooperators are punished and defectors are rewarded. Also the transition between the Leader/Battle of the Sexes and Harmony/Deadlock plateaus shifts in the negative T direction because there are circumstances when defectors in the Harmony/Deadlock plateau are rewarded. When both R and P are negative the agents in the Harmony/Deadlock plateau become bipartisan and do not cooperate 100% of the time as when R is positive and P is negative. This is because in this plateau when P is negative there is a equilibrium attractor solution.

The impact of initial cooperating probability, neighborhood size, learning factors, and grid size were evaluated. The impact of changing the initial cooperating probability is mainly on the percentage of cooperators in the Leader/Battle of the Sexes plateau. The percentage of cooperators in this plateau is approximately equal to the initial cooperating probability.

The plateaus and transitions are similar for different neighborhood sizes. There is a slightly steeper transition for smaller neighborhood sizes between the Prisoners' Dilemma and Leader/Battle of the Sexes plateaus because the payoff function values are fewer and more discrete in nature.

Changing the learning factors significantly changes the location of the transition between the Prisoners' Dilemma and Harmony/Deadlock plateaus, but the location is always consistent with where the roots to the analytical solution become complex. The percentage of cooperators in the Leader/Battle of the Sexes plateau can also be affected since higher learning factors can cause an agent to reach a cooperating or defecting state with certainty more quickly.

Lastly, it was found that smaller grid sizes make the simulation results more irregular in the Prisoners' Dilemma and Leader/Battle of the Sexes plateaus. This is due to the fact that these plateaus involve a stochastic element and more regular results will occur with larger grid sizes due to the Law of Large Numbers.

References

Ausloos, M., Dawid, H., & Merlone, U. (2015). Spatial interactions in agent-based modeling. In P. Commendatore, S. Kayam, & I. Kubin (Eds.), *Complexity and geographical economics: Topics and tools* (pp. 353–377). Cham: Springer International Publishing.

Axelrod, R., & Keohane, R. O. (1985). Achieving cooperation under anarchy: Strategies and institutions. *World Politics, 38*(1), 226–254.

Bush, R. R., & Mosteller, F. (1955). *Stochastic models of learning.* New York, NY: Wiley.

Davidsson, P. (2001). Multi agent based simulation: Beyond social simulation. In S. Moss & P. Davidsson (Eds.), *Multi-agent-based simulation* (pp. 141–155). Berlin/Heidelberg, Cambridge, MA: Springer.

Davidsson, P. (2002). Agent based simulation: A computer science view. *Journal of Artificial Societies and Social Simulations, 5*(1). http://jass.soc.surry.ac.uk/5/1/7.html.

Epstein, J., & Axtell, R. (1996). *Growing artificial societies: Social science from the bottom up.* Cambridge, MA: MIT Press.

Flache, A., & Macy, M. W. (2002). Stochastic collusion and the power law of learning. *Journal of Conflict Resolution, 45*(5), 629–653.

Hauert, C. (2001). Fundamental clusters in spatial 2×2 games. *Proceedings of the Royal Society of London B: Biological Sciences, 268*(1468), 761–769. https://doi.org/10.1098/rspb.2000.1424.

Hauert, C. (2002). Effects of space in 2×2 games. *International Journal of Bifurcation and Chaos, 12*(07), 1531–1548. https://doi.org/10.1142/S0218127402005273.

Macy, M. W., & Flache, A. (2002). Learning dynamics in social dilemmas. *Proceedings of the National Academy of Sciences USA, 99*, 7229–7236.

McAdams, R. H. (2009). Beyond the Prisoners' Dilemma: Coordination, game theory, and the law. *Southern California Law Review, 82*, 209–225.

Merlone, U., Sandbank, D. R., & Szidarovszky, F. (2013). Equilibria analysis in social dilemma games with Skinnerian agents. *Mind & Society, 12*(2), 219–233.

Merlone, U., Sandbank, D. R., & Szidarovszky, F. (2018). Applicability of the analytical solution to n-person social dilemma games. *Frontiers in Applied Mathematics and Statistics, 4*, 15. https://doi.org/10.3389/fams.2018.00015. https://www.frontiersin.org/article/10.3389/fams.2018.00015.

Posch, M., Pichler, A., & Sigmund, K. (1999). The efficiency of adapting aspiration levels. *Proceedings of the Royal Society of London B: Biological Sciences, 266*, 1427–1435.

Poundstone, W. (1992). *Prisoner's Dilemma*. New York: Doubleday.

Power, C. (2009). A spatial agent-based model of n-person Prisoner's Dilemma cooperation in a socio-geographic community. *Journal of Artificial Societies and Social Simulation, 12*(1), 8. http://jasss.soc.surrey.ac.uk/12/1/8.html.

Szilagyi, M. N. (2003). An investigation of N-person Prisoners' Dilemmas. *Complex Systems, 14*(2), 155–174.

Szilagyi, M. N. (2009). Cars or buses: Computer simulation of a social and economic dilemma. *International Journal of Internet and Enterprise Management, 6*(1), 23–30.

Thorndike, E. L. (1911). *Animal intelligence*. Darien, CT: Hafner.

Zhao, J., Szidarovszky, F., & Szilagyi, M. N. (2007). Finite neighborhood binary games: A structural study. *Journal of Artificial Societies and Social Simulation, 10*(3), 3. http://jasss.soc.surrey.ac.uk/10/3/3.html.

Zhao, J., Szilagyi, M. N., & Szidarovszky, F. (2008). An n-person battle of sexes game. *Physica A: Statistical Mechanics and its Applications, 387*(14), 3669–3677. https://doi.org/10.1016/j.physa.2007.09.053. http://www.sciencedirect.com/science/article/pii/S0378437108002082.

Related Topics

Optimizing Imperfect Preventive Maintenance Policy for a Multi-unit System with Different Virtual Ages

Maryam Hamidi, Reza Maihami and Behnam Rahimikelarijani

Abstract A special game against nature is examined in this paper, in which nature controls timings of repairable and non-repairable failures of a system of equipment, and preventive maintenance, repairs and preventive replacement policies are the countermeasures. We optimize the imperfect preventive maintenance policy for a multi-unit system with different initial virtual age units. We develop a binary integer programming problem, where the management decides on the optimal preventive maintenance policy to minimize the total expected maintenance cost not to exceed a given budget. The mathematical formulation is developed for a multi-unit system, considering different preventive maintenance levels for each unit. Numerical examples with sensitivity analysis are performed to illustrate the performance and efficiency of the proposed model. Also, the model is examined for a real case study from railway industry. The results determine the optimal preventive maintenance policy and provide managerial insights based on computational analysis.

Keywords Multi-unit system · Imperfect preventive maintenance · Initial virtual age · Selective maintenance · Optimization

1 Introduction

The study of games against nature is an important field of game theory. The payoff of the active player might depend on weather conditions, rainfall, currency exchange rates, timings of breakdowns of systems of equipment to mention only a few. All these uncertain factors can be predicted with mathematical statistical methods and so some probabilistic characterizations become available making these factors random

M. Hamidi (✉) · B. Rahimikelarijani
Department of Industrial Engineering, Lamar University, Beaumont, TX, USA
e-mail: mhamidi@lamar.edu

B. Rahimikelarijani
e-mail: brahimikelar@lamar.edu

R. Maihami
School of Business and Leadership, Our Lady of the Lake University, Houston, TX, USA
e-mail: r.maihami@ollusa.edu

© Springer Nature Singapore Pte Ltd. 2020
F. Szidarovszky and G. I. Bischi (eds.), *Games and Dynamics in Economics*,
https://doi.org/10.1007/978-981-15-3623-6_13

233

variables. There are usually two different ways to deal with such situations. In one case the probabilities of unfavorable outcomes are limited or minimized considering the situation as a zero-sum two-person game. In the other approach the expected payoff is optimized. This second type of approach is the usual method in reliability and quality engineering. In this paper we will also follow this approach.

In competing industries, units and machines are subject to failure by usage and time (Wang 2002). Performing perfect maintenance is not always possible for all the units due to the limitation in resources, budgets and time (Liu and Huang 2010). Maintenance can be performed imperfectly at different levels to return the system to somewhere between as good as new and as bad as old. Different models of imperfect maintenance have been developed by academic researchers, such as virtual age models. A comprehensive review is presented in (Pham and Wang 1996).

Virtual age models are developed by Kijima (1989). In one model, the virtual age of a system after n_{th} maintenance is $y_n = y_{n-1} + \alpha X_n$, where y_{n-1} is the virtual age of the system before n_{th} maintenance, α is the level of maintenance ($0 \leq \alpha \leq 1$) and X_n is the n_{th} time to failure. In another model, the n_{th} repair decreases all the accumulated damage up to n_{th} failure, $y_n = \alpha(y_{n-1} + X_n)$. Two extensions of the Kijima model; proportional age reduction (PAR) and proportional age setback (PAS), have been studied by different researchers (Sanchez et al. 2009; Martorell et al. 1999; Zhou et al. 2007). Ferreira et al. (2015) presented a Weibull-based generalized renewal process using mixed virtual age model. Tanwar ct al. (2014) provided a survey for imperfect repair models for repairable systems using the concepts of Generalized Renewal Process (GRP), arithmetic reduction of age (ARA), and arithmetic reduction of intensity (ARI).

Due to limitations in budget, resources, and time, maintenance might be performed at different levels and managers should make the decision according to actual condition of each unit. This sort of maintenance action is called selective maintenance (Cao et al. 2018; Rice 1999; Cassady et al. 2001), which is widely used in industry. Cassady et al. (2001) developed a mathematical programming model to select a subset of maintenance actions for making selective maintenance decisions, where component life length followed Weibull distribution. Cassady et al. (2001) established a framework for modeling and optimizing selective maintenance, considering different models and concluded different models resulted in different optimal selective maintenance decisions. Lüx et al. (2012) proposed a non-linear binary mathematical model for selective maintenance considering cannibalization and multiple maintenance actions. Pandey et al. (2013) addressed a selective maintenance model for a binary system under imperfect maintenance. They consider age reduction and hazard adjustment to make the model assumption more realistic. Cao et al. (2017) proposed a simulation method for selective maintenance model to maximize system availability. Time and budget are the most frequent constraints in selective maintenance models, which could be negligible, certain or uncertain (Ali et al. 2011). We consider the budget as a challenging constraint to select the best subset of maintenance actions.

Optimizing maintenance model of a system is more complex when the system consists of many components. Multi-unit maintenance models are focused on optimal maintenance policies for a system with several units (Nicolai and Dekker 2008;

Cho and Parlar 1991). A multi-unit system might be affected by competing risks, which is modeled in (Zhang and Yang 2015). The authors considered a repairable multi-component system, where maintenance policy restores the entire system to as-good-as-new state after maintenance. Such assumptions are not realistic. Combining multi-unit system assumption and virtual age is used in (Liu and Huang 2010; Dao and Zuo 2017) to select optimal maintenance strategies. Liu et al. (2018) developed a selective maintenance model to choose a subset of maintenance actions, where maintenance time is stochastic. The authors applied the proposed model on a three-unit system and proposed their model application in industry. A systematic review of the selective maintenance models in multi-unit systems is presented in (Cao et al. 2018).

In this paper, we introduce a model for a selective maintenance policy of a multi-unit system with different initial virtual ages and different maintenance levels. The objective is to find the optimal preventive maintenance level for each unit to minimize the total maintenance cost subject to budget constraints. The imperfect preventive maintenance cost, replacement cost (for non-repairable failures) and minimal mainte-nance cost (for repairable failures) are included in the maintenance costs. We develop a binary integer programming model to analyze the proposed problem. To the best of our knowledge, this is the first study that optimizes the maintenance program for a multi-unit system with considering initial virtual ages, imperfect maintenance, and various maintenance levels.

This paper is organized as follows. In Sect. 2, we present the problem definition and problem formulation. Numerical examples are given in Sect. 3 to illustrate the model and its efficiency. Sensitivity analysis is performed in this section as well. Section 3 includes a real case study from the railway industry. Finally, Sect. 4 concludes the paper and provides future research directions.

2 Problem Definition

The basic problem is finding the optimal preventive maintenance policy for a multi-unit system with different initial virtual ages. In this problem, we assume that there are different preventive maintenance levels for each unit. First, we briefly describe the maintenance level concept.

In many industrial environments, there are different maintenance levels for differ-ent machines and units. As an example, based on the information of Mobility Work website (Giorgio and Pulcini 2018), the maintenance levels could be categorized in 5 different levels. Level 1 of maintenance includes simple maintenance actions that are necessary for the operations, such as condition monitoring rounds and daily lubrication. A maintenance Level 2 is the simple procedures performed for the equip-ment that is usually implemented by a qualified worker with a brief training, such as controlling the operation parameters in equipment, breaking and safety devices control. Level 3 consists of operations which need complex procedures and a quali-fied technician with detailed procedure. Level 4 maintenance, performed by a team, takes care of operations whose procedures use specific techniques or technologies, such as measuring and analyzing the machine vibration and Level 5 maintenance, which is named renovation or reconstruction operations, includes operations whose procedures apply a particular know-how and need special techniques, technologies or processes, like complete inspection on dismantled machines. In here, first different maintenance levels for each unit needs to be determined and next the following formulation can be used to determine the optimal maintenance policy.

2.1 Problem Formulation

The following notations are used throughout the paper.

$k = (1, 2, \ldots, N)$	Set of all units
M_k	Number of possible preventive maintenance levels for each unit k
$i (0 \leq i \leq M_k)$	Preventive maintenance level i, which decreases the virtual age of each unit k from T_k to $\alpha_{ki} T_k$ where $\alpha_{ki} \in [0, 1]$
$C_{ki}^{(m)}$	Preventive maintenance cost for unit k at maintenance level i
R_k	Number of repairable failure types for each unit k
$\rho_{kj}(t)$ for $j = 1, 2, \ldots, R_k$	The failure rate of each repairable failure type j for each unit k
$c_{kj}^{(r)}$	Cost of minimal repair for repairable failure type j for each unit k
$F_k(t)$	The CDF of time to the first non-repairable failure for each unit k from zero virtual age
$C_k^{(R)}$	Cost of failure replacement for unit k
$B^{(m)}$	Preventive maintenance budget
$B^{(R)}$	Replacement budget
$B^{(r)}$	Minimal corrective repair budget

Consider a multi-unit system with N units and initial virtual ages $T_1, T_2, \ldots, T_N$. The management wants to decide on the optimal preventive maintenance plan, which would minimize total expected cost. The preventive maintenance is performed at time zero and the planning horizon is the next T time periods. For each unit k, preventive maintenance with level i decreases the virtual age of the unit from T_k to $\alpha_{ki} T_k$, where $\alpha_{ki} \in [0, 1]$. There are M_k possible preventive maintenance levels, for each unit, i.e. $0 \leq i \leq M_k$. The preventive maintenance cost $C_{ki}^{(m)}$ depends on the unit k and the maintenance level i. For each unit k, maintenance level 0 means that no preventive maintenance is performed with factor $\alpha_{k0} = 1$ and cost $C_{k0}^{(m)} = 0$, while the value $\alpha_{ki} = 0$ refers to preventive replacement.

Each unit is subjected to both non-repairable failure and R_k types of repairable failures, with failure rate $\rho_{kj}(t)$ for $j = 1, 2, \ldots, R_k$. Let $c_{kj}^{(r)}$ and $C_k^{(R)}$ be the costs of minimal repairs for repairable failure type j and that of the failure replacement including possible damages. So generally three types of maintenance is considered for each unit: preventive maintenance, minimum repair for repairable failures and replacement for non-repairable failures.

During any time period of length X, the expected number of type j repairable failures is clearly

$$
\int_{\alpha_{ki} T_k}^{\alpha_{ki} T_k + X} \rho_{kj}(t) dt \tag{1}
$$

So the total expected repair cost becomes:

$$
\sum_{j=1}^{R_k} c_{kj}^{(r)} \int_{\alpha_{ki} T_k}^{\alpha_{ki} T_k + X} \rho_{kj}(t) dt \tag{2}
$$

Let t denote the time of the first non-repairable failure, then the conditional CDF considering the initial virtual age and imperfect preventive maintenance is given as

$$
F_{ki}(t) = \frac{F_k(t + \alpha_{ki} T_k) - F_k(\alpha_{ki} T_k)}{1 - F_k(\alpha_{ki} T_k)} \tag{3}
$$

For simplicity, we assume that at most one non-repairable failure might occur during the considered time period of length T. If it occurs at time $X \in (0, T)$, then the unit becomes as new, so the expected number of repairable failures until the end of the planning horizon is

$$
\int_0^{T-X} \rho_{kj}(t) dt \tag{4}
$$

for failure type j, so the expected total repair cost for all types of repairable failures after the non-repairable failure is

$$
\sum_{j=1}^{R_k} c_{kj}^{(r)} \int_0^{T-X} \rho_{kj}(\tau) d\tau \tag{5}
$$

If the time of the non-repairable failure $X \in (0, T)$ were known, then the total minimal repair cost for repairable failures would have the following form:

$$\Gamma_{ki}(X) = \sum_{j=1}^{R_k} c_{kj}^{(r)} \left[\int_{\alpha_{ki} T_k}^{\alpha_{ki} T_k + X} \rho_{kj}(t)dt + \int_0^{T-X} \rho_{kj}(t)dt \right] \tag{6}$$

Let $f_{ki}(t) = F'_{ki}(t)$. The total expected cost of preventive maintenance, minimal repairs and possible replacement during time period T is given as

$$\psi_{ki}(T) = \int_0^T \Gamma_{ki}(X) f_{ki}(X)dX + \Gamma_{ki}(T)(1 - F_{ki}(T)) + C_k^{(R)} F_{ki}(T) + C_{ki}^{(m)}$$

$$= \sum_{j=1}^{R_k} c_{kj}^{(r)} \int_0^T \left[\int_{\alpha_{ki} T_k}^{\alpha_{ki} T_k + X} \rho_{kj}(t)dt + \int_0^{T-X} \rho_{kj}(t)dt \right] f_{ki}(x)dx$$

$$+ \sum_{j=1}^{R_k} c_{kj}^{(r)} \left[\int_{\alpha_{ki} T_k}^{\alpha_{ki} T_k + T} \rho_{kj}(t)dt \right] (1 - F_{ki}(T)) + C_k^{(R)} F_{ki}(T) + C_{ki}^{(m)} \tag{7}$$

Now, we can formulate the optimization problem:

$$\min \sum_{k=1}^{N} \sum_{i=0}^{M_k} x_{ki} \psi_{ki}(T) \tag{8}$$

Subject to:

$$\sum_{i=0}^{M_k} x_{ki} = 1, \quad \forall k \tag{9}$$

$$\sum_{k=1}^{N} \sum_{i=0}^{M_k} x_{ki} C_{ki}^{(m)} \le B^{(m)} \tag{10}$$

$$\sum_{k=1}^{N} \sum_{i=0}^{M_k} x_{ki} \left(\int_0^T \Gamma_{ki}(X) f_{ki}(X)dX + \Gamma_{ki}(T)(1 - F_{ki}(T)) \right) \le B^{(r)} \tag{11}$$

$$\sum_{k=1}^{N} \sum_{i=0}^{M_k} C_k^{(R)} x_{ki} F_{ki}(\alpha_{ki} T_k + T) \le B^{(R)} \tag{12}$$

In the proposed model, the decision variables are as follows:

$$x_{ki} = \begin{cases} 1 \text{ if maintenance level } i \text{ is chosen for unit } k \\ 0 \qquad\qquad\qquad\qquad\qquad\qquad\quad \text{otherwise} \end{cases}$$

Equation (8) is the objective function which is the overall expected maintenance cost. Equation (9) implies that for each unit only one maintenance level should be considered and constraint (10) shows the limitation in total preventive maintenance cost. The budget limitation for total corrective repair cost and replacement cost are required by inequalities (11) and (12), respectively.

It should be noted that if the unit failure rate follows Weibull distribution, then we have

$$\rho_{kj}(t) = \frac{\beta_{kj}}{\eta_{kj}} \left(\frac{t}{\eta_{kj}} \right)^{\beta_{kj}-1}, \forall k, j \tag{13}$$

$$F_k(t) = 1 - e^{-\left(\frac{t}{\eta_k} \right)^{\beta_k}}, \forall k \tag{14}$$

$$F_{ki}(X) = \frac{F_k(X + \alpha_{ki} T_k) - F_k(\alpha_{ki} T_k)}{1 - F_k(\alpha_{ki} T_k)}, \forall k, i \tag{15}$$

$$f_{ki}(X) = F'_{ki}(X) = \frac{f_k(X + \alpha_{ki} T_k)}{1 - F_k(\alpha_{ki} T_k)}$$

$$= \frac{1}{1 - F_k(\alpha_{ki} T_k)} \frac{\beta_{kj}}{\eta_{kj}} \left(\frac{X + \alpha_{ki} T_k}{\eta_{kj}} \right)^{\beta_{kj}-1} e^{-\left(\frac{X + \alpha_{ki} T_k}{\eta_{kj}} \right)^{\beta_{kj}}} \quad \forall k, i \tag{16}$$

So Eq. (6) can be rewritten as:

$$\Gamma_{ki}(X) = \sum_{j=1}^{R_k} c_{kj}^{(r)} \left[\int_{\alpha_{ki} T_k}^{\alpha_{ki} T_k + X} \rho_{kj}(t)dt + \int_0^{T-X} \rho_{kj}(t)dt \right]$$

$$= \sum_{j=1}^{R_k} c_{kj}^{(r)} \left(\frac{1}{\eta_{kj}} \right)^{\beta_{kj}} \left[(\alpha_{ki} T_k + X)^{\beta_{kj}} - (\alpha_{ki} T_k)^{\beta_{kj}} + (T - X)^{\beta_{kj}} \right]$$

Then, we can obtain Eq. (7) as follows:

$$\psi_{ki}(T) = \int_0^T \Gamma_{ki}(X) f_{ki}(X)dX + \Gamma_{ki}(T)(1 - F_{ki}(T)) + C_k^{(R)} F_{ki}(T) + C_{ki}^{(m)}$$

$$= \sum_{j=1}^{R_k} c_{kj}^{(r)} \left(\frac{1}{\eta_{kj}} \right)^{\beta_{kj}} \int_0^T [(\alpha_{ki} T_k + X)^{\beta_{kj}} - (\alpha_{ki} T_k)^{\beta_{kj}}$$

$$+ (T - X)^{\beta_{kj}}] f_{ki}(X)dX$$

$$+ \sum_{j=1}^{R_k} c_{kj}^{(r)} \left(\frac{1}{\eta_{kj}} \right)^{\beta_{kj}} \left[(\alpha_{ki} T_k + T)^{\beta_{kj}} - (\alpha_{ki} T_k)^{\beta_{kj}} \right] (1 - F_{ki}(T))$$

$$+ C_k^{(R)} F_{ki}(T) + C_{ki}^{(m)} \tag{17}$$

We can compute the equations of constraints accordingly. We use Weibull distribution as the unit failure rate in the following numerical examples.

3 Numerical Examples

In this section, we solve two simple numerical examples and one real world case study by utilizing CPLEX software to illustrate the efficiency of the proposed model.

3.1 Example 1

We consider a 3 unit system $K = (1, 2, 3)$ with initial virtual ages $T_k = (1, 1, 1.5)$ years. Each unit has 3 preventive maintenance levels and 4 repairable failure types, $j = (1, 2, 3, 4)$. The planning horizon is $T = 3$ years, and the budget for preventive maintenance, minimal repair and replacements are $B^{(m)} = \$40$, $B^{(r)} = \$1000$, $B^{(R)} = \$200$, respectively. The Weibull scale parameter and shape parameter for each repairable failure for each unit is presented in Tables 1 and 2, respectively. Cost of replacement is $C_k^{(R)} = (5, 5, 5)$ and cost of minimal repair and preventive maintenance is presented in Tables 3 and 4, respectively. Effect of each preventive maintenance level for each unit can be seen in Table 5.

Applying the above parameters, we first compute the total expected maintenance costs, $\psi_{ki}(T)$, and the results are presented in Table 6. The total maintenance cost includes preventive maintenance, minimal repair for all repairable failure types and replacement cost for non-repairable failure.

Table 1 Value of scale parameter η_{kj} of Weibull distribution for Example 1

		η_{kj} (scale parameter of repairable failure of type j)			
		1	2	3	4
Units, k	1	1	0.9	1	1
	2	0.8	0.6	0.4	0.9
	3	0.6	0.7	0.5	0.3

Table 2 Value of shape parameter β_{kj} of Weibull distribution for Example 1

		β_{kj} (shape parameter of repairable failure of type j)			
		1	2	3	4
Units, k	1	1.1	1.2	1.3	1.4
	2	1.4	1.3	1.2	1.1
	3	1.6	1.6	1.6	1.6

Table 3 Cost $c_{kj}^{(r)}$ of minimal repair of failure of type j for each machine k for Example 1

		$c_{kj}^{(r)}$ (cost of minimal repair for repairable failure of type j)			
		1	2	3	4
Units, k	1	2	2.1	2.2	2.3
	2	1.5	1.4	1.3	1.2
	3	1.7	1.7	1.7	1.7

Table 4 Preventive maintenance cost $C_{ki}^{(m)}$ for unit k at maintenance level i for Example 1

		$C_{ki}^{(m)}$ (preventive maintenance cost at maintenance level i)		
		1	2	3
Units, k	1	1	1.5	2
	2	2	2.5	3
	3	2	2.5	3

Table 5 Age reduction coefficient α_{ki} in virtual age model for Example 1

		α_{ki} (age reduction coefficient, level i)		
		1	2	3
Units, k	1	0.9	0.8	0.7
	2	0.5	0.4	0.3
	3	0.7	0.6	0.5

Table 6 The total expected cost $\psi_{ki}(T)$ of preventive maintenance, minimal repairs and possible replacement for Example 1

		$\psi_{ki}(T)$ (total expected maintenance cost at level i)		
		Level 1	Level 2	Level 3
Units, k	Unit 1	40.705	40.882	41.223
	Unit 2	45.003	45.517	46.240
	Unit 3	139.032	136.161	137.086

Next, we optimize function (14) and obtain \$222.318 as the optimal objective value. The optimal decision variables can be seen in Table 7. It shows that the management should consider preventive maintenance level 1 for unit 1 and unit 2, and level 2 for unit 3. It is clear that any other combination of maintenance levels for the units would lead to larger maintenance costs.

Table 7 The optimal preventive maintenance policy for Example 1

		Maintenance level, i		
		Level 1	Level 2	Level 3
Units, k	Unit 1	1	0	0
	Unit 2	1	0	0
	Unit 3	0	1	0

3.1.1 Discussion on Weibull Distribution Parameters

The Weibull parameters are the critical factors in determining the optimal solutions. We are now examining how the value of shape and scale parameters affect the optimal solution. First, we solve the proposed example for random values of shape parameter β_{kj}. The results are shown in Table 8. It is pretty clear that even a small change in shape parameter would lead to change in optimal solution. Likewise, we perform

Table 8 Optimal solution corresponding to various values of shape parameter β_{kj}

	β_{kj} (shape parameter of each repairable failure type j for each unit k)	Optimal maintenance level
Case 1	$\beta_{kj} = \begin{bmatrix} 1.2\ 1.3\ 1.4\ 1.5 \\ 1.5\ 1.4\ 1.3\ 1.2 \\ 1.7\ 1.7\ 1.7\ 1.7 \end{bmatrix}$	$x_{ki} = \begin{bmatrix} 0\ 1\ 0 \\ 1\ 0\ 0 \\ 0\ 1\ 0 \end{bmatrix}$
Case 2	$\beta_{kj} = \begin{bmatrix} 1.3\ 1.4\ 1.5\ 1.6 \\ 1.6\ 1.5\ 1.4\ 1.3 \\ 1.8\ 1.8\ 1.8\ 1.8 \end{bmatrix}$	$x_{ki} = \begin{bmatrix} 0\ 1\ 0 \\ 1\ 0\ 0 \\ 0\ 1\ 0 \end{bmatrix}$
Case 3	$\beta_{kj} = \begin{bmatrix} 1.4\ 1.5\ 1.6\ 1.7 \\ 1.7\ 1.6\ 1.5\ 1.4 \\ 1.9\ 1.9\ 1.9\ 1.9 \end{bmatrix}$	$x_{ki} = \begin{bmatrix} 0\ 0\ 1 \\ 1\ 0\ 0 \\ 0\ 1\ 0 \end{bmatrix}$
Case 4	$\beta_{kj} = \begin{bmatrix} 1\ \ 1.1\ 1.2\ 1.3 \\ 1.3\ 1.2\ 1.1\ 1 \\ 1.5\ 1.5\ 1.5\ 1.5 \end{bmatrix}$	$x_{ki} = \begin{bmatrix} 1\ 0\ 0 \\ 1\ 0\ 0 \\ 0\ 1\ 0 \end{bmatrix}$
Case 5	$\beta_{kj} = \begin{bmatrix} 0.9\ 1\ \ 1.1\ 1.2 \\ 0.9\ 1.2\ 1.8\ 0.8 \\ 1.1\ 1.5\ 0.9\ 2 \end{bmatrix}$	$x_{ki} = \begin{bmatrix} 1\ 0\ 0 \\ 1\ 0\ 0 \\ 0\ 1\ 0 \end{bmatrix}$
Case 6	$\beta_{kj} = \begin{bmatrix} 0.8\ 0.9\ 1\ \ 1.1 \\ 0.8\ 1.7\ 0.6\ 1.2 \\ 1\ \ 0.8\ 1.9\ 1.4 \end{bmatrix}$	$x_{ki} = \begin{bmatrix} 1\ 0\ 0 \\ 1\ 0\ 0 \\ 0\ 1\ 0 \end{bmatrix}$

Table 9 Optimal solution corresponding to various value of scale parameter η_{kj}

	η_{kj} (scale parameter of each repairable failure type j for each unit k)	Optimal maintenance level
Case 1	$\eta_{kj} = \begin{bmatrix} 1.5 & 1.2 & 0.8 & 1.8 \\ 0.3 & 1.2 & 1 & 0.6 \\ 0.9 & 1.7 & 1.5 & 0.9 \end{bmatrix}$	$x_{ki} = \begin{bmatrix} 1 & 0 & 0 \\ 0 & 0 & 1 \\ 0 & 0 & 1 \end{bmatrix}$
Case 2	$\eta_{kj} = \begin{bmatrix} 1 & 1.5 & 2 & 2.5 \\ 0.3 & 0.4 & 0.5 & 0.6 \\ 4 & 3 & 2 & 1 \end{bmatrix}$	$x_{ki} = \begin{bmatrix} 1 & 0 & 0 \\ 0 & 0 & 1 \\ 0 & 0 & 1 \end{bmatrix}$
Case 3	$\eta_{kj} = \begin{bmatrix} 0.7 & 1.7 & 1.2 & 0.4 \\ 1.3 & 0.2 & 1.9 & 1.6 \\ 1.9 & 0.7 & 0.5 & 1.8 \end{bmatrix}$	$x_{ki} = \begin{bmatrix} 1 & 0 & 1 \\ 1 & 0 & 0 \\ 0 & 1 & 0 \end{bmatrix}$
Case 4	$\eta_{kj} = \begin{bmatrix} 1.2 & 1.2 & 1.2 & 1.2 \\ 1 & 1 & 1 & 1 \\ 0.8 & 0.8 & 0.8 & 0.8 \end{bmatrix}$	$x_{ki} = \begin{bmatrix} 1 & 0 & 0 \\ 1 & 0 & 0 \\ 0 & 0 & 1 \end{bmatrix}$
Case 5	$\eta_{kj} = \begin{bmatrix} 1 & 0.9 & 0.6 & 0.3 \\ 2 & 1.6 & 1.2 & 0.8 \\ 3 & 2 & 1 & 0.1 \end{bmatrix}$	$x_{ki} = \begin{bmatrix} 0 & 1 & 0 \\ 1 & 0 & 0 \\ 0 & 0 & 1 \end{bmatrix}$
Case 6	$\eta_{kj} = \begin{bmatrix} 0.3 & 0.9 & 1.4 & 0.3 \\ 1.2 & 0.6 & 1.2 & 0.6 \\ 0.5 & 2 & 0.3 & 0.1 \end{bmatrix}$	$x_{ki} = \begin{bmatrix} 0 & 1 & 0 \\ 1 & 0 & 0 \\ 1 & 0 & 0 \end{bmatrix}$

the optimization problem for various value of scale parameter η_{kj} (Table 9). Like as shape parameter, any deviation in scale parameter value remarkably changes the optimal solution. Therefore, it is very important for decision makers to indicate the precise and correct value of Weibull parameters if they want to obtain the real optimal solution.

There are different methods to estimate the distribution parameters. Generally, these methods categorized into two groups: (1) the graphically method such as probability plotting and hazard plotting, and (2) the analytically methods such as method of moment (MOM) least square method (LSM), maximum likelihood estimation (MLE) and density power method (DPM). All these methods depend on the data quality that are used to estimate the parameters. Many of data analysis and statistical packages easily compute the Weibull parameters based on the given data.

3.2 Example 2

In this example, we consider a system with 10 units, 5 preventive maintenance levels for each unit, and 8 types of possible repairable failures. The planning horizon is $T = 5$ years and we assume the maintenance budgets as $B^{(m)} = \$4000$, $B^{(R)} = \$20000$, and $B^{(r)} = \$100000$. The rest of the parameters are presented in Tables 10, 11, 12, 13, 14 and 15.

We next compute the total expected maintenance costs for all units in all possible preventive maintenance levels, and the results can be seen in Table 16.

We then solve the optimum problem to find the optimal solution. The results show that minimum total maintenance cost for all units is \$64768.95 The corresponding optimal decision variable are given in Table 17. Based on the results, we conclude that the management should apply maintenance level 1 for all units except for units 7 and 10, where maintenance level 2 is optimal.

Next, we perform sensitivity analysis based on the Example 2 information. First, we vary the value β_{11} of Weibull distribution shape parameter of unit 1 at failure type 1. The results are shown in Fig. 1.

It is clear that when the shape parameter is increased, the total maintenance cost is increased, as well. The same process for the Weibull scale parameter η_{11} of failure type 1 can be done.

Next we repeat the analysis of the maintenance costs based on different values of T for all units in maintenance level 1. The results given in Fig. 2 show that increase in the total life cycle for each unit leads to an increase in the maintenance cost. Thus, when the considered unit age becomes larger, the maintenance cost increases as well. From managerial point of view, it is implied that the unit replacement strategy is justified when the unit age becomes old.

3.3 Case Study

In this section we present a real world application of the model for railroad tracks. Track repairable failure types are categorized into two main categories, structural and geometrical failures. While structural defects are created by structural conditions of the track, including rail, sleeper, fastening, sub-grade and drainage system, geometry failures are related to bad condition of the rail geometry parameters, such as profile and alignment (He et al. 2015). In this study, we consider three of the major repairable geometry failure types as follow:

- The first is the surface failure type, which measures any non-uniformity of the top surface of a single rail. As can be seen in Fig. 3, the surface measurement can be positive or negative when there is a hump or a dip, respectively.
- The second repairable failure type, demonstrated in Fig. 4, is DIP, which measures a fall or a rise in the centerline of the track.

Table 10 Value η_{kj} of scale parameter of Weibull distribution for Example 2

Units, k	η_{kj} (scale parameter of repairable failure of type j)							
	1	2	3	4	5	6	7	8
1	1	0.9	0.8	0.7	0.6	0.3	0.4	0.3
2	1	0.9	0.8	0.7	0.6	0.5	0.4	0.3
3	1	0.9	0.8	0.7	0.6	0.5	0.4	0.3
4	0.5	0.9	0.8	0.7	0.9	0.5	0.4	0.3
5	1	0.4	0.8	0.8	0.6	0.5	0.4	0.9
6	1	0.9	0.2	0.7	0.6	0.5	0.4	0.3
7	0.5	0.9	0.8	0.5	0.6	0.5	0.4	0.3
8	1	0.9	0.8	0.7	0.6	0.5	0.4	0.3
9	1	0.9	0.8	0.7	0.4	0.5	0.4	0.3
10	0.4	0.9	0.8	0.3	0.6	0.9	0.4	0.3

Table 11 Value β_{kj} of shape parameter of Weibull distribution for Example 2

| Units, k | β_{kj} (shape parameter of reparable failure of type j) | | | | | | | |
	1	2	3	4	5	6	7	8
1	1.1	1.2	1.3	1.4	1.5	1.6	1.7	1.8
2	1.9	1.8	1.3	1.4	1.2	1.6	1.7	1.8
3	1.1	1.2	1.3	1.4	1.5	1.6	1.7	1.8
4	1.1	1.9	1.3	1.3	1.5	1.6	1.7	1.8
5	1.1	1.2	1.3	1.4	1.4	1.6	1.7	1.8
6	1.1	1.4	1.3	1.4	1.5	1.6	1.7	1.8
7	1.1	1.2	1.9	1.4	1.5	1.6	1.7	1.8
8	1.1	1.2	1.7	1.4	1.6	1.6	1.7	1.8
9	1.1	1.1	1.3	1.4	1.5	1.6	1.7	1.8
10	1.4	1.2	1.2	1.4	1.6	1.6	1.7	1.8

Table 12 Cost $c_{kj}^{(r)}$ of minimal repair of repairable failure of type j for unit k for Example 2

Units, k	$c_{kj}^{(r)}$ (cost of minimal repair of failure of type j)							
	1	2	3	4	5	6	7	8
1	10	12	17	16	18	20	22	24
2	10	12	14	15	18	20	22	24
3	19	12	14	16	18	27	13	24
4	10	12	14	16	18	20	22	24
5	22	20	19	17	15	27	22	23
6	22	20	19	17	15	27	21	23
7	10	12	14	16	18	20	21	24
8	10	12	14	16	18	20	22	24
9	10	9	14	16	18	13	22	29
10	10	12	14	14	18	20	22	24

Table 13 Preventive maintenance cost $C_{ki}^{(m)}$ for unit k at maintenance level i for Example 2

		$C_{ki}^{(m)}$ (preventive maintenance cost at maintenance level i)				
		1	2	3	4	5
Units, k	1	18	16	14	12	10
	2	20	18	16	14	12
	3	22	20	18	16	14
	4	24	22	20	18	16
	5	26	24	22	20	18
	6	10	8	6	4	2
	7	12	10	8	6	4
	8	14	12	10	8	6
	9	16	14	12	10	8
	10	18	16	14	12	10

Table 14 Age reduction coefficient α_{ki} in virtual age model for Example 2

		α_{ki} (age reduction coefficient at maintenance level i)				
		1	2	3	4	5
Units, k	1	0.5	0.6	0.7	0.8	0.9
	2	0.4	0.5	0.6	0.7	0.8
	3	0.3	0.4	0.5	0.6	0.7
	4	0.2	0.3	0.4	0.5	0.6
	5	0.1	0.2	0.3	0.4	0.5
	6	0.5	0.6	0.7	0.8	0.9
	7	0.4	0.5	0.6	0.7	0.8
	8	0.3	0.4	0.5	0.6	0.7
	9	0.2	0.3	0.4	0.5	0.6
	10	0.1	0.2	0.3	0.4	0.5

- The third is the cross level (X-level) failure type, which measures the difference in elevation of top surface of two rails at any specific point of the railroad track. The cross level measurement is mostly performed under load since the rails can move up or down under a load. Figure 5 presents cross level defect.

Geometry cars equipped with sensors, GPS and measurement devices, periodically inspect tracks and record different track geometries such as track alignment, elevation, curvature and track surface. Part of the data that geometric cars gather are segment number, milepost, defect amplitude, and class. A brief definition of these variables is as follows.

- Segment number: Segment is like tracks connecting two cities
- Milepost: Point on the track segment

Table 15 Values of $C_k^{(R)}$ and T_k

Units, k	1	2	3	4	5	6	7	8	9	10
$C_k^{(R)}$ (Replacement cost)	100	110	120	115	90	90	95	100	140	70
T_k (Virtual initial age)	2	2.2	1	2.3	1.5	1.5	1	1	1.5	1.8

Table 16 The total expected cost $\psi_{ki}(T)$ of preventive maintenance, minimal repairs and replacements for Example 2

		$\psi_{ki}(T)$ (total expected maintenance cost at level i)				
		Level 1	Level 2	Level 3	Level 4	Level 5
Units, k	Unit 1	7434.66	7595.41	7770.12	7936.26	8077.1
	Unit 2	6797.55	6939.69	6863.79	7027.83	7026.21
	Unit 3	5903.92	6039.37	6200.2	6370.71	6536.61
	Unit 4	6677.03	6723.92	6686.54	6831.1	6870.19
	Unit 5	3908.57	4275.88	4112.81	4174.67	4296.98
	Unit 6	7578.84	7826.99	8422.37	8079.31	8233.5
	Unit 7	6995.12	6706.4	7100.34	7271.06	7251.42
	Unit 8	6339.08	6490.41	6813.02	6861.25	7087.52
	Unit 9	6873.41	7029.82	7283.39	7502.28	7892.27
	Unit 10	7061.84	6549.49	6724.75	7376.53	7224.52

Table 17 Optimal solutions for Example 2

		Maintenance level, i				
		Level 1	Level 2	Level 3	Level 4	Level 5
Units, k	Unit 1	1	0	0	0	0
	Unit 2	1	0	0	0	0
	Unit 3	1	0	0	0	0
	Unit 4	1	0	0	0	0
	Unit 5	1	0	0	0	0
	Unit 6	1	0	0	0	0
	Unit 7	0	1	0	0	0
	Unit 8	1	0	0	0	0
	Unit 9	1	0	0	0	0
	Unit 10	0	1	0	0	0

- Defect type: Geometry defect types
- Defect amplitude: Size of defect in inches or degrees
- Class: All tracks get a number between one and five. Each class represents operating speed limits for passenger and freight traffic. Class one has the lowest speed limit and class five has the highest speed limit.

Federal Railroad Administration (FRA) defines the defect amplitude threshold of each failure type and a defect amplitude recorded by geometry cars is considered a failure if greater than the threshold. Such defects violate FRA safety standards and need immediate maintenance. The failure threshold for each failure type is presented in Table 18.

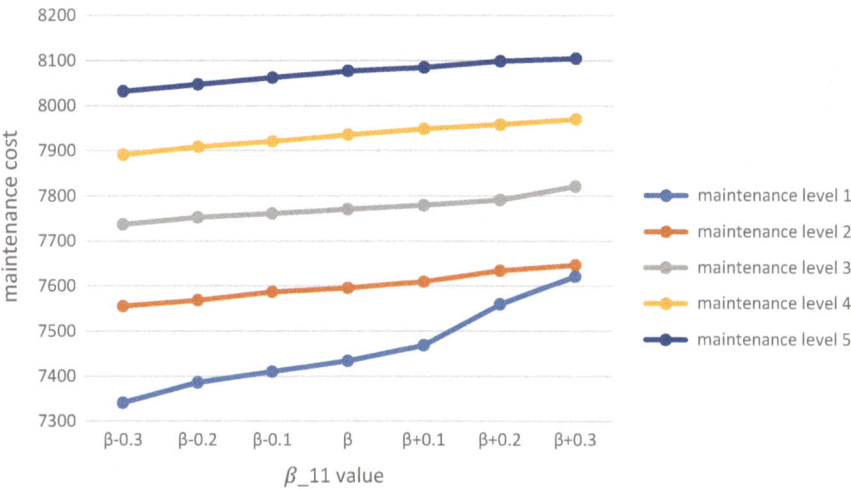

Fig. 1 Total expected maintenance costs for unit 1 with varying Weibull shape parameter β_{11} of failure type 1

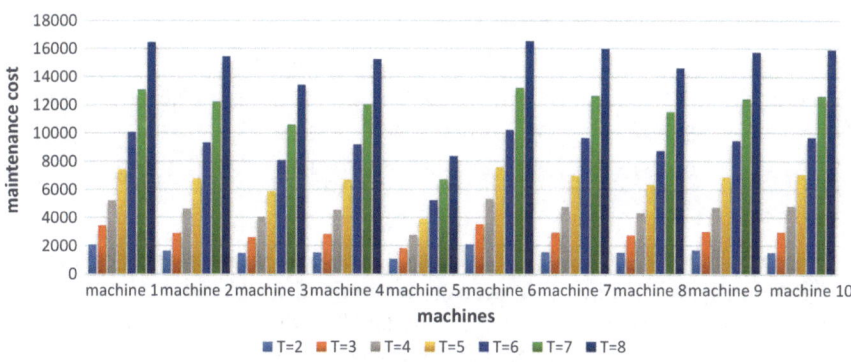

Fig. 2 Total expected maintenance costs for all units at maintenance level 1 regard to various T

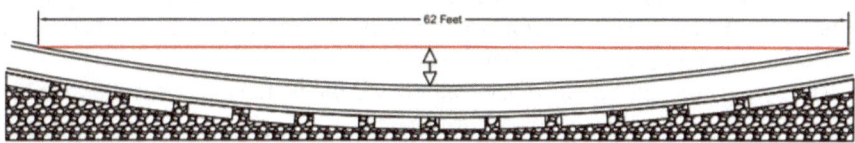

Fig. 3 Graphical representation of surface failure

In this study we consider segments as different units of the system, where each segment can have three types of repairable failure; DIP, surface and X-level. A unit

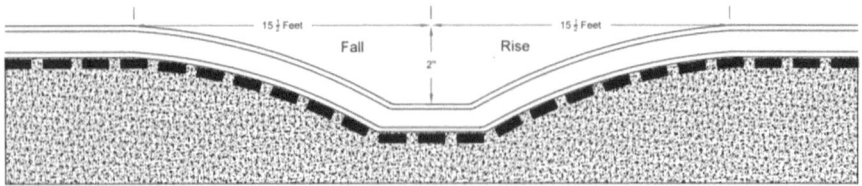

Fig. 4 Graphical representation of DIP failure

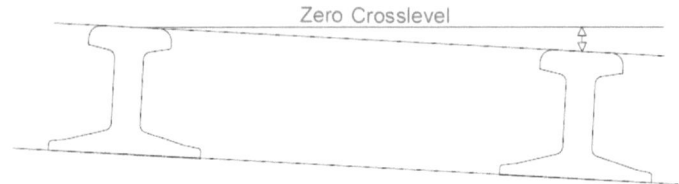

Fig. 5 Graphical representation of cross level failure

Table 18 Failure amplitude threshold for each failure type in different rail classes (inch)

		Class of rails				
		Class 1	Class 2	Class 3	Class 4	Class 5
Failure type	DIP	3	2.75	2.25	1.75	1.5
	Surface	3	2.75	2.25	2	1.25
	Cross level	3	2	1.75	1.25	1

is considered failed when the defect amplitude of at least one milepost is greater than the FRA threshold. That milepost needs to be minimally repaired.

To elaborate the real application of the model, the required data has been obtained from Burlington Northern and Santa Fe (BNSF) Railway Company. BNSF Railway is one of the major freight railroad networks in North America and is one of the seven class I railroads in US. We consider the track geometry failures from 2007 to 2013. We consider a track with three segments. For each segment, we analyze the failure time data recorded by geometry cars and estimate a Weibull distribution for each failure type in class 5 rails. Table 19 shows the results.

The cost of minimal repair for each unit based on the failure type is presented in Table 20. The minimal repair cost is the same for each segment.

There are two maintenance levels to preventively maintain any track segment; tamping and stone blowing. In tamping process, a tamping machine raise the sleepers and ballast the stone under them, while in the stone blowing process the ballast rests and the stone will be blown under them. The preventive maintenance cost for each segment is shown in Table 21. The age reduction coefficient α_{ki} for all segments in tamping process is 0.6 and in stone blowing maintenance is 0.8.

Table 19 Estimated Weibull parameters for different modes of failure in rail segments

		Segment, k	Shape parameter, β_{kj}	Scale parameter, η_{kj}
Failure type	DIP	3	1.5	146
	Surface	2	1.2	212
		3	1.4	181
	Cross level	1	1.3	211
		2	1.3	238
		3	1.4	211

Table 20 Cost $c_{kj}^{(r)}$ of minimal repair of failure type j in segment k

		Cost of minimal repair for each failure type		
		DIP	Surface	Cross level
Segment, k	1	$1125	$1125	$1534
	2	$1125	$1125	$1534
	3	$1125	$1125	$1534

Table 21 Preventive maintenance cost $C_{ki}^{(m)}$ for segment k at maintenance level i

		Preventive maintenance level, i	
		Level 1, stone blowing	Level 2, tamping
Segment, k	1	$125460	$139400
	2	$157658	$175175
	3	$117551	$130613

The initial virtual age of each segment is $T_k = (4, 5, 3)$ years respectively. The replacement cost for all segments is same and is equal to $480000. We assume the time horizon of $T = 5$ years and maintenance budgets as $B^{(m)} = \$500000$, $B^{(r)} = \$1000000$, $B^{(R)} = \$600000$.

Using the above data, the total maintenance cost for each segment is shown in Table 22.

Table 22 The total expected cost, $\psi_{ki}(T)$ of maintenance, repairs and possible replacement

		Maintenance level, i	
		Level 1	Level 2
Segment, k	1	$129189	$142472
	2	$160854	$178247
	3	$120121	$133675

Table 23 Optimal solutions
for case study

		Maintenance level, i	
		Level 1	Level 2
Segment, k	1	1	0
	2	1	0
	3	1	0

The optimal solution shows that to minimize the total maintenance cost, all segment should consider the maintenance level 1 (Table 23). The minimal total maintenance cost for the whole system is $410167.

4 Conclusion

In this paper, we developed a new model to study imperfect maintenance of a multi-unit system with different maintenance levels and different initial virtual ages. The mathematical formulation was given and numerical examples were presented. A real world case study of rail tracks was presented and the optimal maintenance level for each unit and the corresponding total maintenance costs for each maintenance policy were presented in the results. Moreover, sensitivity analysis was performed to study the impact of changing some model parameter values. This result can assist the management to realize the importance of the correct estimation of the model parameters.

The model introduced in this paper can be extended in several ways. The preventive maintenance time can be optimized in addition to the current maintenance level. Instead of expected cost we could also consider expected cost per unit time, when the cycle ends either at time T or at the time of non-repairable failure, which occurs first. These model variants will be the subject of our next project.

References

Ali, I., Khan, M. F., Raghav, Y. S., & Bari, A. (2011). Allocation of repairable and replaceable components for a system availability using selective maintenance with probabilistic maintenance time constraints. *American Journal of Operations Research, 1*(3), 147.

Cao, W., Jia, X., Hu, Q., Song, W., & Ge, H. (2017). Selective maintenance for maximising system availability: A simulation approach. *International Journal of Innovative Computing and Applications, 8*(1), 12–20.

Cao, W., Jia, X., Hu, Q., Zhao, J., & Wu, Y. (2018). A literature review on selective maintenance for multi-unit systems. *Quality and Reliability Engineering International, 34*(5), 824–845.

Cassady, C. R., Murdock, W. P., Jr., & Pohl, E. A. (2001a). Selective maintenance for support equipment involving multiple maintenance actions. *European Journal of Operational Research, 129*(2), 252–258.

Cassady, C. R., Pohl, E. A., & Paul Murdock, W. (2001). Selective maintenance modeling for industrial systems. *Journal of Quality in Maintenance Engineering, 7*(2), 104–117.

Cho, D. I., & Parlar, M. (1991). A survey of maintenance models for multi-unit systems. *European Journal of Operational Research, 51*(1), 1–23.

Dao, C. D., & Zuo, M. J. (2017). Selective maintenance of multi-state systems with structural dependence. *Reliability Engineering & System Safety, 159,* 184–195.

Ferreira, R. J., Firmino, P. R. A., & Cristino, C. T. (2015). A mixed kijima model using the weibull-based generalized renewal processes. *PLoS ONE, 10*(7), e0133772.

Giorgio, M., & Pulcini, G. (2018). A new state-dependent degradation process and related model misidentification problems. *European Journal of Operational Research, 267*(3), 1027–1038.

He, Q., Li, H., Bhattacharjya, D., Parikh, D. P., & Hampapur, A. (2015). Track geometry defect rectification based on track deterioration modelling and derailment risk assessment. *Journal of the Operational Research Society, 66*(3), 392–404.

Kijima, M. (1989). Some results for repairable systems with general repair. *Journal of Applied Probability, 26*(1), 89–102.

Liu, Y., Chen, Y., & Jiang, T. (2018). On sequence planning for selective maintenance of multi-state systems under stochastic maintenance durations. *European Journal of Operational Research, 268*(1), 113–127.

Liu, Y., & Huang, H.-Z. (2010). Optimal selective maintenance strategy for multi-state systems under imperfect maintenance. *IEEE Transactions on Reliability, 59*(2), 356–367.

LüX, Z., Yu, Y. L., & Zhang, L. (2012). Selective maintenance model considering cannibalization and multiple maintenance actions. *Acta Armamentarii, 33*(3), 360–366.

Martorell, S., Sanchez, A., & Serradell, V. (1999). Age-dependent reliability model considering effects of maintenance and working conditions. *Reliability Engineering & System Safety, 64*(1), 19–31.

Nicolai, R. P., & Dekker, R. (2008). Optimal maintenance of multi-component systems: A review. In *Complex system maintenance handbook* (pp. 263–286). Springer.

Pandey, M., Zuo, M. J., Moghaddass, R., & Tiwari, M. K. (2013). Selective maintenance for binary systems under imperfect repair. *Reliability Engineering & System Safety, 113,* 42–51.

Pham, H., & Wang, H. (1996). Imperfect maintenance. *European Journal of Operational Research, 94*(3), 425–438.

Rice, W. F. (1999). Optimal selective maintenance decisions for series systems (Mississippi State University. Department of Industrial Engineering).

Sanchez, A., Carlos, S., Martorell, S., & Villanueva, J. F. (2009). Addressing imperfect maintenance modelling uncertainty in unavailability and cost based optimization. *Reliability Engineering & System Safety, 94*(1), 22–32.

Tanwar, M., Rai, R. N., & Bolia, N. (2014). Imperfect repair modeling using Kijima type generalized renewal process. *Reliability Engineering & System Safety, 124,* 24–31.

Wang, H. (2002). A survey of maintenance policies of deteriorating systems. *European Journal of Operational Research, 139*(3), 469–489.

Zhang, N., & Yang, Q. (2015). Optimal maintenance planning for repairable multi-component systems subject to dependent competing risks. *IIE Transactions, 47*(5), 521–532.

Zhou, X., Xi, L., & Lee, J. (2007). Reliability-centered predictive maintenance scheduling for a continuously monitored system subject to degradation. *Reliability Engineering & System Safety, 92*(4), 530–534.

Newton-Type Solvers Using Outer Inverses for Singular Equations

Ioannis K. Argyros and Stepan Shakhno

Abstract We are motivated by a seminal paper of Nashed and Chen on Newton-type solvers for Banach space valued operators equations. The novelty of our paper lies in the fact that we present a more flexible, finer semi-local convergence analysis and without additional hypotheses. We also study the local convergence analysis not given in the aforementioned paper.

1 Introduction

Let B_1, B_2 stand for Banach spaces, $\Omega \subseteq B_1$ be convex, open and nonempty, $L(B_1, B_2) := \{Q : B_1 \to B_2 \text{ is linear and continuous}\}$ and $U(x, \rho) := \{y \in B_1 : \|y - x\| < \rho, \rho > 0\}$. Consider $H : \Omega \to B_2$ to be a continuous operator. One of the most challenging and important tasks is to find a solution x^* of equation

$$RH(x) = 0, \tag{1}$$

where $R \in L(B_2, B_1)$.

Numerous problems from diverse branches such as Mathematical: Biology, Chemistry, Economics, Medicine, Physics and also and Engineering to mention a few are reduced to solving equation (1) using mathematical modelling (Argyros and Magrenán 2017, 2018; Argyros and Shakhno 2019a, b; Argyros et al. 2019; Ben-Israel 1966, 1968; Ben-Israel and Greville 1974; Chen and Yamamoto 1989; Deuflhard and Heindl 1979; Nashed and Chen 1993; Häußler 1986; Akilov 1981; Nashed 1976, 1987; Potra and Ptak 1984; Ortega and Rheinboldt 1970; Traub 1964; Shakhno 2014, 2010, 2009; Shakhno et al. 2014; Yamamoto 1986, 1987, 1989). The desired

I. K. Argyros (✉)
Department of Mathematics, Cameron University, Lawton 73505, USA
e-mail: iargyros@cameron.edu

S. Shakhno
Department of Theory of Optimal Processes, Ivan Franko National University of Lviv, Lviv 79000, Ukraine
e-mail: stepan.shakhno@lnu.edu.ua

© Springer Nature Singapore Pte Ltd. 2020
F. Szidarovszky and G. I. Bischi (eds.), *Games and Dynamics in Economics*,
https://doi.org/10.1007/978-981-15-3623-6_14

closed form solution can only be found in rare cases. Therefore, iterative solvers are introduced of the form

$$x_{n+1} = x_n - T(x_n)^{\#} H(x_n), \, n = 0, 1, 2, \ldots, \tag{2}$$

where $T(x_n)^{\#}$ stands for a bounded outer inverse operator (BOIO). That is it satisfies

$$T(x_n)^{\#} T(x_n) T(x_n)^{\#} = T(x_n)^{\#}. \tag{3}$$

In the seminal paper by Nashed and Chen (1993) a semi-local convergence analysis of solver (2) was given based on Yamamoto type Ben-Israel (1968) conditions for Newton-type solvers

$$y_{n+1} = y_n - T(x_n)^{-1} H(x_n) \tag{4}$$

but extended to BOIO from T^{-1}.

The convergence analysis in Nashed and Chen (1993) extended and generalized earlier results by Häußler (1986), Deuflhard and Heindl (1979) and others. The problem with all these results is that the sufficient convergence criteria are strong leading to a small convergence region in general. Moreover, the error bounds and the information on the solution are not optimal. We address all these problems in this paper, and introduce a finer convergence analysis which is not involving additional hypotheses. Furthermore, our choice of the BOIO is more flexible also allowing the study of equations with a non-differentiable term. We also study the local convergence of solver (2) not given in Nashed and Chen (1993) with similar advantages.

The layout of the rest of the paper involves: mathematical background semi-local, local convergence and conclusion in Sect. 2–Sect. 4, respectively.

2   Mathematical Background

In order to make the paper as self continued as possible, we reproduce some results from Ben-Israel (1968), Nashed and Chen (1993), Häußler (1986), Nashed (1987), Yamamoto (1989).

Let $N(Q)$, $R(Q)$ stand for the null-space, and the range respectively of a linear operator Q.

We need the following auxiliary results.

Lemma 1 *If $Q^{\#}$ is a BOIO of $Q \in L(B_1, B_2)$, then $B_1 = R(Q^{\#}) \oplus N(Q^{\#} Q)$, $B_2 = N(Q^{\#}) \oplus R(Q Q^{\#})$.*

Lemma 2 *Let $Q^{\#}$ be an outer inverse of $Q \in L(B_1, B_2)$ satisfying $\|Q^{\#}(Q_1 - Q)\| < 1$ for $Q_1 \in L(B_1, B_2)$. Then, $Q_1^{\#} := (I + Q^{\#}(Q_1 - Q))^{-1} Q^{\#}$ is a BOIO of Q_1, $N(Q_1^{\#}) = N(Q)$, $R(Q_1^{\#}) = R(Q)$,*

$$\|(Q_1^{\#} - Q^{\#})\| \leq \frac{\|Q^{\#}(Q_1 - Q)Q_1^{\#}\|}{1 - \|Q^{\#}(Q_1 - Q)\|} \leq \frac{\|Q^{\#}(Q_1 - Q)\|\|Q^{\#}\|}{1 - \|Q^{\#}(Q_1 - Q)\|}$$

and

$$\|Q_1^{\#}Q\| \leq \frac{1}{1 - \|Q^{\#}(Q_1 - Q)\|}.$$

Lemma 3 *Let $Q, Q_1 \in L(B_1, B_2)$ with $Q^{\#}, Q_1^{\#}$ denoting outer inverses of Q, and Q_1, respectively. Then, the following implication holds*

$$Q_1^{\#}(I - QQ_1^{\#}) = 0 \Leftrightarrow N(Q^{\#}) \subset N(Q_1^{\#}).$$

3 Semilocal Convergence

The semi-local convergence is based on the conditions (A):

(A_1) $T(x) \in L(B_1, B_2)$. There exist an open convex subset Ω_0 of Ω, BOIO $T^{\#}$ of T with $T(x_0) = T$, and constant $\eta > 0$ such that $\|T^{\#}H(x_0)\| \leq \eta_0$.

(A_2) For $L_0, l_0 \geq 0$, $x \in \Omega_0$

$\|T^{\#}(T(x) - T(o))\| \leq L_0\|x - x_0\| + l_0$, and $l_0 = 0$, if $x = x_0$.

Set $\Omega_1 = \Omega \cap U\left(x_0, \dfrac{1 - l_0}{L_0}\right)$ for $L_0 \neq 0$, and $l_0 \in [0, 1]$.

(A_3) There exist $K_0 > 0$, $M_0, \mu_0 \geq 0$ such that for each $x, y \in \Omega_1$

$$\|T^{\#}(H(y) - H(x) - T(x)(y - x))\| \leq \frac{K_0}{2}\|y - x\|^2 + (M_0\|x - x_0\| + \mu_0)\|y - x\|.$$

(A_4) $b_0 := \mu_0 + l_0 < 1$.

(A_5) For $\delta_0 := \max\{K_0, M_0 + L_0\}$

$$h_0 = \delta_0\eta_0 \leq \frac{1}{2}(1 - b_0)^2$$

and

$$\bar{U}(x_0, u^*) \subset \Omega_0,$$

where $u^* = \dfrac{1 - b_0 - \sqrt{(1 - b_0)^2 - 2h_0}}{\delta_0}$. It is also relevant to define scalar functions

$$f_0(t) = \frac{\delta_0}{2}t^2 - (1 - b_0)t + \eta_0,$$

$$g_0(t) = 1 - L_0t - l_0,$$

as well as a sequence $\{u_n\}$ by

$$u_0 = 0, \quad u_{n+1} = u_n + \frac{f_0(u_n)}{g_0(u_n)}.$$

Clearly, by the inequality in (A_5), $f_0(t)$ has roots u^* and u^{**} with $0 < u^* \leq u^{**}$, and $u_n \leq u_{n+1}$, so $\lim_{n \to \infty} u_n = u^*$, where

$$u^{**} = \frac{1 - b_0 + \sqrt{(1 - b_0)^2 - 2h_0}}{\delta_0}.$$

Next, the preceding conditions and notation suffice to present the semi-local convergence.

Theorem 1 *Under the conditions (A), the following items hold:*

(a) The sequence generated by solver (2) stays in $U(x_0, u^*)$, and converges to a solution $x^* \in \bar{U}(x_0, u^*)$ of equation $T^\# H(x) = 0$, provided that

$$T(x_n)^\# = (I + T^\#(T(x_n) - T))^{-1} T^\#.$$

(b) The point x^* is the only solution of equation $T^\# H(x) = 0$ in $\Omega_2 = \tilde{U}_0 \cap \{R(T^\#) + x_0\}$, where

$$\tilde{U}_0 = \begin{cases} \bar{U}_0(x_0, u^*) \cap \Omega_0, & h_0 = \dfrac{1}{2}(1 - b_0)^2 \\ U_0(x_0, u^{**}) \cap \Omega_0, & h_0 < \dfrac{1}{2}(1 - b_0)^2, \end{cases}$$

and

$$R(T^\#) + x_0 = \{x + x_0 : x \in R(T^\#)\}.$$

Proof We use mathematical induction to show:

$$\|x_{n+1} - x_n\| \leq u_{n+1} - u_n. \tag{5}$$

By (A_1), we have

$$\|x_1 - x_0\| \leq \eta = u_1 - u_0, \tag{6}$$

so (5) is true for $n = 0$. Using (A_2), the definition of u^*, and (6), we get

$$\|T^\#(T(x_1) - T\| \leq L_0\|x_1 - x_0\| + l_0 \leq L_0 u_1 + l_0 \leq L_0 u^* + l_0 < 1. \tag{7}$$

It follows by Lemma 2 that

$$(T(x_1)^\# := (I + T(T(x_1) - T))^{-1} T^\#$$

is an outer inverse of $T(x_1)$,

$$\|(T(x_1)^\# T\| \leq \frac{1}{1 - L_0\|x_1 - x_0\| - l_0} \leq \frac{1}{1 - L_0 u_1 - l_0}, \qquad (8)$$

and $N(T(x_1)^\#) = N(T^\#)$.

Suppose that

$$\|x_m - x_{m-1}\| \leq u_m - u_{m-1} \qquad (9)$$

and

$$N(T(x_{m-1})^\#) = N(T^\#) \qquad (10)$$

for each $m = 1, 2, ..., n$.

Then, we have

$$\|x_m - x_0\| \leq \|x_m - x_{m-1}\| + \|x_{m-1} - x_{m-2}\| + ... + \|x_1 - x_0\|$$
$$\leq (u_m - u_{m-1}) + (u_{m-1} - u_{m-2}) + ... + (u_1 - u_0) = u_m - u_0 = u_m \leq u^* \qquad (11)$$

and

$$N(T(x_m)^\#) = N(T(x_{m-1})^\#) = N(T^\#). \qquad (12)$$

Using Lemma 3, we get

$$T(x_n)^\#(I - T(x_{n-1})T(x_{n-1})^\#) = 0. \qquad (13)$$

Moreover, by solver (2), we get in turn that

$$x_{n+1} - x_n = -T(x_n)^\# H(x_n)$$
$$= -T(x_n)^\#(H(x_n) - T(x_{n-1})(x_n - x_{n-1}) - T(x_{n-1})T(x_{n-1})^\# H(x_{n-1}) \qquad (14)$$
$$= -[T(x_n)^\# T][T^\#(H(x_n) - H(x_{n-1}) - T(x_{n-1})(x_n - x_{n-1}))].$$

Notice that by Lemma 3, and $N(T(x_n)^\#) = N(T^\#)$, we have

$$T(x_n)^\#(I - T(x_n)T^\#) = 0. \qquad (15)$$

In view of (A_3), (8), the definition of sequence $\{u_n\}$, and (14), we obtain in turn that

$$\|x_{n+1} - x_n\| = \|[T(x_n)^\# T][T^\#(H(x_n) - H(x_{n-1}) - T(x_{n-1})(x_n - x_{n-1})]$$

$$\leq \frac{1}{1 - L_0 u_n - l_0}[\frac{K_0}{2}\|x_n - x_{n-1}\|^2 + (M_0\|x_{n-1} - x_0\| + \mu_0)\|x_n - x_{n-1}\|]$$

$$\leq \frac{1}{1 - L_0 u_n - l_0}[\frac{\delta_0}{2}(u_n - u_{n-1})^2 + (M_0 u_{n-1} + \mu_0)(u_n - u_{n-1})]$$

$$= \frac{1}{g_0(u_n)}[\frac{\delta_0}{2}(u_n - u_{n-1})^2 + (M_0(u_n - u_{n-1})u_{n-1} + \tag{16}$$

$$+ f_0(u_{n-1}) + \mu_0(u_n - u_{n-1}) - g_0(u_{n-1})(u_n - u_{n-1})]$$

$$= \frac{1}{g_0(u_n)}[\frac{\delta_0}{2}u_n^2 - (1 - b_0)u_n + \eta_0 + (\delta_0 - M_0 - L_0)u_{n-1}(u_n - u_{n-1})]$$

$$\leq \frac{f_0(u_n)}{g_0(u_n)} = u_{n+1} - u_n,$$

which terminates the induction for (9). Then, so far we have for each n

$$\|T^\#(T(x_{n+1}) - T)\| \leq L_0\|x_{n+1}) - x_0\| + l_0 \leq L_0 u_{n+1} + l_0 \leq L_0 u^* + l_0 < 1 \tag{17}$$

$$\|x_n - x_0\| \leq u_n \leq u^* \tag{18}$$

and

$$T(x_{n+1})^\# := (I + T^\#(T(x_{n+1}) - T))^{-1}T^\# \tag{19}$$

is an outer inverse of $T(x)$. By (9), sequence $\{x_n\}$ is fundamental, and as such it converges to some $x^* \in \bar{U}(x_0, u^*)$. Moreover, by (19), we get that

$$T^\#(H(u^*)) = \lim_{n\to\infty} T^\# H(x_n) = \lim_{n\to\infty}(I + T^\#(T(x_n) - T))(x_n - x_{n-1}) = 0,$$

so $T^\# H(x^*) = 0$.

(b) By Lemma 1, we can write $R(T(x_n)^\#) = R(T^\#)$ for each n, so

$$x_{n+1} - x_n = -T(x_n)^\# H(x_n) \in R(T(x_n)^\#) = R(T^\#).$$

Then, by Lemma 2, we get $R(T^\#) = R(T^\# T)$ leading to $x_{n+1} \in x_n + R(T^\#)$ and $x_n \in x_0 + R(T^\#)$ for each n. Let $y^* \in \bar{U}_0 \cap \{x_0 + R(T^\#)\}$ be a solution of equation $T^\# H(x) = 0$, and $y^* - x^* \in R(T^\#)$. Then, we can write for each n

$$T^\# T(y^* - x_n) = T^\# T(y^* - x_0) + T^\# T(x_n - x_0) = y^* - x_n. \tag{20}$$

Hence, by (A_1), (A_3) and (20)

$$\|y^* - x_1\| = \|y^* - x_0 + T^\# H(x_0) - T^\# H(y^*)\|$$

$$= \|T^\# T(y^* - x_0) + T^\#(H(x_0) - H(y^*))\|$$
$$= \|T^\#(H(x_0) - H(y^*)) + T(y^* - x_0)\|$$
$$= \|\|T^\#(H(y^*)) - H(x_0) - T(y^* - x_0)\| \le (\frac{\delta_0}{2}\|y^* - x_0\| + l_0 + \mu_0)\|\|y^* - x_0\| = p(\rho)$$

for $\rho = \|y^* - x_0\|$. It then follows from

$$\|y^* - x_0\| \le \|y^* - x_1\| + \|x_1 - x_0\| \le p(\rho) + \eta_0 = f_0(\rho) + \eta_0,$$

so $p(\rho) \ge 0$, and consequently $y^* \in \bar{U}(x_0, u^*)$. We must show by induction that

$$\|y^* - x_n\| \le u^* - u_n. \tag{21}$$

By $y^* \in \bar{U}(x_0, u^*)$, (21) holds for $n = 0$. Suppose (21) holds for all $i = 0, 1, 2, ..., n$. In view of (A_2), (A_3), (8) and (21), we obtain in turn that

$$\|y^* - x_{j+1}\| = \|y^* - x_j + T(x_j)^\# H(x_j) - T(x_j)^\# H(y^*)\|$$

$$= \|[T(x_j)^\# T(x_j)][T(x_j)^\#(T(x_j)(y^* - x_j))] + H(x_j) - H(y^*)\|$$

$$\le \frac{1}{1 - L_0 u_j - l_0}[\frac{K_0}{2}\|y^* - x_j\|^2 + (M_0\|x_j - x_0\| + \mu_0)\|y^* - x_j\|]$$

$$\le \frac{1}{g_0(u_j)}[\frac{\delta_0}{2}(u^* - u_j)^2 + (M_0 u_j + \mu_0)(u^* - u_j)]$$

$$= \frac{1}{g_0(u_j)}[\frac{\delta_0}{2}(u^*)^2 + \mu_0 u^* - (\delta_0 - M_0)u_j)(u^* - u_j)]$$

$$= \frac{1}{g_0(u_j)}[\frac{\delta_0}{2}(u^* - \eta - l_0 u^* - (\delta_0 - M_0)u_j(u^* - u_j) - \frac{\delta_0}{2}u_j^2 - \mu_0 u_j]$$

$$= u^* - u_j + \frac{1}{g_0(u_j)}[-(u^* - u_j)g_0(x_j) + u^* - \eta$$

$$-l_0 u^* - \frac{\delta_0}{2}u_j^2 - \mu_0 u_j - (\delta_0 - M_0)u_j(u^* - u_j)]$$

$$= u^* - u_j - \frac{1}{g_0(u_j)}[\frac{\delta_0}{2}u_j^2 - (1 - \mu_0 - l_0)u_j + \eta + (\delta_0 - M_0 - L_0)u_j(u^* - u_j)]$$

$$\le u^* - u_{j+1},$$

$$\tag{22}$$

which completes the induction for (21). It then follows from (21) that $\lim_{n\to\infty} x_n = y^*$. But we showed $\lim_{n\to\infty} x_n = x^*$, so $x^* = y^*$.

Remark 1 (*a*) The proof of Theorem 1 extends the corresponding one in Nashed and Chen (1993). In particular, set $H = H_1$ and $H_2 = 0$. Then, the corresponding conditions in Nashed and Chen (1993) are:

$\|T^{\#}H(x_0)\| \leq \eta$,

$\|T^{\#}(T(x) - T(x_0))\| \leq L\|x - x_0\| + l$ for each $x \in \Omega$,

$\|T^{\#}(H_1(x) - H_1(x))\| \leq K\|y - x\|$ for each $x, y \in \Omega$,

$\|T^{\#}(H_1(x) - T(x))\| \leq M\|x - x_0\| + \mu$ for each $x \in \Omega$,

$b = \mu + l < 1$,

$h = \delta\eta \leq \dfrac{1}{2}(1 - b)^2, \quad \delta = \max\{K, M + L\}$,

$v^* = \dfrac{1 - b - \sqrt{(1 - b)^2 - 2h}}{\delta}$,

$v^{**} = \dfrac{1 - b + \sqrt{(1 - b)^2 - 2h}}{\delta}$,

$f(t) = \dfrac{\delta}{2}t^2 - (1 - b)t + \eta, \quad g(t) = 1 - Lt - l$,

$v_0 = 0, \quad v_{n+1} = v_n + \dfrac{f(v_n)}{g(v_n)}$,

and

$D_1 = \tilde{U}_0^1 \cap \{R(T^*) + x_0\}$,

$$\tilde{U}_0^1 = \begin{cases} \bar{U}_0(x_0, v^*) \cap \Omega_0, & h = \dfrac{1}{2}(1 - b)^2 \\[2mm] U_0(x_0, v^{**}) \cap \Omega_0, & h < \dfrac{1}{2}(1 - b)^2. \end{cases}$$

Then, by $\Omega_1 \subseteq \Omega$, we have

$$\eta_0 = \eta,$$

$$L_0 \leq L,$$

$$l_0 \leq l,$$

$$K_0 \leq K,$$

$$M_0 \leq M,$$

$$\mu_0 \leq \mu,$$

$$b_0 \leq b,$$

$$\delta_0 \leq \delta,$$

$$h \leq \frac{1}{2}(1 - b)^2 \implies h_0 \leq \frac{1}{2}(1 - b_0)^2 \tag{23}$$

but not vice versa unless, if $\delta_0 = \delta$ and $b_0 = b$. Hence, we have extended the applicability of solver (2) without adding hypotheses, since in practice the computation of $L, l, K, M, \mu, b, \delta$ requires that of $L_0, l_0, K_0, M_0, \mu_0, b_0, \delta_0$ as special cases.

(b) If $T^{\#} = T^{-1}$, Theorem 5 reduces to the one in Yamamoto (1987) for Newton-like solvers.

(c) Let us specialize further for the case of Newton's solver, i.e. for $T = H_1'$. Then, we obtain $\eta_0 = \eta$, $L = l_0 = M = M_0 = \mu = \mu_0 = b = b_0 = 0$, $\delta_0 = \max\{K_0, L\}$, $\delta = K$. Moreover, we have

$$h = K\eta \leq \frac{1}{2} \implies h_0 = \delta_0\eta \leq \frac{1}{2}, \tag{24}$$

and from the uniqueness part of the proof

$$\|x^* - x_{n+1}\| \leq \frac{K_0}{2(1 - L_0 u^*)} \|x^* - x_n\|^2 \tag{25}$$

instead of the old

$$\|x^* - x_{n+1}\| \leq \frac{K}{2(1 - K v^*)} \|x^* - x_n\|^2 \tag{26}$$

$$u^* = \frac{1 - \sqrt{1 - \delta_0 \eta}}{\delta_0} \leq v^* = \frac{1 - \sqrt{1 - K\eta}}{\eta}. \tag{27}$$

It is also worth noticing that if we replace $\bar{U}(x_0, u^*) \subset \Omega_0$ by $U(x_1, u^* - \eta) \subset \Omega$ in the proof of Theorem 1 (similarly for the old case), then under the assumptions of Theorem 1 we obtain the estimates as in Yamamoto (1986, 1987)

$$\|x^* - x_n\| \leq e_n^0 \leq \frac{u^* - u_n}{u_{n+1} - u_n} \alpha_n \leq \frac{u^* - u_n}{u_n - u_{n-1}} \alpha_{n-1} \leq u^* - u_n,$$

instead of

$$\|x^* - x_n\| \leq e_n \leq \frac{v^* - v_n}{v_{n+1} - v_n} \alpha_n \leq \frac{v^* - v_n}{v_n - v_{n-1}} \alpha_{n-1} \leq v^* - v_n,$$

so

$$e_n^0 \leq e_n, \tag{28}$$

where

$$e_n^0 := \frac{2\alpha_n}{1 + \sqrt{1 - \frac{2\delta_0 \alpha_n}{1 - \delta_0 \Delta_n}}},$$

$$e_n := \frac{2\alpha_n}{1 + \sqrt{1 - \frac{2K\alpha_n}{1 - K\Delta_n}}},$$

$\Delta_n := \|x_n - x_0\|$, and $\alpha_n := \|x_{n+1} - x_n\|$. Hence, again not only the sufficient convergence criteria (23) and (24) are weakened but the location of the solution is more precise (see 27) and the error estimations are tighter (see 28).

(d) The convergence conditions can become even more general, if (A_2) and (A_3) are replaced, respectively by

$(A_2)'$

$$\|S^{\#}(T - S(x_0))\| \leq L_0 \|x - x_0\| + l_0,$$

and $l_0 = 0$, if $x = x_0$, where $S(x) \in L(B_1, B_2)$, and there exist BOIO $S^{\#}$ of S with $S = S(x_0) = T(x_0)$

and

$(A_3)'$ there exist $K_0 > 0$, $M_0, \mu_0 \geq 0$ such that for each $x, y \in \Omega_1$

$$\|S^{\#}(H(y) - H(x) - T(x)(y - x))\| \leq \frac{K_0}{2} \|y - x\|^2 + (M_0 \|x - x_0\| + \mu_0) \|y - x\|.$$

Then, the conclusions of Theorem 1 hold true, if $(A_2)'$ and $(A_3)'$ reduce (A_2) and (A_3), if $S = T$. Condition (A_2) is used to show the existence of $T(x)^{\#}$. The same goal however is achieved by the more flexible $(A_2)'$. Concerning the proof in this setting $T(x)$ is simply replaced $T^{\#}SS^{\#}$. Clearly this setting allows a greater flexibility in the choice of T. These results allow us to see equations containing a nondifferentiable term in a new, more usefull and more flexible setting. Moreover, earlier methods can be compared in a unified setting.

Clearly, if appropriately specialized and along the same lines the results of other solvers such as Secant, Steffensen, Stirling's, Newton-Secant, Newton-Kurchatov, Aitken and other solvers can be extended along the same lines, and the same conditions (A). Therefore these methods are compared to each other under the same set of conditions.

4   Local Convergence

The local convergence is based on the conditions (C):

(C_1) $H : \Omega \rightarrow B_2$ is continuous, $T(x) \in L(B_1, B_2)$. There exist an open convex subset Ω_0 of Ω, BOIO $T^{\#}$ of T, x^* a solution of $T^{\#}H(x) = 0$, such that $T = T(x^*)$.

(C_2) For $L_0, l_0 \geq 0$, $x \in \Omega_0$

$$\|T^{\#}(T(x) - T)\| \leq L_0 \|x - x^*\| + l_0$$

and $l_0 = 0$, if $x = x^*$.

Set $\Omega_1 = \Omega \cap \bar{U}(x^*, \frac{1 - l_0}{L_0})$, $L_0 \neq 0$.

(C_3) There exist $K_0 > 0$, M_0, $\mu_0 \geq 0$ such that for each $x \in \Omega_1$

$$\|T^\#(H(x^*) - H(x) - T(x)(x^* - x))\| \leq \frac{K_0}{2}\|x^* - x\|^2 + (M_0\|x^* - x\| + \mu_0)\|\|x^* - x\|.$$

(C_4) = (A_4).
(C_5) $\bar{U}(x^*, r^*) \subset \Omega_0$, where

$$r^* = \frac{1 - l_0 - \mu_0}{\frac{K_0}{2} + M_0 + L_0}. \tag{29}$$

Next, we present the local convergence of solver (2) utilizing (C), and the preceding notation.

Theorem 2 *Under the conditions (C), choose $x_0 \in U(x^*, r^*)$. Then, the following items hold*

$$\{x_n\} \subset U(x^*, r^*). \tag{30}$$

$$\lim_{n \to \infty} x_n = x^*, \tag{31}$$

$$\|x_{n+1} - x^*\| \leq q_n\|x_n - x^*\| \leq \|x_n - x^*\| \leq r^*, \tag{32}$$

and x^* is the only solution of equation $T^\#H(x) = 0$ in $\Omega_4 = \Omega \cap \bar{U}(x^*, r^*)$, where

$$q_n = \frac{\frac{K_0}{2}\|x^* - x_n\| + M_0\|x^* - x_n\| + \mu_0}{1 - L_0\|x_n - x^*\| - l_0} \in [0, 1]. \tag{33}$$

Proof in view of the uniqueness part in Theorem 1 (see 22 for $x_0 = x^*$), but using the (C) instead of the (H) conditions, we obtain the conclusions of Theorem 2.

Remark 2 (a) Local results were not given in Nashed and Chen (1993). But if they were, the radius would have been

$$\bar{r}^* = \frac{1 - l - \mu}{\frac{K}{2} + M + L},$$

where the constants are related as before.
(b) In the case of Newton's solver

$$r^* = \frac{2}{2L_0 + K_0},$$

where as

$$\bar{r}^* = \frac{2}{3K},$$

was given independently by Rheinboldt Ortega and Rheinboldt (1970) and Traub Traub (1964). Notice that

$$\bar{r}^* \leq r^*.$$

(c) As in the semi-local case consider instead of (C_2) and (C_3), respectively
$(C_2)'$

$$\|S^{\#}(T(x) - S)\| \leq L_0\|x - x^*\| + l_0,$$

and $l_0 = 0$, if $x = x^*$, where $S(x) \in L(B_1, B_2)$, and there exist BOIO $S^{\#}$ of S with $S = S(x^*) = T(x^*)$.
$(C_3)'$ There exist $K_0 > 0$, M_0, $\mu_0 \geq 0$ such that for each $x \in \Omega_1$

$$\|S^{\#}(H(^*x) - H(x) - T(x)(x^* - x))\| \leq \frac{K_0}{2}\|x^* - x\|^2 + (M_0\|x^* - x\| + \mu_0)\|\|x^* - x\|.$$

The rest of the comments are similar to the ones in Remark 1.

Corresponding error estimations are as in Remark 1, and our information on the location of solution is more accurate. Hence, we have a wider choice of initial guesses x_0, and fewer iterations are needed to obtain a desired error tolerance, and the new information on the location of the solution is more accurate. Concrete examples, where the new constants are smaller than the old ones (so the advantages are obtained) can be found in Argyros and Magrenán (2017), Argyros and Magrenán (2018), Argyros and Shakhno (2019a), Argyros and Shakhno (2019b), Argyros et al. (2019).

5 Conclusion

We presented a finer semi-local as well as a local convergence solver (2) using more general conditions than before. Despite this fact, in the

semi-local case: We obtained weaker sufficient convergence criteria, tighter error estimations, and a better information on the location of the solution;

local case: We deliver a larger radius of convergence and rest same as above.

These extensions are also obtained using the same computational effort, since the new constants are special cases of the old ones. Hence, we extended the applicability of solver (2), and without additional hypotheses.

References

Argyros, I. K., & Magrenán, Á. A. (2017). *Iterative methods and their dynamics with applications: A contemporary study*. CRC Press.
Argyros, I. K., & Magrenán, Á. A. (2018). *A contemporary study of iterative methods*. New York, NY, USA: Elsevier (Academic Press).

Argyros, I. K., & Shakhno, S. (2019). Extended local convergence for the combined newton-kurchatov method under the generalized lipschitz conditions. *Mathematics, 7*(2), 207. https://doi.org/10.3390/math7020207.

Argyros, I. K., & Shakhno, S. (2019). Extending the applicability of two-step solvers for solving equations. *Mathematics, 7*(1), 62. https://doi.org/10.3390/math7010062.

Argyros, I. K., Shakhno, S., & Yarmola, H. (2019). Two-step solver for nonlinear equations. *Symmetry, 11*(2), 128. https://doi.org/10.3390/sym11020128.

Ben-Israel, A. (1968). On applications of generalized inverses in nonlinear analysis. In T. L. Boullion, & P. P. Odell (Ed.), *Theory and application of generalized inverses of matrices* (pp. 183–202). Lubbock: Texas Tech University Press.

Ben-Israel, A. (1966). Newton-Raphson method for the solution of equations. *Journal of Mathematical Analysis and Applications, 15*, 243–253.

Ben-Israel, A., & Greville, T. N. E. (1974). *Generalized inverses: Theory and applications*. New York: Wiley and Sons.

Chen, X., & Yamamoto, T. (1989). Convergence domains of certain iterative methods for solving nonlinear equations. *Numererical Functional Analysis and Optimization, 10*, 37–48.

Deuflhard, P., & Heindl, G. (1979). Affine invariant convergence theorems for Newton's method and extensions to related methods. *SIAM Journal of Numerical Analysis, 16*, 1–10.

Häußler, W.M. (1986). A Kantorovich-type convergence analysis for the Gauss-Newton-method. *Numerische Mathematik, 48*, 119–125.

Kantorovich. L. V., Akilov, G. (1981). Functional analysis in normal spaces. Fizmathiz, Moscow (1959); English translation (2nd edn.). Pergamon Press, London.

Nashed, M. Z. (1976). *Generalized inverses and applications*. New York: Academic Press.

Nashed, M. Z. (1987). Inner, outer, and generalized inverses in Banach and Hilbert spaces. *Numerical Functional Analysis and Optimization, 9*, 261–325.

Nashed, M. Z., & Chen, X. (1993). Convergence of Newton-like methods for singular operator with outer inverses. *Numerische Mathematik, 66*, 235–257.

Ortega, J. M., & Rheinboldt, W. C. (1970). *Iterative solution of nonlinear equations in several variables*. New York: Academic Press.

Potra, F. A., Ptak, V. (1984). Nondiscrete induction and iterative processes. Research Notes in Mathematics, vol. 103. Pitman, Boston.

Shakhno, S. M. (2009). On an iterative algorithm with superquadratic convergence for solving nonlinear operator equations. *Journal of Computational and Applied Mathematics, 231*, 222–235.

Shakhno, S. M. (2010). On a two-step iterative process under generalized Lipschitz conditions for first-order divided differences. *Journal of Mathematical Sciences, 168*, 576–584.

Shakhno, S. M. (2014). Convergence of the two-step combined method and uniqueness of the solution of nonlinear operator equations. *Journal of Computational and Applied Mathematics, 261*, 378–386.

Shakhno, S. M., Mel'nyk, I. V., & Yarmola, H. P. (2014). Analysis of the convergence of a combined method for the solution of nonlinear equations. *Journal of Mathematical Sciences, 201*, 32–43.

Traub, J. F. (1964). *Iterative methods for the solution of equations*. Englewood Cliffs, New York: Prentice-Hall Inc.

Yamamoto, T. (1989). Uniqueness of the solution in a Kantorovich-type theorem of Häußler for the Gauss-Newton method. *Japan Journal of Industrial and Applied Mathematics, 6*, 77–81.

Yamamoto, T. (1986). A method for finding sharp error bounds for Newtons method under the Kantorovich assumptions. *Numerische Mathematik, 49*, 203–320.

Yamamoto, T. (1987). A convergence theorems for Newton-like methods in Banach spaces. *Numerische Mathematik, 51*, 545–557.

Properties of Linear Time-Dependent Systems

Sandor Molnar and Mark Molnar

Abstract In economic modelling time-dependent linear systems are frequently used to analyse a multitude of interesting situations. One interesting area of application is reachability which can drive the state of the system to a given desired state in any future time. In this study we elaborate a generalisation of the Kalman-type rank conditions which provide the sufficient and necessary conditions for reachability of such systems.

1 Introduction

For economic models it is very important to study market behaviour and interactions which are typically represented by a dynamic price function and the dynamics of market share (Bischi et al. 2010). These processes can be represented by state and time-dependent systems (Matsumoto and Szidarovszky 2018; Szidarovszky and Yen 1993). This study focuses on the latter approach and provides a general discussion of reachability.

Under system qualities reachability from 0, controllability, observability and reconsctructability and a sort of stability of input-output systems is meant. We don't discuss the latter comprehensively as the classical Lyapunov-type methods and the characterisation with Ricatti-equations essentially solve the problem in this system class (see e.g. Pontryagin et al. 1964; Kaplansky 1976).

S. Molnar (✉)
Institute of Mechanics and Machinery, Szent István University, Páter K. U. 1, Gödöllő 2103, Hungary
e-mail: molnar.sandor@gek.szie.hu

M. Molnar
Department of Macroeconomics, Institute of Economics, Szent István University, Páter K. U. 1, Gödöllő 2103, Hungary
e-mail: molnar.mark@gtk.szie.hu

© Springer Nature Singapore Pte Ltd. 2020
F. Szidarovszky and G. I. Bischi (eds.), *Games and Dynamics in Economics*,
https://doi.org/10.1007/978-981-15-3623-6_15

271

For the classical canonical form of time dependent parametric systems

$$\begin{aligned}\dot{x}(t) &= A(t)x(t) + B(t)u(t) \\ y(t) &= C(t)x(t) + D(t)u(t)\end{aligned} \tag{1}$$

R. Kalman solved all fundamental issues (Szigeti 1992; Molnár et al. 1993; Molnár and Szidarovszky 1994; Molnár 2001; Serre 1992). He proved the dualities for the basic qualities he defined, e.g. controllability, reachability, observability and reconstructibility. Furthermore he proved the equivalence for the reachability and controllability and observability and reconstructability pairs for continuous time systems (of form (1)). For example price and demand change in state and time due to the change of preferences of consumers, market saturation, technological development, etc. Complying with this we will only deal with one quality, reachability. The classical results concerning the main features of linear systems are summarised in Szidarovszky and Bahill (1998).

2 Preliminaries

Let's consider the reachability of the system on a given fixed $[0,T]$ interval, where we assume that functions

$$\begin{aligned}A &: [0, T] \to \mathbb{R}^{n \times n}, \; B : [0, T] \to \mathbb{R}^{n \times k}, \\ C &: [0, T] \to \mathbb{R}^{l \times n}, \; D : [0, T] \to \mathbb{R}^{l \times k},\end{aligned}$$

are at least continuous.

R. Kalman characterised reachability with the invertibility of the so called Kalman-Gram-type matrix. For this we need to define the fundamental matrix of the system. Let us consider the initial value problem

$$\dot{x}(t) = A(t)x(t), \; x(\tau) = I \tag{2}$$

on $\mathbb{R}^{n \times n}$. In case of a continuous coefficient matrix this has a singular solution on the entire $[0, T]$ interval

$$t \mapsto \Phi(t, \tau) \in \mathbb{R}^{n \times n}$$

which is continuously differentiable as a bivariate (t, τ)-function for which $\Phi(t, \tau)$ can be inverted for all (t, τ) pairs. Consider for this the solution $t \mapsto \Psi(t, \tau)$ of the initial value problem

$$\dot{Y}(t) = -Y(t)A(t), \; Y(\tau) = I$$

which has a domain on the entire $[0, T]$ interval. Then

$$\frac{d}{dt}(\Psi(t,\tau)\Phi(t,\tau)) = \dot\Psi(t,\tau)\Phi(t,\tau) + \Psi(t,\tau)\dot\Phi(t,\tau)$$
$$= (-\Psi(t,\tau)A(t)\Phi(t,\tau) + \Psi(t,\tau)(A(t)\Phi(t,\tau))) = 0,$$

that is, $\Psi(t,\tau)\Phi(t,\tau) = I$, which is sufficient on $\mathbb{R}^n$ for having $\Psi(t,\tau) = \Phi(t,\tau)^{-1}$.

Furthermore $\Phi(t,\tau) = \Phi(t,0)\Phi(\tau,0)^{-1}$, since $t \mapsto \Phi(t,0)\Phi(\tau,0)^{-1}$ is a solution of the equation $\dot x(t) = A(t)x(t)$ and $\Phi(\tau,0)\Phi(\tau,0)^{-1} = I$. Switching the role of variables t and τ

$$\Phi(\tau,t) = \Phi(\tau,0)\,\Phi(t,0)^{-1} = \Phi(\tau,0)\,\Psi(t,0),$$

that is

$$\frac{d}{dt}\Phi(\tau,t) = \Phi(\tau,0)\frac{d}{dt}\Psi(t,0) = \Phi(\tau,0)(-\Psi(t,0)A(t))$$
$$= -\Phi(\tau,0)\Phi(t,0)^{-1}A(t) = -\Phi(\tau,t)A(t),\ \Phi(\tau,\tau) = I.$$

Thus

$$\Phi(t,\tau)^{-1} = \Phi(\tau,t).$$

After these preparations following up on R. Kalman we can define the Kalman-Gram type reachability matrix:

$$R[0,T] = \int_0^T \Phi(T,t)B(t)B(t)^*\Phi(T,t)^*dt$$

2.1 The Kalman-Type Reachability Theorem

The system in (1) is reachable from state 0 if and only if the Kalman-Gram type reachability matrix is invertible, or, equivalently if it is positive definite (Kalman and Falb 1969). We note here that similar theorem stands for controllability. For this define the Kalman-Gram-type controllability matrix as follows:

$$C[0,T] = \int_0^T \Phi(0,t)B(t)B(t)^*\Phi(0,t)^*dt.$$

If we define the dual system of (1)

$$\begin{aligned}
\dot{x}(t) &= A(t)^* x(t) + C(t)^* u(t) \\
y(t) &= B(t)^* x(t) + D(t)^* u(t),
\end{aligned} \tag{3}$$

we could denote input with y and output with u thus indicating that their roles switch.

2.2 The Kalman-Type Duality Theorem

Important system properties like controllability, observability and reconstructability is defined in many sources, see e.g. Szidarovszky and Bahill (1998) and Okoguchi and Szidarovszky (1997). The system described in (1) is controllable if and only if (3) can be reconstructed and (1) is reachable if and only if (3) is observable (Kalman and Falb 1969).

Since the dual system's dual pair is the original system therefore it is also true that (1) is observable if and only if (3) is reachable and (1) is reconstructible if and only if (3) is controllable.

The observability Kalman-Gram-type matrix is

$$O[0, T] = \int_0^T \Phi(T, t)^* C(t)^* C(t) \Phi(T, t) dt,$$

while reconstructability is equivalent with the invertibility and positive definiteness of the Kalman-Gram-type matrix

$$R_e[0, T] = \int_0^T \Phi(0, t)^* C(t)^* C(t) \Phi(0, t) dt$$

In the followings we focus on the reachability of system (1) and the reachability of the more general canonical form systems

$$\begin{aligned}
\dot{x}(t) &= A(t)x(t) + \sum_j B_j(t) u^{(j)} \\
y(t) &= C(t)x(t) + \sum_j D_j(t) u^{(j)}
\end{aligned} \tag{4}$$

Multiple studies examine the system described in (1) see e.g. (Molnár 1993; Molnár 2001; Molnár 1993; Molnár and Szidarovszky 1994; Molnár and Szigeti 1994; Molnár and Szigeti 1994). Starting from these we examine the general canonical form systems with special regards to the theoretical constructibility of persistent excitation conditions.

3   Lie-Algebras

We will need a few new terms, which we define herewith. Let L be a vector space on $\mathbb{R}$, where we also define a productive operation: the so-called Lie-product or Lie-brackets: if $l_1, l_2 \in L$ then $[l_1, l_2] \in L$, $l_1 \mapsto [l_1, l_2]$ and $l_2 \mapsto [l_1, l_2]$ are linear mappings and

(1) $[l, l] = 0$ for all $l \in L$,
(2) $[l_1, l_2] + [l_2, l_1] = 0$ for all $l_1, l_2 \in L$,
(3) $[l_1, [l_2, l_3]] + [l_2, [l_3, l_1]] + [l_3, [l_1, l_2]] = 0$, for all $l_1, l_2, l_3 \in L$.

According to the second attribute $[l_1, l_2] = -[l_2, l_1]$ is an expression of anti-commutativity, while 3 is the extent of non-associativity.

In fact

$$[l_1, [l_2, l_3]] = -[l_2, [l_3, l_1]] - [l_3, [l_1, l_2]]$$
$$= [[l_1, l_2], l_3] - [l_2, [l_3, l_1]],$$

since if $[l_2, [l_3, l_1]] = 0$, then the remaining

$$[l_1, [l_2, l_3]] = [[l_1, l_2], l_3]$$

equation exactly means associativity.

Examples

(1) Let $L = \mathbb{R}^{n \times n}$, $[A, B] = AB - BA$. The Lie-product defined in this manner renders $\left(\mathbb{R}^{n \times n}, [., .]\right)$ a Lie-algebra.
(2) Consider on the $\Omega \subset \mathbb{R}^n$ open set the vector space of the analytic functions $f : \Omega \to \mathbb{R}^n$ over $\mathbb{R}$. Define the Lie-bracket in the following manner:

$$[f, g](x) = f'(x)g(x) - g'(x)f(x). \tag{5}$$

Denote with $A(\Omega)$ the vectorspace of all the analytical $\mathbb{R}^n$ -valued functions (the analytic vectorfields on Ω) over $\mathbb{R}$ with the Lie-product. Then we get the $(A(\Omega), [\cdot, \cdot])$ Lie-algebra.
(3) Consider the vector fields C^∞ on the open set domain of $\Omega \subset \mathbb{R}^n$. Proceeding in a similar manner define the Lie-product using (5) over the vector space $C^\infty(\Omega)\mathbb{R}$. Then the $(C^\infty(\Omega), [\cdot, \cdot])$ Lie-algebra is defined.

3.1   Application for Solving System States

Let us now return to our system. Consider the partial Lie-algebra $L \subset \mathbb{R}^{n \times n}$, $(L, [., .])$ generated by the subset

$$\{A(t) : t \in [0, T]\} \subset \mathbb{R}^{n \times n}$$

that is, the smallest Lie-algebra for which $\{A(t) : t \in [0, T]\} \subset L$ stands. Such exists, since the set of containing partial Lie-algebras is not empty, obviously it contains $\mathbb{R}^{n \times n}$, therefore there is a minimal element of it, like for example the section of all such partial Lie-algebras. We call this the $L = L(A(t)) \subset \mathbb{R}^{n \times n}$ Lie-algebra generated by $\{A(t)\}$-s. Since the dimension of $\mathbb{R}^{n \times n}$ is n^2, it is finite, therefore the $L \subset \mathbb{R}^{n \times n}$ Lie-algebra's partial Lie-algebra has also finite dimension. Consider an $A_1, A_2, \ldots, A_l \in L$ basis of L.

In this basis

$$A(t) = \sum_{i=1}^{l} a_i(t) A_i$$

Express the Lie-product $[A_i, A_j] \in L$ in the basis $A_1, A_2, \ldots, A_l$:

$$[A_i, A_j] = \sum_{k=1}^{l} \Gamma_{ij}^k A_k.$$

Since $X \mapsto [A_i, X] = Ad A_i(X)$ is a linear mapping on the L vectorspace (it is also a Lie-algebra), therefore the matrix representation of $Ad A_i$ in the base $A_1, A_2, \ldots, A_l$ can be expressed with the Γ_{ij}^k numbers.

Let $X = \sum_{j=1}^{l} x_j A_j$; then

$$[A_i, X] = \left[A_i, \sum_{j=1}^{l} x_j A_j \right] = \sum_{j} x_j \left(\sum_{h=1}^{l} \Gamma_{ij}^h \right) A_h,$$

where $\sum_{j} \Gamma_{ij}^h x_j$ is the hth component of the following matrix product:

$$\begin{pmatrix} \Gamma_{i1}^1 & \Gamma_{i2}^1 & \cdots & \Gamma_{il}^1 \\ \Gamma_{i1}^2 & \Gamma_{i2}^2 & \cdots & \Gamma_{il}^2 \\ \vdots & \vdots & \vdots & \vdots \\ \Gamma_{i1}^l & \Gamma_{i2}^l & \cdots & \Gamma_{il}^l \end{pmatrix} \begin{pmatrix} x_1 \\ x_2 \\ \vdots \\ x_l \end{pmatrix} = \Gamma_i \mathbf{x}$$

That is, due to the $Ad A_i X \leftrightarrow \Gamma_i \mathbf{x}$ and $X \leftrightarrow \mathbf{x}$ compliances Γ_i is the matrix representation of $Ad A_i$ in the basis $A_1 A_2 \ldots A_l \in L$. We know that based on the Cauchy-formula the solution of system (1) satisfying the initial condition $x(0) = \xi$ is

$$x(t) = \Phi(t, 0)\xi + \int_0^t \Phi(t, \tau)B(\tau)u(\tau)d\tau,$$

moreover, a similar solution holds for the generalised canonical form system

$$x(t) = \Phi(t, 0)\xi + \int_0^t \Phi(t, \tau)\left(\sum_j B_j(\tau)u^j(\tau)\right)d\tau.$$

This is handsome, but we can calculate the fundamental matrix $\Phi(t, \tau)$ only in few cases. For the constant coefficient systems the basic system is the solution of

$$\dot{x}(t) = \mathbf{A}x(t), \ x(\tau) = I$$

that is

$$\Phi(t) = \exp A(t - \tau).$$

Moreover if the system's structure matrix $\mathbf{A}(t)$ has the form $\mathbf{A}(t) = a(t)\mathbf{A}$ then the basic system is

$$\Phi(t, \tau) = \exp \mathbf{A} \int_\tau^t a(s)ds.$$

For a general time dependent structure system

$$\dot{x}(t) = \sum_{i=1} a_i(t)\mathbf{A}_i x + \sum_j \mathbf{B}_j(t)u^{(j)}(t)$$

the basic system has a

$$\Phi(t, \tau) = (\exp \mathbf{A}_1 g_1(t, \tau)), (\exp \mathbf{A}_2 g_2(t, \tau)), \ldots, (\exp \mathbf{A}_l g_i(t, \tau))$$

form. We assume here also that in the **L** Lie-algebra generated by $\mathbf{A}(t)$'s $\mathbf{A}_1, \mathbf{A}_2, \ldots, \mathbf{A}_l$ is a basis and the matrix representation of $Ad A_i$ is the matrix $\Gamma_i \in \mathbb{R}^{l \times l}$. Then the existence of the above representation is provided by the Wei-Norman theorem.

Theorem. Wei-Norman-theorem *Let $\gamma(t) = g(t, \tau) \in \mathbb{R}^k$ be the solution of*

$$\left(\sum_{i=1}^{I} (\exp \Gamma_1 \gamma_1), (\exp \Gamma_2 \gamma_2), \ldots, (\exp \Gamma_{i-1} \gamma_{i-1}) E_{ii} \right)^{-1} \dot{\gamma} = a$$

$$\gamma(\tau) = 0$$

so called Wei-Norman differential equation. (The solution for this equation exists locally since due to the inital condition $\gamma(\tau) = 0$ the matrix to be inverted in $\tau = 0$ is the identity matrix, thus it is invertible in an appropriate neighbourhood of τ therefore it can be made explicit.)
    Then

$$\Phi(t, \tau) = \exp A_1 g_1(t, \tau), \exp A_2 g_2(t, \tau), \ldots, \exp A_I g_i(t, \tau)$$

has the exponential product form.

For the proof, see (Wei and Norman 1964).
We consider knowing that

$$\exp A_i g_i(t, \tau) = \sum_{j=0}^{n-1} q_{ij} \big(g_{ij}(t, \tau) \big) A_i^j$$

is polynomial in A_i with a maximum degree of $n - 1$, quasi-polynomial in $g(t, \tau)$ that it is polynomial in $g_i(t, \tau)$, $\sin \alpha_l g_i(t, \tau)$, $\cos \beta_l g_i(t, \tau)$ and $\exp \lambda_l g_i(t, \tau)$ where λ_l is the real part of the eigenvalues of A_i while α_l, β_l are the imaginary parts. The related fundamental results can be found in classical monographies of matrix theory and ordinary differential equations, see e.g. (Coddington and Levinson 1955).
    Substituting the formulation in the exponential product we get

$$\Phi(t, \tau) = \sum_{\mathbf{n}} Q_{\mathbf{n}}(\mathbf{g}(t, \tau)) A_1^{n_1} A_2^{n_2}, \ldots, A_I^{n_I},$$

where $Q_{\mathbf{n}}$ are the quasi polynomials of $\mathbf{g}(t, \tau) = (g_1(t, \tau), g_2(t, \tau), \ldots, g_I(t, \tau))$, (certain products of the $g_{ij}(g_i(t, \tau))$ quasi-polynomials).

3.2   Determination of Terminal State

Let's consider now the terminal state of the generalised time dependent coefficient linear system (4) at time $T = 0$ under the initial condition $x(0) = 0$. Based on the Cauchy-formula

$$x(T) = \int\limits_0^T \Phi(T, t)\left(\sum_j B_j(t)u^{(j)}(t)\right)dt.$$

We make this formula more practical with the application of an integration by parts procedure. If the highest order u-derived term is $u^{(J)}(t)$, then we assume that for all $j = 1,2, ..., J - 1$ the boundary conditions $u^{(j)}(0) = 0, u^{(j)}(T) = 0$ hold. This condition has no influence on the subspace of reachable terminal states on $\mathbb{R}^n$:

$$x(T) = \int\limits_0^T \Phi(T, t)B_0(t)u(t)dt + \sum_{j=1}^T \int\limits_0^T \Phi(T, t)B_j(t)u^{(j)}(t)dt$$

$$= \int\limits_0^T \Phi(T, t)B_0 u(t)dt + \sum_{j\geq 1}[\Phi(T, t)B_j(t)u^{j-1}(t)]_0^T$$

$$-\sum_{j=1}\int\limits_0^T \frac{d}{dt}\left(\Phi(T, t)B_j(t)\right)u^{(j-1)}(t)dt = \int\limits_0^T \Phi(T, t)B_0(t)u(t)dt$$

$$+\int\limits_0^T \Phi(T, t)\left(A(t)B_1(t) - B_1'(t)\right)u(t)dt$$

$$+\sum_{j\geq 2}\int\limits_0^T \Phi(T, t)(A(t)B_j(t) - B_j'(t))u^{(j-1)}(t)dt.$$

Repeating this for the last element we get the equations of the next step:

$$\sum_{j\geq 2}\int\limits_0^T \Phi(T, t)(A(t)B_j(t) - B_j'(t))u^{(j-1)}(t)dt$$

$$= \sum_{j\geq 2}[\Phi(T, t)(A(t)B_j(t) - B_j'(t))u^{(j-2)}(t)]_0^T$$

$$-\int\limits_0^T \frac{d}{dt}(\Phi(T, t)(A(t)B_2(t) - B_2'(t)))u(t)dt$$

$$-\sum_{j\geq 3}\int\limits_0^T \frac{d}{dt}(\Phi(T, t)(A(t)B_j(t) - B_j'(t)))u^{(j-2)}(t)dt$$

$$= \int\limits_0^T \Phi(T, t)\left(A(t)^2 B_2(t) - 2A(t)B_2'(t) - A'(t)B_2(t) + B''(t)\right)u(t)dt$$

$$+ \sum_{j \geq 3} \int_0^T \Phi(T,\ t)\Big(A(t)^2 B_j(t) - 2A(t)B_j'(t) - A'(t)B_j(t) + B_j''(t)\Big)u^{(j-2)}(t)dt.$$

Repeating the integration by parts for the last element we get a similar equation:

$$\sum_{j \geq 3} \int_0^T \Phi(T,t)(A(t)^2 B_j(t) - 2A(t)B_j'(t) - A'(t)B_j(t) + B_j''(t))u^{(j-2)}(t)dt$$

$$= \int_0^T \Phi(T,t)(A(t)^3 B_3(t) - 2A(t)A(t)'B_3(t) - A'(t)A(t)B_3(t) - 3A(t)^2 B_3'(t)$$

$$+ 3A'(t)B_3'(t) + 3A(t)B_3''(t) - B_3'''(t))u(t)dt$$

$$+ \sum_{j \geq 4} \int_0^T \Phi(T,t)(A(t)^3 B_j(t) - 3A(t)^2 B_j'(t) - 2A(t)A'((t))B_j(t)$$

$$+ 3A(t)B_j''(t) - A'(t)A(t)B_j(t) + 3A'(t)B_j'(t) + A''(t)B_j(t)$$

$$- B_j'''(t))u^{(j-3)}(t)dt.$$

Continuing, after the Jth step the derivative of $u(t)$ will not be in the integral. We can summarise the results of our calculations after introducing some abbreviating notations. Let $0 \leq \alpha_1, \alpha_2, \ldots, \alpha_\gamma,\ 1 \leq \beta_1, \beta_2, \ldots, \beta_\gamma$ be integer numbers. Then introducing a vector notation $\boldsymbol{\alpha} = (\alpha_1, \alpha_2, \ldots, \alpha_\gamma)$, $\boldsymbol{\beta} = (\beta_1, \beta_2, \ldots, \beta_\gamma)$ not highlighting dimension γ. We also introduce the $|\boldsymbol{\alpha}| = \sum |\alpha_i|$ notation of norm.

For the product $\big(A(t)^{(\alpha_1)}\big)^{\beta_1}, \big(A(t)^{(\alpha_2)}\big)^{\beta_2}, \ldots, \big(A(t)^{(\alpha_\gamma)}\big)^{\beta_\gamma}$ we introduce the abbreviation $\big(A^{(\boldsymbol{\alpha})}(t)\big)^{\boldsymbol{\beta}}$. Thus

$$x(T) = \int_0^T \Phi(T,t)B_0(t)u(T)dt + \int_0^T \Phi(T,t)(A(t)B_1(t) - B_1'(t))u(t)dt$$

$$+ \int_0^T \Phi(T,t)(A(t)^2 - A'(t)B_2(t) - 2A(t)B_2'(t) + B_2''(t))u(t)dt$$

$$+ \int_0^T \Phi(T,t)((A(t)^3 - 2A(t)A(t)' - A(t)'A(t))B_3(t)$$

$$- (3A(t)^2 - 3A'(t))B_3'(t) + 3A(t)B_3'' - B_3''')u(t)dt$$

$$+ \ldots$$

$$= \int_0^T \Phi(T,t) \left[\sum_{j=0}^J \left(\sum_{|\alpha+\beta|=j-\widehat{j}} C_{\alpha,\beta} \left(A^{(\alpha)}(t) \right)^\beta \right) B_j^{(\widehat{j})}(t) \right] u(t)dt.$$

For example in the third element for the coefficient of $B_3(t)$ the possible α, β indices are $(0, 3)$, $(1, 1) + (0, 1)$, for which $|\alpha + \beta| = 3 - 0$.

The coefficient of $B_3'(t)$ complying with indices $(0, 2)$, $(1, 1)$ fulfills the $|\alpha+\beta| = 3 - 1 = 2$ condition, the coefficients $C_{\alpha,\beta}$ are always integer.

4 General Reachability Conditions

In accordance with the classical Kalman-type discussion we can introduce the Kalman-Gram-type reachability matrix for the generalised systems:

$$R[0, T] =$$

$$\int_0^T \Phi(T,t) \left[\sum_{j=0}^J \left(\sum_{|\alpha+\beta|=j-\widehat{j}} C_{\alpha,\beta} \left(A^{(\alpha)}(t) \right)^\beta \right) B_j^{(\widehat{j})}(t) \right] \cdot$$

$$\cdot \left[\sum_{j=0}^J \left(\sum_{|\alpha+\beta|=j-\widehat{j}} C_{\alpha,\beta} \left(A^{(\alpha)}(t) \right)^\beta \right) B_j^{(\widehat{j})}(t) \right]^T \Phi(T,t)^T dt,$$

and the following theorem can be stated.

Theorem 1 The generalised linear system with time dependent coefficients is completely reachable on the $[0,T]$ interval if and only if the $R[0, T]$ Kalman-Gram-type matrix is invertible, or equivalently, positive definite.

Proof The proof is similar to the classical time dependent linear case.

Let L be the Lie-algebra generated by $\{A(t); \ t \in [0, T]\} \subset \mathbb{R}^{n \times n}$ and $A_1, A_2, \ldots, A_I \in L$ is one basis of this, as mentioned earlier. In this basis

$$A(t) = \sum_{i=1}^I a_i(t) A_i.$$

Let's proceed in a similar manner with the matrices $B_0(t)$, $B_1(t)$, ..., $B_J(t)$. If $V = V\{B_0(t), B_1(t), \ldots, B_J(t) : t \in [0, T]\} \subset \mathbb{R}^{n \times k}$ is the subspace of the vectorspace $\mathbb{R}^{n \times k}$ spanned by $B_j(t)$'s then we can select a $B_1, B_2, \ldots, B_{\widehat{I}}$ basis in V in which

$$B_j(t) = \sum_{i=1}^{\widehat{I}} b_{j\widehat{i}}(t) B_{\widehat{i}}.$$

Since $\Phi(T, t)$ can be written as a polynomial of $A_1, A_2, \ldots, A_I$ as the quasi-polynomial of the Wei-Norman-type equation's $g_1, g_2, \ldots, g_I$ solutions and the $x(T)$ integral representation's core function $\sum_{j=0}^{J} \left(\sum_{|\alpha+\beta|=j-\widehat{j}} C_{\alpha,\beta} \left(A^{(\alpha)}(t) \right)^{\beta} \right) B_j^{(\widehat{j})}(t)$ can

be written as the polynomial of $A_1, A_2, \ldots, A_I$ and $B_1, B_2, \ldots, B_{\widehat{I}}$ as the differential polynomial of $a_i(t)$ and $b_{\widehat{j}i}(t)$ (with integer coefficients and first order $B_{\widehat{i}}$'s). From this, exploiting that the exponents of A_i can be easily arranged in the natural order $(A_1^{m_1}, A_2^{m_2}, \ldots, A_I^{m_I}$, where for all inequalities $0 \le m_i < n$ or $0 \le \mathbf{m} < \mathbf{n}$ holds) by

exchanging the neighbouring elements with the aid of $A_{i_1} A_{i_2} = A_{i_2} A_{i_1} + \sum_{h=1}^{I} \Gamma_{i_1 i_2} A_h$,

and $A_i^n = \sum_{\widehat{i}=0} C_{\widehat{i} i} A_i^{\widehat{i}}$ equalities, assuming that the characteristic polynomial of A_i's

has the form $\lambda^n = \sum_{i=0}^{n-1} C_{\widehat{i} i} \lambda^i$.

The above yields the following:

$$x(T) = \sum_{0 \le \mathbf{m} < \mathbf{n}} \sum_{\widehat{i}=0}^{\widehat{i}} A_1^{m_1}, A_2^{m_2}, \ldots, A_I^{m_I} B_{\widehat{i}} \int_0^T P_{\mathbf{m}, \widehat{i}} \left(g(T, t), a^{[\infty)}(t), b_1^{[\infty)}(t), \ldots, b_J^{[\infty)}(t) \right) u(t) dt.$$

Here the $P_{\mathbf{m}, \widehat{i}} \left(g(T, t), \mathbf{a}^{[\infty)}(t), \ldots, b_j^{[\infty)}(t), \ldots \right)$ expressions are quasi-polynomials in $g_1(T, t), g_2(T, t), \ldots, g_I(T, t)$ -and differential polynomials in $a_1(t), a_2(t), \ldots, a_I(t)$, $b_{11}(t), b_{12}(t), \ldots, b_{1\widehat{I}}(t)$, $b_{21}(t), \ldots, b_{2\widehat{I}}(t), \ldots, b_{J1}(t), b_{J2}(t), \ldots, b_{J\widehat{I}}(t)$'s.

From this it is visible that the reachable subspace on $[0, T]$ of the generalised system has to be in the image space

$$\mathrm{Im} \left\{ \ldots, A_1^{m_1}, A_2^{m_2}, \ldots, A_I^{m_I} B_{\widehat{i}}, \ldots \right\}$$

which is the same as for the classical canonical system, seemingly not extending the reachability subspace with the derivatives acting as additional inputs.

In fact, the extension can be reviewed as follows:

Let $V_0 = V\{B_0(t); t \in [0, T]\} \subset \mathbb{R}^{n \times k}$ and $V_J = V_J\{B_0(t), B_1(t), \ldots, B_J(t); t \in [0, T]\} \subset \mathbb{R}^{n \times k'}$ be the vectorspaces generated by the respective $B_J(t)$ matrices. Choose the basis of V_J so that the first $\widehat{I}_0$ elements will be the basis of V_0:

$$V_0 = V \left\{ B_1, B_2, \ldots, B_{\widehat{I}_0} \right\},$$

$$V_J = V\left\{B_1, B_2, \ldots, B_{\widehat{I_0}}, B_{\widehat{I_0+1}}, B_{\widehat{I_0+2}}, \ldots, B_{\widehat{I_J}}\right\}$$

From this it is obvious that for the relevant generalised Kalman-Gram-type matrix (from now shortly referred to as Kalman-type matrices) for the generalised system it holds that their image space contains the image space of the classical system's Kalman-type matrix' image space.

Proceeding with these results proven for the canonical system, it can be easily deduced what is the excitation condition for having the reachability subspace of the generalised system match the image space of the general Kalman-type matrix assigned to the system. Assume that ξ is a vector which is in the image space of the Kalman-type matrix, denoted by $\mathrm{Im}\left\{\ldots, A_1^{m_1}, A_2^{m_2}, \ldots, A_I^{m_I} B_{\widehat{i}}, \ldots\right\} = K_{gen}$.

This image space is not identical with the generalised system's reachability subspace over $[0, T]$ if and only if there exists $\xi \neq 0, \xi \in K_{gen}$ for which $\langle \xi, x(T) \rangle = 0$ for all possible $u(t)$ inputs.

This means, that

$$0 = \langle \xi, x(T) \rangle =$$

$$= \left\langle \xi, \sum_{0 \leq m < n} \sum_{\widehat{i}=0}^{\widehat{I}} A_1^{m_1} A_2^{m_2}, \ldots, A_I^{m_I} B_{\widehat{i}} \int_0^T P_{\mathbf{m}, \widehat{i}}\left(g(T, t), \mathbf{a}^{[\infty)}(t), \ldots, b_j^{[\infty)}, \ldots\right) u(t) dt \right\rangle$$

$$= \int_0^T \left\langle \sum_{0 \leq m < n} \sum_{\widehat{i}=0}^{\widehat{I}} P_{\mathbf{m}, \widehat{i}}\left(g(T, t), \mathbf{a}^{[\infty)}(t), \ldots, b_j^{[\infty)}, \ldots\right) \right.$$

$$\left. \left(A_I^T\right)^{m_I}, \left(A_{I-1}^T\right)^{m_{I-1}}, \ldots, \left(A_1^T\right)^{m_1} B_{\widehat{i}}^T \xi, u(t) \right\rangle dt.$$

From the classical Lagrange-type lemma it follows that if this holds for all „good" functions (e.g. continuous functions) then

$$\sum_{0 \leq m < n} \sum_{\widehat{i}=0}^{\widehat{I}} P_{\mathbf{m}, \widehat{i}}\left(g(T, t), \mathbf{a}^{[\infty)}(t), \ldots, b_j^{[\infty)}, \ldots\right)\left(A_I^T\right)^{m_I}, \left(A_{I-1}^T\right)^{m_{I-1}}, \ldots, \left(A_1^T\right)^{m_1} B_{\widehat{i}}^T \xi = 0.$$

$$(10)$$

To replace the analytical functions of $\exp \lambda g$, $\cos \alpha g$, $\sin \alpha g$ type we can introduce the new variables, denote these with $\bar{g} = \exp \lambda g$, $\widehat{g} = \cos \alpha g$, $g = \sin \alpha g$. The respective differential equations can be easily derived:

$$\dot{\bar{g}} = \lambda \dot{g} \exp \lambda g = \lambda \dot{g} \bar{g},$$

$$\dot{\widehat{g}} = -\alpha \dot{g} \sin \alpha g = -\alpha \dot{g} g,$$

$$\dot{g} = \alpha \dot{g} \cos \alpha g = \alpha \dot{g} \widehat{g}.$$

$$(11)$$

It is visible that these are not explicit differential equations, both sides contain derivatives.

Consider now the Wei-Norman-type differential equation:

$$\left(\sum_{i=1}^{I} \exp \Gamma_1 g_1, \exp \Gamma_2 g_2, \ldots, \exp \Gamma_{i-1} g_{i-1} E_{ii} \right) \dot{g} = \mathbf{a}, \ \ \mathbf{g}(0) = \mathbf{0},$$

where

$$E_{ii} = (\delta_{ij}) = \begin{pmatrix} 0 & & & & 0 \\ & \ddots & & \cdot^{\cdot} & \\ & & 1 & & \\ & \cdot^{\cdot} & & \ddots & \\ 0 & & & & 0 \end{pmatrix} \in \mathbb{R}^{I \times I}.$$

This also contains exponential products, but these are the $\Gamma_1, \Gamma_2, \ldots, \Gamma_{I-1}$ Lie-algebra's product table's exponentials (the Cristoffel symbols) (Serre 1992). In these, similarly to the previous steps the $\exp \lambda g$, $\cos \alpha g$, $\sin \alpha g$ non-polynomial forms can be introduced as new variables which yet again will bring the addition of more—already polynomial—differential equations and the Wei-Norman equation will also become polynomial but not an explicit equation.

From here, if we want $\dot{g}$ can be made explicit with inverting in the original derivatives:

$$\dot{g} = \left(\sum_{i=1}^{I} \exp \Gamma_1 g_1, \exp \Gamma_2 g_2, \ldots, \exp \Gamma_{i-1} g_{i-1} E_{ii} \right)^{-1} \mathbf{a},$$

which will make the equations also explicit but with a fraction denominator:

$$\det \left(\sum_{i=1}^{I} \exp \Gamma_1 g_1, \exp \Gamma_2 g_2, \ldots, \exp \Gamma_{i-1} g_{i-1} E_{ii} \right).$$

Multiplying the explicit system of equations with these eventually we will get a polynomial differential equation implicit on $\mathbf{g}, \bar{\mathbf{g}}, \hat{\mathbf{g}}, \breve{\mathbf{g}}$ variables which will contain one derivative in each equation. This means that this is a regular differential equation which can be made explicit in the derivatives (with fractional right hand sides).

As a result, instead of the quasi-polynomials $P_{\mathbf{m},\,\hat{i}}\left(\mathbf{g}(T, t), \mathbf{a}^{[\infty)}(t), \mathbf{b}_1^{[\infty)}(t), \mathbf{b}_2^{[\infty)}(t), \ldots, \mathbf{b}_I^{[\infty)}(t) \right)$ we will get polynomials

$$\overline{P}_{\mathbf{m},\,\hat{i}}\left(\mathbf{g}(T, t), \bar{\mathbf{g}}(T, t), \hat{g}(T, t), g(T, t), \mathbf{a}^{[\infty)}(t), \mathbf{b}_1^{[\infty)}(t), \mathbf{b}_2^{[\infty)}(t), \ldots, \mathbf{b}_I^{[\infty)}(t) \right)$$

in variables $\mathbf{g}, \bar{\mathbf{g}}, \hat{\mathbf{g}}, \breve{\mathbf{g}}$ and differential polynomials in $\mathbf{a}(t), \mathbf{b}_0(t), \mathbf{b}_1(t), \ldots, \mathbf{b}_J(t)$ functions.

We will rewrite the implicit polynomial equation developed above introducing vector variables $\mathbf{x} = (x_1, x_2, \ldots, x_N)$ for $\mathbf{g}, \bar{\mathbf{g}}, \hat{\mathbf{g}}, \mathbf{g}$ and $\mathbf{u} = (u_1, u_2, \ldots, u_K)$ for $\mathbf{a}, \mathbf{b}_1, \mathbf{b}_2, \ldots, \mathbf{b}_J$, then

$$F(\mathbf{x}, \dot{\mathbf{x}}, \mathbf{u}, \dot{\mathbf{u}}, \ldots) = 0.$$

Equation (10) can also be rewritten with $\mathbf{x}, \mathbf{u}$ as

$$\sum_{0 \le m < n} \sum_{\hat{i}=0}^{I} \overline{P}_{bm,\hat{i}}(\mathbf{x}, \dot{\mathbf{x}}, \mathbf{u}, \dot{\mathbf{u}}, \ldots)\left(A_I^T\right)^{m_I}, \left(A_{I-1}^T\right)^{m_{I-1}}, \ldots, \left(A_1^T\right)^{m_1} B_{\hat{i}}^T \xi = 0.$$

From this define the output equation

$$y = \sum_{0 \le m < n} \sum_{ki_2=0}^{I} \overline{P}_{m,ki_2}(\mathbf{x}, \mathbf{u}, \dot{\mathbf{u}}, \ldots,)(A_I^T)^{m_I}, (A_{I-1}^T)^{m_{I-1}}, \ldots, (A_1^T)^{m_1} B_{ki_2}^T \xi$$

$$= G(\mathbf{x}, \mathbf{u}, \dot{\mathbf{u}}, \ldots, \xi)$$

This yields the

$$\begin{aligned} F(\mathbf{x}, \dot{\mathbf{x}}, \mathbf{u}, \dot{\mathbf{u}}, \ldots, \xi) &= 0 \\ y &= G(\mathbf{x}, \mathbf{u}, \dot{\mathbf{u}}, \ldots, \xi) \end{aligned} \tag{12}$$

input-output system, which is polynomial and implicit in the derivatives of $\dot{x}$, with a regularity condition $\partial_{\dot{x}} F(\mathbf{x}, \dot{\mathbf{x}}, \mathbf{u}, \dot{\mathbf{u}}, \ddot{\mathbf{u}}, \ldots, \xi) \ne 0$. Here $\mathbf{u}$ are inputs, $\mathbf{x}$ are states, and $\mathbf{y}$ are outputs.

Consider another representation with perhaps different states but with the same inputs and outputs:

$$\begin{aligned} \overline{F}\left(\overline{\mathbf{x}}, \dot{\overline{\mathbf{x}}}, \mathbf{u}, \dot{\mathbf{u}}, \ldots, \xi\right) &= 0 \\ y &= \overline{G}(\overline{\mathbf{x}}, \mathbf{u}, \dot{\mathbf{u}}, \ldots, \xi). \end{aligned} \tag{13}$$

Let us consider (12) and (13) as input-output systems. They are equivalent if for all $(\mathbf{u}, \mathbf{y})$ input-output pair it holds that (12) has a solution $\mathbf{x}$ if and only if when (13) has an $\overline{\mathbf{x}}$ solution. This means that the two systems function identically.

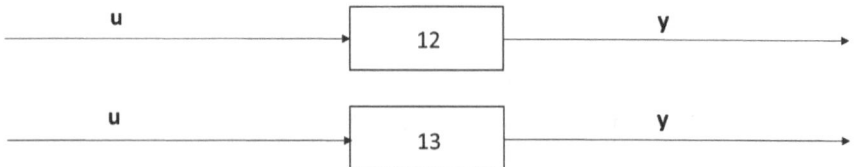

Systems (12) and (13) can be written in a more compact form, allowing for the y outputs' derivatives,

$$J(\mathbf{x}, \dot{\mathbf{x}}, \mathbf{u}, \dot{\mathbf{u}}, \ddot{\mathbf{u}}, \ldots, \mathbf{y}, \dot{\mathbf{y}}, \ddot{\mathbf{y}}, \ldots, \xi) = 0, \tag{14}$$

and

$$\overline{J}(\bar{\mathbf{x}}, \dot{\bar{\mathbf{x}}}, \mathbf{u}, \dot{\mathbf{u}}, \ddot{\mathbf{u}}, \ldots, \mathbf{y}, \dot{\mathbf{y}}, \ddot{\mathbf{y}}, \ldots, \xi) = 0.$$

Diop prooved (1991) that there exists a finite purely algebraic algorithm which can provide differential polynomials

$$\widehat{J}(\mathbf{u}, \dot{\mathbf{u}}, \ddot{\mathbf{u}}, \ldots, \mathbf{y}, \dot{\mathbf{y}}, \ddot{\mathbf{y}}, \ldots, \xi)$$
$$\widehat{G}(\mathbf{u}, \dot{\mathbf{u}}, \ddot{\mathbf{u}}, \ldots, \mathbf{y}, \dot{\mathbf{y}}, \ddot{\mathbf{y}}, \ldots, \xi)$$

for which if we consider the implicit equation and non-equality condition

$$\widehat{J}(\mathbf{u}, \dot{\mathbf{u}}, \ddot{\mathbf{u}}, \ldots, \mathbf{y}, \dot{\mathbf{y}}, \ddot{\mathbf{y}}, \ldots, \xi) = 0$$
$$\widehat{G}(\mathbf{u}, \dot{\mathbf{u}}, \ddot{\mathbf{u}}, \ldots, \mathbf{y}, \dot{\mathbf{y}}, \ddot{\mathbf{y}}, \ldots, \xi) \neq 0 \tag{15}$$

then based on the above system-equivalence the input output system $\mathbf{u} \mapsto \mathbf{y}$ definable from these is equivalent with system (14).

It is visible that the latter does not have a state variable $\mathbf{x}$ that's the reason we can say that the Diop-type algorithm is a state elimination algorithm. We repeat the definition of equivalence. The correspondences (14) and (15) define an equivalent input-output system if for a $(\mathbf{u}, \mathbf{y})$ input-output pair system (14) has a solution for state $\mathbf{x}$ (in other words the triplet $(\mathbf{x}, \mathbf{u}, \mathbf{y})$ is a solution for (14)) if $(\mathbf{u}, \mathbf{y})$ is a solution of the polynomial equation

$$\widehat{J}(\mathbf{u}, \dot{\mathbf{u}}, \ddot{\mathbf{u}}, \ldots, \mathbf{y}, \dot{\mathbf{y}}, \ddot{\mathbf{y}}, \ldots, \xi) = 0$$

and it holds that

$$\widehat{G}(\mathbf{u}, \dot{\mathbf{u}}, \ddot{\mathbf{u}}, \ldots, \mathbf{y}, \dot{\mathbf{y}}, \ddot{\mathbf{y}}, \ldots, \xi) \neq 0.$$

This latter comes from the stepwise division in the algorithm by a differential polynomial, which naturally can not be 0, thus this assertion has to be made. The products of these provide the $\widehat{G}(\mathbf{u}, \dot{\mathbf{u}}, \ddot{\mathbf{u}}, \ldots, \mathbf{y}, \dot{\mathbf{y}}, \ddot{\mathbf{y}}, \ldots, \xi)$ differential polynomials. If this product is not 0, then none of its factors can be 0.

Returning to our originally acquired input-output system, regarding the linewise-continuously ordered $\mathbf{u} = (\mathbf{a}, \mathbf{b}_0, \mathbf{b}_1, \ldots, \mathbf{b}_j)$ input we get

$$F(\mathbf{x}, \dot{\mathbf{x}}, (\mathbf{a}, \mathbf{b}_0, \mathbf{b}_1, \ldots, \mathbf{b}_j), (\dot{\mathbf{a}}, \dot{\mathbf{b}}_0, \dot{\mathbf{b}}_1, \ldots, \dot{\mathbf{b}}_j), \ldots, \xi) = 0,$$
$$G(\mathbf{x}, (\mathbf{a}, \mathbf{b}_0, \mathbf{b}_1, \ldots, \mathbf{b}_j), (\dot{\mathbf{a}}, \dot{\mathbf{b}}_0, \dot{\mathbf{b}}_1, \ldots, \dot{\mathbf{b}}_j), \ldots, \xi) = 0.$$

Writing in the state eliminated system derived from system (12) the correspondences following from the system equivalence.

Entering the correspondences from the system equivalence

$$\widehat{J}\big((\mathbf{a}, \mathbf{b}_0, \mathbf{b}_1, \ldots, \mathbf{b}_J), (\dot{\mathbf{a}}, \dot{\mathbf{b}}_0, \dot{\mathbf{b}}_1, \ldots, \dot{\mathbf{b}}_J), \ldots, 0, 0, 0, \ldots, \xi\big) = 0$$
$$\widehat{G}\big((\mathbf{a}, \mathbf{b}_0, \mathbf{b}_1, \ldots, \mathbf{b}_J), (\dot{\mathbf{a}}, \dot{\mathbf{b}}_0, \dot{\mathbf{b}}_1, \ldots, \dot{\mathbf{b}}_J), \ldots, 0, 0, 0, \ldots, \xi\big) \neq 0$$

(16)

into the state eliminated system for the input-output pair $((\mathbf{a}, \mathbf{b}_0, \mathbf{b}_1, \ldots, \mathbf{b}_J), 0)$ we can consider these as the sufficient condition for all $\mathbf{u}$ input the $\mathbf{x}(T)$ final state to be perpendicular to the

$$\xi \in \mathrm{Im}\big\{\ldots, A_1^{m_1}, A_2^{m_2}, \ldots, A_I^{m_I} B_j, \ldots\big\}$$

(17)

vector.

Definition We say that the time dependent coefficients $\mathbf{a}, \mathbf{b}_0, \mathbf{b}_1, \ldots, \mathbf{b}_J$ excite our system persistently if the subspace of reachable states matches the image space of the generalised Kalman-type rank condition's matrix. In other words it will be the greatest dimensional subspace possible. This means, based on our equations that if the conditions of (14) hold then state ξ must be 0.

The most interesting is the special case when the image space of the generalised Kalman-type rank condition's matrix is the complete $\mathbb{R}^n$. Then it holds that coefficients $\mathbf{a}, \mathbf{b}_0, \mathbf{b}_1, \ldots, \mathbf{b}_J$ excite the system persistently if and only if the system is completely reachable on the $[0, T]$ interval.

So if the coefficients do not excite the system persistently then

$$\widehat{J}\big((\mathbf{a}, \mathbf{b}_0, \mathbf{b}_1, \ldots, \mathbf{b}_J), (\dot{\mathbf{a}}, \dot{\mathbf{b}}_0, \dot{\mathbf{b}}_1, \ldots, \dot{\mathbf{b}}_J), \ldots, 0, 0, 0, 0, \ldots, \xi\big) = 0$$
$$\widehat{G}\big((\mathbf{a}, \mathbf{b}_0, \mathbf{b}_1, \ldots, \mathbf{b}_J), (\dot{\mathbf{a}}, \dot{\mathbf{b}}_0, \dot{\mathbf{b}}_1, \ldots, \dot{\mathbf{b}}_J), \ldots, 0, 0, 0, 0, \ldots, \xi\big) \neq 0$$
$$\xi \neq 0$$

can be solved. Considering the equation as an implicit function of ξ it can be solved for ξ,

$$\xi = \widehat{f}\left((\mathbf{a}, \mathbf{b}_0, \mathbf{b}_1, \ldots, \mathbf{b}_J), (\dot{\mathbf{a}}, \dot{\mathbf{b}}_0, \dot{\mathbf{b}}_1, \ldots, \dot{\mathbf{b}}_J), \ldots, \right), \tag{18}$$

Substituting this into the two inequalities we get that the condition of „persistent non-excitation" is that the following two inequalities hold simultaneously:

$$0 \neq \widehat{G}((\mathbf{a}, \mathbf{b}_0, \mathbf{b}_1, \ldots, \mathbf{b}_J), (\dot{\mathbf{a}}, \dot{\mathbf{b}}_0, \dot{\mathbf{b}}_1, \ldots, \dot{\mathbf{b}}_J), \ldots, 0, 0, \ldots,$$
$$\widehat{f}\left((\mathbf{a}, \mathbf{b}_0, \mathbf{b}_1, \ldots, \mathbf{b}_J), (\dot{\mathbf{a}}, \dot{\mathbf{b}}_0, \dot{\mathbf{b}}_1, \ldots, \dot{\mathbf{b}}_J), \ldots\right)$$
$$0 \neq \widehat{f}(\mathbf{a}, \mathbf{b}_0, \mathbf{b}_1, \ldots, \mathbf{b}_J), (\dot{\mathbf{a}}, \dot{\mathbf{b}}_0, \dot{\mathbf{b}}_1, \ldots, \dot{\mathbf{b}}_J) \ldots)$$

The contradiction of this is the condition of persistent excitation:

$$0 = \widehat{G}((\mathbf{a}, \mathbf{b}_0, \mathbf{b}_1, \ldots, \mathbf{b}_J), (\dot{\mathbf{a}}, \dot{\mathbf{b}}_0, \dot{\mathbf{b}}_1, \ldots, \dot{\mathbf{b}}_J), \ldots, 0, 0, \ldots,$$
$$\widehat{f}\left((\mathbf{a}, \mathbf{b}_0, \mathbf{b}_1, \ldots, \mathbf{b}_J), (\dot{\mathbf{a}}, \dot{\mathbf{b}}_0, \dot{\mathbf{b}}_1, \ldots, \dot{\mathbf{b}}_J), \ldots\right)$$

or

$$0 = \widehat{f}((\mathbf{a}, \mathbf{b}_0, \mathbf{b}_1, \ldots, \mathbf{b}_J), (\dot{\mathbf{a}}, \dot{\mathbf{b}}_0, \dot{\mathbf{b}}_1, \ldots, \dot{\mathbf{b}}_J) \ldots).$$

We return to the solvability of the implicit function of ξ in order to get (18).

We can also apply here the Diop-type elimination theorem (algorithm). For this consider vector ξ a state vector which can be eliminated. For this we would need an equation for the system state, a dynamics which is a differential equation of ξ. For this simply consider that ξ is constant, that is, the dynamics simply is $\dot{\xi} = 0$.

Results discussed above starting from Theorem 1. are concluded herewith in Theorem 2 as the main result. Specify system (4) in the following manner.

Let $A_1, A_2, \ldots, A_I \in \mathbb{R}^{n \times n}$ be given structure matrices and $B \in \mathbb{R}^n$ be the input matrix of 1 dimension. Consider

$$\dot{x}(t) = \sum_{i=1}^{I} a_i(t) A_i x + b(t) B u \tag{19}$$

LTV-system. For this system the above discussion and results are summarised in the following theorem.

Theorem 2 If the generalised Kalman-type rank condition holds for system in (19), that is (9) is the complete $\mathbb{R}^n$, and the excitation condition from the Diop-type elimination holds then (19) is reachable.

5 Conclusions

Similar theorems hold for state-dependent parametric linear systems as above. There, the difficulty is the lack of Diop-type elimination theorem for the partial differential algebras. Therefore in that case a different approach has to be applied to find the results for the sufficient and necessary conditions to hold. The Diop-type elimination procedure can be represented in such cases by excitation conditions.

References

Bischi, G. I., Chiarella, C., Kopel, M., & Szidarovszky, F. (2010). *Nonlinear oligopolies: Stability and bifurcations*. Berlin/Heidelberg: Springer.

Coddington, E. A., & Levinson, N. (1955). Theory of ordinary differential equations. New York: McGraw-Hill; New Delhi: Tata McGraw-Hill.

Diop, S. (1991). Elimination in control theory. *Mathematics of Control, Signals, and Systems, 4*(1), 17–32.

Kalman, R. E., Falb, P.L., & Arbib M.A. (1969). Topics in mathematical system theory (New York, San Francisco, St. Louis, Toronto, London, Sydney: McGraw-Hill Book Company).

Kaplansky, I. (1976). *An introduction to differential algebra*. Paris: Hermann.

Matsumoto, A., & Szidarovszky, F. (2018). *Dynamic oligopolies with time delays*. Tokyo: Springer.

Molnár, S. (1993a). Kalman's rank conditions for time dependent linear systems. *Pure Mathematics and Applications, 4*(3), 353–361.

Molnár, S. (1993b). Stabilization of verticum-type systems. *Pure Mathematics and Applications, 4*(4), 493–499.

Molnár, S. (2001). Időtől függő vertikum-típusú lineáris rendszerekről (On linear time dependent verticum type systems, *in Hungarian*), in MTA Közgyűlési előadások, (May 2000, Vol II, pp. 645–657, MTA). ISSN: 1585-1915.

Molnár, S., & Szidarovszky, F. (1994a). A note on the coverability problem in input-output systems. *Pure Mathematics and Applications, 5*(4), 425–429.

Molnár, S., & Szidarovszky, F. (1994b). A dinamikus termelői-fogyasztói modell irányíthatóságáról. *SZIGMA, 26*(1–2), 49–54.

Molnár, S., & Szigeti, F. (1994a). On time varying discrete-time linear systems: reachability, distinguishability and identifiability. *Pure Mathematics and Applications, 5*(1), 415–424.

Molnár, S., & Szigeti, F. (1994b). On "Verticum"-type linear systems with time-dependent linkage. *Applied Mathematics and Computation, 60*, 89–102.

Molnár, S., Szigeti, F., Vera, C. E. (1993). Kalman-féle rangfeltételek az időtől függő lineáris rendszerekre, Alkalmazott Matematikai Lapok (Vol. 17, pp. 3–4. sz., 279–286. old).

Okoguchi, K., & Szidarovszky, F. (1997). *The theory of oligopoly with multi-product firms* (2nd ed.). Boston, Heidelberg, New York: Springer.

Pontryagin, L. S., Boltyanskii, V. G., Gamkrelidze, R. S., & Mischenko, E. F. (1964). *Mathematical theory of optimal processes*. New York: Pergamon Press, The Macmillan Co.

Serre, J.-P. (1992). *Lie algebras and Lie groups* (2nd ed.). New York: Springer-Verlag.

Szidarovszky, F., & Bahill, A. T. (1998). *Linear systems theory* (2nd ed.). Boca Raton, Boston, New York, Washington DC, London: CRC Press.

Szidarovszky, F., & Yen, J. (1993). On the controllability of discrete dynamic oligopolies under adaptive expectations. *Applied Mathematics and Computation, 56*, 49–57.

Szigeti, F. (1992). A differential-algebraic condition for controllability and observability of time varying linear systems. In *Proceedings of the 31st IEEE Conference on Decision and Control* (Vol. 4, pp. 3088–3090). Arizona: Tucson.

Wei, J., & Norman, E. (1964). On global representation of the solutions of linear differential equations as a product of exponentials. *Proceedings of the American Mathematical Society, 15*(12), 327–334.